THE BEEKEEPING BIBLE

A Beekeeper's Essential Guide to Honey Harvest, Beeswax Processing, and the Art of Building a Sustainable Hive

Harrison Miller

© 2025 The Beekeeper's Bible

All rights reserved.

This book is for informational use only.

The publisher is not liable for any damages from its use or misuse.

Unauthorized copying, sharing, or transmitting of this book, entirely or partially, is forbidden.

The book is provided "as is" without warranties of any kind. All trademarks and brand names are owned by their respective holders.

Table of Contents

Part 1: Introduction to the World of Bees 12

Chapter 1: The Importance of Bees in the Ecosystem 13
- Bees' Role in Pollination and Food Supply 14
- Environmental Benefits of Bees 16
- Decline of Bees: Impact on Ecosystems and Agriculture 16
- Global Initiatives for Bee Protection 18

Chapter 2: A Brief History of Beekeeping 21
- Ancient Beekeeping Practices 22
- Medieval Beekeeping Innovations 22
- Modern Beekeeping: Langstroth Hive Innovation 23

Chapter 3: The Fascination with Bees 25
- Bee Myths and Cultural Stories 25
- Symbolism of Bees in Ancient Cultures 26
- Bees in Art, Literature, and Science 27
- The Enduring Appeal of Bees 28

Part 2: Bee Biology and Behavior 30

Chapter 4: Anatomy of Bees 31
- Bee Anatomy: Head, Thorax, and Abdomen Functions 31
- Pheromones and Hive Communication 32
- Specialized Bee Organs for Survival and Hive Duties 33
- Bee Adaptations for Efficient Nectar and Pollen Collection 34

Chapter 5: Bee Society 36
- Hive Social Hierarchy: Queen, Workers, and Drones 36
- Selecting and Raising a New Queen Bee 37
- Age-Based Roles of Worker Bees 38
- Hive Cooperation and Communication 39

Chapter 6: The Life Cycle of Bees 41
- Bee Development: From Egg to Adult 41
- Lifespan of Worker Bees, Drones, and the Queen 42

    Factors Influencing the Bee Life Cycle 43

Chapter 7: Bee Communication 45
    The Waggle Dance and Bee Communication 46
    Pheromones in Hive Communication 46
    Bee Communication: Sounds and Vibrations 47

Part 3: Types of Bees 49

Chapter 8: Honeybees 50
    Key Characteristics of Honeybees 51
    European vs. Africanized Honeybee Differences 52
    Popular Honeybee Breeds in Beekeeping 53
    Adaptability of Honeybees to Climates and Environments 54

Chapter 9: Solitary Bees 56
    Solitary Bees vs. Honeybees: Lifestyle and Behavior 56
    Solitary Bees and Their Ecological Roles 57
    Solitary Bees: Key Pollinators of Crops and Wild Plants 58

Chapter 10: Other Bee Species 60
    Stingless Bees: Characteristics and Habitats 61
    Wild Bees: Biodiversity and Plant Reproduction 61
    Threats to Lesser-Known Bee Species 62
    Bee Conservation Initiatives Worldwide 63

Chapter 11: Interactions Among Species 65
    Bee and Pollinator Resource Competition 65
    Bee Species and Ecological Balance 66
    Effects of Beekeeping on Wild Pollinators 67
    Bee Species Interactions and Conservation 68

Part 4: The Importance of Biodiversity 70

Chapter 12: Bees and Plants 71
    Bee and Flower Co-Evolution 72
    Flowers Attractive and Beneficial to Bees 73
    Bees' Impact on Agricultural Productivity 74

Chapter 13: Threats to Biodiversity 75
    Impact of Habitat Destruction on Bees 76
    Pesticides' Impact on Bee Health and Behavior 77

- Climate Change Impact on Bee Habitats ... 78
- Diseases and Parasites Affecting Bees ... 79

Chapter 14: The Beekeeper's Role in Conservation ... 82
- Responsible Beekeeping and Biodiversity ... 82
- Creating Bee-Friendly Environments ... 83
- Engaging Communities in Bee Conservation ... 84
- Urban Beekeeping: Boosting Biodiversity and Pollination ... 85

Chapter 15: Promoting Biodiversity ... 87
- Bee-Friendly Gardens and Landscapes ... 88
- Green Corridors for Pollinator Habitats ... 88
- Importance of Pollinators ... 89

Part 5: Preparing for Beekeeping ... 91

Chapter 16: Essential Beekeeping Tools for Beginners ... 92
- Types of Hives and Their Unique Features ... 92
- Personal Protective Equipment for Beekeeping Safety ... 93
- Hive Management Tools ... 94

Chapter 17: Choosing the Ideal Hive Location ... 96
- Environmental Factors for Hive Placement ... 97
- Nectar and Water Sources for Hives ... 97
- Avoiding Risky Zones for Beekeeping ... 98
- Integrating Hives into Urban and Rural Settings ... 99

Chapter 18: Acquiring Your First Bee Colonies ... 101
- Reliable Sources for Buying Bee Colonies ... 102
- Introducing Bees to a New Hive Safely ... 102
- Differences: Natural Swarms vs. Bred Queens ... 103

Chapter 20: Economic Aspects of Beekeeping ... 105
- Beekeeping Costs: Setup and Ongoing Expenses ... 106
- Income Streams from Honey, Wax, and Propolis ... 107
- Government Incentives for Beginner Beekeepers ... 108
- Managing Financial Risks in Beekeeping ... 108

Part 6: Building and Managing Hives ... 110

Chapter 21: Types of Hives ... 111
- Langstroth Hives: Pros and Cons ... 112

- Top-Bar Hives: Sustainable Beekeeping Choice ... 113
- Warré Hives: Tradition Meets Innovation ... 114

Chapter 22: DIY Hive Construction Techniques ... 116
- Durable and Eco-Friendly Hive Materials ... 117
- Step-by-Step Guide to Assembling a Langstroth Hive ... 118
- Designing Flexible Top-Bar Hives ... 119
- Maintaining Handmade Hives ... 120

Chapter 23: Proper Hive Placement ... 122
- Maximizing Hive Success on Your Property ... 123
- Deterring Predators from Hives ... 124
- Protecting Hives from Harsh Weather ... 124
- Ventilation for Hive Health and Moisture Control ... 125

Chapter 24: Monitoring Hive Health ... 127
- Signs of a Healthy Hive ... 128
- Detecting Foulbrood and Chalkbrood Diseases ... 129
- Integrated Pest Management for Varroa Mites ... 130
- Identifying Hive Stressors and Triggers ... 132

Chapter 25: Expanding the Hive ... 134
- Adding Frames and Supers for Hive Growth ... 134
- Splitting Hives for New Colonies and Overcrowding Prevention ... 135
- Introducing New Queens for Colony Stability ... 136

Part 7: Seasons and Beekeeping ... 138

Chapter 26: Spring ... 139
- Preparing Hives for Active Season ... 140
- Post-Winter Hive Health Inspections ... 141
- Early Nutrition for Honey Production and Growth ... 142

Chapter 27: Summer Beekeeping Essentials ... 144
- Monitoring Hive Activity and Managing Heat Stress ... 145
- Harvesting Honey for Quality and Quantity ... 145
- Managing Summer Pest Issues in Beekeeping ... 146

Chapter 28: Autumn Beekeeping ... 148
- Assessing Honey Reserves for Winter Survival ... 148
- Supplementary Feeding with Sugar Syrup ... 149

  Queen Bee Health and Replacement..150

Chapter 29: Winter Beekeeping Strategies...152
  Insulating Hives for Winter Stability..152
  Monitoring Hive Activity Without Disturbance..153
  Preventing Moisture Buildup in Hives...154

Part 8: Harvesting Hive Products..156

Chapter 31: Honey..157
  Types of Honey: Clover, Acacia, Wildflower...158
  Honey Extraction Techniques..159
  Storing Honey: Best Practices...160
  Labeling and Marketing Honey for Consumer Appeal..161

Chapter 32: Royal Jelly...162
  Health Benefits of Royal Jelly...162
  Harvesting Royal Jelly Safely..163
  Storage Techniques for Freshness and Quality...164

Chapter 33: Propolis...166
  Bee Propolis: Production and Hive Uses..167
  Extracting Propolis Safely and Efficiently...167
  Propolis in Natural Medicine and Wellness..168
  Propolis Processing Techniques...169

Chapter 34: Beeswax..171
  Collecting and Purifying Beeswax...172
  Crafting with Beeswax...173
  Melting and Molding Beeswax...174
  Selling Beeswax Products for Extra Income..175

Chapter 35: Pollen...176
  Nutritional Benefits of Pollen...176
  Pollen Collection Without Hive Stress..177
  Storing and Using Pollen in Recipes..178

Part 9: Cooking with Honey...180

Chapter 36: Sweet Recipes..181
  Honey Cakes: History, Culture, and Baking Techniques...182
  Honey Cookies and Pies: Traditional and Modern...182

- Honey Mousse and Custards 183
- Pairing Honey with Traditional Desserts 183

Chapter 37: Savory Honey Recipes 185
- Honey Marinades and Glazes for Meats and Vegetables 185
- Balancing Honey's Sweetness in Dressings 186
- Honey in Main Dishes 187
- Enhancing Soups with Honey 187

Chapter 38: Honey Beverages 189
- The Art of Making Mead 189
- Sweetening Teas with Honey 190
- Honey-Based Cocktails 191
- Natural Energy Drinks with Honey 192

Chapter 39: Preserving Recipes 193
- Preserving Honey's Natural Flavor in Cooking 193
- Storing Honey-Based Dishes 194
- Personal Honey Recipe Book Ideas 195

Chapter 40: Gastronomic Pairings 197
- Honey and Cheese Pairings 198
- Pairing Honey with Fruits for Balanced Flavors 198

Part 10: Crafting with Hive Products 200

Chapter 41: Beeswax Candles 201
- Traditional Beeswax Candle Techniques 201
- Decorative Candle Ideas 202
- Creating Therapeutic Scented Candles 203

Chapter 42: Natural Cosmetics 204
- Creams and Lotions with Beeswax 205
- Nourishing Lip Balms with Natural Ingredients 205
- Homemade Soap with Beeswax and Hive Products 206

Chapter 43: Decorative Objects 208
- Artistic Uses of Beeswax 208
- Making Reusable Food Wraps 209
- Care and Storage of Decorative Items 210

Chapter 44: Starting a Small Craft Business 211

- Market Demand for Beeswax and Propolis Crafts 212
- Selling Crafts: Online, Markets, Direct Channels 212

Chapter 45: Sustainability in Crafting 214
- Ethical and Sustainable Hive Resource Use 214
- Eco-Friendly Crafting Materials 215

Part 11: Health and Wellness with Bee Products 217

Chapter 46: Medicinal Properties of Honey 218
- Honey for Wounds and Burns 218
- Boosting Immunity with Hive Products 219
- Honey's Health Benefits: Scientific Research 220

Chapter 47: Propolis for Health 222
- Propolis: Boosting Immunity and Wellness 223
- Using Propolis in Daily Health Routines 224
- Propolis in Respiratory Health and Infection Control 225

Chapter 48: Royal Jelly and Vitality 226
- Royal Jelly: Nutritional Energy Booster 227
- Incorporating Royal Jelly into Diets and Skincare 228
- Myths and Realities of Royal Jelly Properties 228

Chapter 49: Bee Pollen and Nutrition 231
- Bee Pollen: A Nutrient-Dense Superfood 232
- Bee Pollen Recipes and Meal Ideas 232
- Allergy Precautions for Pollen in Diets 233

Chapter 50: Holistic Bee Product Approaches 235

Part 12: Sustainability and Bees 236

Chapter 51: Challenges Bees Face Today 237
- Pesticides: Long-term Consequences for Bees 238
- Climate Change Impact on Bees 240
- Diseases and Pests Threatening Hives 241

Chapter 52: Sustainable Beekeeping Practices 243
- Natural Pest Control in Beekeeping 244
- Sustainable Harvesting for Colony Health 245

Chapter 53: Supporting Local Pollinators 247

- Creating Pollinator-Friendly Spaces ... 248
- Partnering with Farmers for Pollinator Networks ... 248
- Chapter 54: Education and Advocacy ... 250
  - The Role of Bees in Ecosystems ... 250
  - Teaching Kids About Pollinators ... 251

Part 13: Advanced Beekeeping Techniques ... 253

- Chapter 55: Queen Rearing ... 254
  - Queen's Role in Hive Health and Productivity ... 254
  - Raising Queen Bees: Grafting to Natural Selection ... 255
  - Evaluating Queen Quality ... 256
  - Introducing a New Queen to a Hive ... 257
- Chapter 56: Hive Splitting ... 259
  - Step-by-Step Instructions for Successfully Splitting a Hive ... 260
  - Avoiding Stress-Induced Absconding ... 261
- Chapter 57: Overwintering Hives ... 263
  - Winterizing Hives: Insulation and Ventilation ... 264
  - Feeding Strategies for Low Activity Periods ... 265
  - Monitoring Hive Health in Winter ... 265
- Chapter 58: Migratory Beekeeping ... 267
  - Benefits of Relocating Hives for Seasonal Blooms ... 268
  - Ensuring Hive Health During Relocation ... 269

Part 14: The Cultural Significance of Bees ... 271

- Chapter 59: Bees in History and Mythology ... 272
- Chapter 60: The Honey Trade Through the Ages ... 273
- Chapter 61: Bees in Modern Culture ... 275
- Chapter 62: Inspiring Change Through Bees ... 277

Part 15: The Future of Beekeeping ... 279

- Chapter 63: Technological Innovations in Beekeeping ... 280
- Chapter 64: Global Trends in Beekeeping ... 282
- Chapter 65: Educating Future Beekeepers ... 284

Part 1: Introduction to the World of Bees

CHAPTER 1: THE IMPORTANCE OF BEES IN THE ECOSYSTEM

Bees play a pivotal role in the ecosystem, far beyond the common perception of them merely as honey producers. Their most crucial contribution is pollination, a process vital for the reproduction of many plants and, consequently, the production of a significant portion of the foods we consume. Pollination occurs when bees collect nectar and pollen from flowers. As they move from flower to flower, they inadvertently transfer pollen grains, facilitating the fertilization of plants. This natural service supports the growth of fruits, vegetables, and nuts, essential components of the human diet, as well as forage crops used to feed livestock.

Beyond agriculture, bees' pollination efforts contribute significantly to maintaining the diversity of plant life in wild ecosystems. This biodiversity is not only crucial for wildlife habitats, providing food and shelter for a myriad of species, but it also plays a fundamental role in human survival. Diverse plant life ensures healthy soils and regulates the water cycle, both of which are indispensable for sustainable agriculture and for preventing natural disasters such as floods and landslides.

The environmental benefits of bees extend to the preservation of genetic diversity among plants. By pollinating a wide variety of plants, bees facilitate cross-pollination, which results in plants having a greater genetic diversity. This diversity allows plant species to be more resilient to diseases and changing environmental conditions, ensuring their survival and continuation. Moreover, many plants that rely on bees for pollination are significant sources of medicine, providing raw materials for the pharmaceutical industry and traditional medicines.

However, the contribution of bees to the ecosystem is under threat due to the alarming decline in bee populations worldwide. Factors contributing to this decline include habitat destruction, pesticide use, climate change, and the spread of diseases and parasites. The loss of bees would not only mean a decrease in honey production but, more importantly, a significant disruption of the ecological balance, affecting food production, plant diversity, and ultimately human survival.

Understanding the intricate relationship between bees and the ecosystem is the first step in recognizing the importance of their conservation.

The decline in bee populations poses a critical challenge to global agriculture and food security. Bees are responsible for the pollination of approximately 75% of the crops used directly for human consumption, including fruits, vegetables, and nuts. The economic value of bees' pollination services is immense, estimated to be billions of dollars annually worldwide. Without bees, many of the foods we rely on would become scarce and significantly more expensive, leading to nutritional deficiencies and impacting global health. In response to the decline, there has been a surge in efforts to protect bee populations and their habitats. Conservation initiatives aim to address the root causes of bee population declines, such as habitat destruction and pesticide use. These efforts include promoting organic farming practices that are less harmful to bees, restoring natural habitats to provide bees with the necessary resources for survival, and implementing policies that protect bees from harmful pesticides and chemicals.

Educating the public about the importance of bees and how to protect them is also a crucial component of conservation efforts. Many people are unaware of the vital role bees play in our ecosystem and how their actions can impact bee populations. By raising awareness, individuals can take steps to support bees, such as planting bee-friendly gardens, avoiding the use of harmful pesticides, and supporting local beekeepers by buying local honey. Beekeeping itself has evolved to support the health and sustainability of bee populations. Modern beekeeping practices focus on the welfare of the bees, using techniques that mimic natural conditions and promote the health of the colony. This includes managing hives in ways that reduce stress on the bees, treating diseases and parasites with natural remedies, and ensuring that bees have access to a diverse range of flowers for nutrition.

Research into bee health and conservation is ongoing, with scientists studying the impacts of environmental changes, pesticides, and diseases on bee populations. These studies are crucial for developing strategies to protect bees and ensure their survival for future generations. The relationship between bees and the ecosystem is a delicate balance that requires attention and action to maintain. By understanding the challenges bees face and taking steps to mitigate these threats, we can protect these vital pollinators and ensure the sustainability of our food systems and natural environments.

Bees' Role in Pollination and Food Supply

Bees' role in pollination is a cornerstone of agricultural productivity and biodiversity. Pollination, the transfer of pollen from the male parts of a flower to the female parts, is essential for plant reproduction. Bees, particularly honeybees, are among the most efficient pollinators due to their size, behavior, and abundance. When a bee visits a flower to collect nectar and pollen for food, it brushes against the flower's reproductive organs, unwittingly transferring pollen in the process. This seemingly simple act is a critical component of the reproductive cycle of most flowering plants, leading to the production of seeds and fruit.

The significance of bees to agriculture cannot be overstated. Many of the crops humans rely on for food require insect pollination to produce fruit and seeds. This includes a wide variety of fruits such as apples, oranges, and blueberries, vegetables like cucumbers, pumpkins, and tomatoes, as well as nuts like almonds and cashews. Even crops that do not directly produce food products, such as cotton and flax, benefit from bee pollination for seed production. The efficiency of bees as pollinators also makes them invaluable in enhancing the yield and quality of crops, directly influencing agricultural economics and the global food supply chain.

Moreover, bee pollination is not limited to agricultural crops; it also includes wild plants that form the basis of natural ecosystems. These plants provide habitat and food for wildlife, contribute to soil health, and help regulate the climate. The diversity of plants pollinated by bees is astonishing and highlights the interconnectedness of bees with both managed and natural ecosystems. By facilitating the reproduction of a broad range of plant species, bees play a vital role in maintaining the genetic diversity of plants. This diversity is crucial for plants to adapt to changes in the environment, resist pests and diseases, and continue to thrive.

The intricate relationship between bees and plants is a result of millions of years of co-evolution. Plants have developed colors, scents, and shapes that attract bees, while bees have evolved to become effective foragers and pollinators of these plants. This mutualistic relationship underscores the importance of bees in sustaining the health of ecosystems and the survival of many plant species.

Despite their critical role in pollination and the global food supply, bees face numerous threats from human activities. Habitat loss, pesticide use, climate change, and diseases are some of the challenges that bee populations encounter, leading to declines in their numbers. The repercussions of these declines are far-reaching, affecting not only the availability and cost of food but also the health of ecosystems worldwide. Addressing these threats requires a concerted effort to understand and mitigate the impacts of human activities on bee populations, ensuring their survival and the continued benefits they provide to the environment and humanity.

The economic implications of bee pollination extend beyond the direct value of the crops they help to produce. Pollination services provided by bees are an essential underpinning of the agricultural industry, contributing significantly to the economy. The estimated value of these services runs into billions of dollars annually, highlighting the economic importance of preserving bee populations for the sustainability of agriculture and food production.

To further appreciate the economic impact of bees, consider the almond industry, which is entirely dependent on bee pollination. Almonds, a crop with a multi-billion dollar value to the agricultural sector, require the pollination services of honeybees to produce nuts. The demand for bee pollination in almond cultivation showcases a direct link between bee health and agricultural profitability. As such, the decline in bee populations poses not just an ecological crisis but a significant economic threat to industries reliant on bee-pollinated crops.

In addressing the challenges faced by bees, innovative solutions and sustainable practices are being developed and implemented. For example, integrated pest management (IPM) strategies aim to reduce the reliance on chemical pesticides, which can harm bees, by promoting biological and mechanical control methods. Additionally, habitat restoration projects seek to replenish the natural environments essential for bee health, providing them with a diverse array of pollen and nectar sources. These projects often involve planting wildflower strips alongside agricultural fields to offer bees nutrition and refuge, thereby supporting their populations and the pollination services they provide.

Urban beekeeping has also emerged as a movement to bolster bee populations while educating city dwellers about the importance of pollinators. By installing hives in urban settings, beekeepers can mitigate some of the threats bees face in agricultural landscapes, such as monocultures and pesticide exposure. Urban environments can offer bees a varied diet from parks, gardens, and roadside plantings, contributing to their health and the pollination of urban plants. Furthermore, research into bee genetics and breeding programs aims to enhance bee resilience to diseases, pests, and environmental pressures. By selecting traits that favor disease resistance and stress tolerance, scientists and beekeepers work together to develop bee populations that can thrive despite the challenges.

The role of policy and legislation cannot be understated in protecting bees and their habitats. Regulations that limit the use of harmful pesticides, protect critical habitats, and support sustainable farming practices are crucial for bee conservation. Public policies that encourage organic farming and habitat conservation can also play a significant role in sustaining bee populations.

Community engagement and education are vital in fostering a culture of conservation and appreciation for bees. Through workshops, school programs, and public awareness campaigns, individuals can learn about the importance of bees and how to support them in their own gardens and communities. By making simple changes, such as planting bee-friendly flowers, avoiding pesticides, and supporting local beekeepers, everyone can contribute to the well-being of bees and, by extension, the health of our ecosystems and food systems.

The interconnectedness of bees with agricultural and natural ecosystems illustrates the profound impact these pollinators have on our world. Their role transcends the boundaries of agriculture, touching upon environmental health, biodiversity, and human well-being. As such, the conservation of bee populations and the promotion of bee-friendly practices are imperative for the sustainability of our planet's food resources and the preservation of ecological balance. Through collective action and a commitment to sustainable practices, we can ensure that bees continue to thrive, supporting the intricate web of life that depends on their pollination services.

Environmental Benefits of Bees

Bees contribute significantly to the **fostering of biodiversity** through their pollination activities, which in turn supports plant growth, healthy ecosystems, and the production of a wide array of agricultural crops. The environmental benefits of bees are multifaceted, encompassing not only the direct pollination of plants but also the indirect support of entire ecosystems.

Biodiversity is crucial for ecosystem resilience and productivity. Bees, by pollinating a diverse set of plants, help to maintain the genetic diversity necessary for plants to adapt to changing environmental conditions and threats. This genetic diversity within plant populations contributes to ecosystems' ability to withstand and recover from a variety of disturbances.

In terms of **plant growth**, bees play a critical role in the reproductive process of many plants by ensuring the transfer of pollen from the male anthers of a flower to the female stigma. Without this transfer, most plants cannot produce fruits and seeds, which are essential for the continuation of plant species. This pollination process is not only vital for the plants themselves but also for the animals and humans who rely on these plants for food, shelter, and other resources.

The **symbiotic relationship** between bees and flowers is a perfect example of nature's balance. Flowers have evolved to attract bees with their colors, shapes, and scents, offering nectar and pollen as food sources. In return, bees provide the essential service of pollination, which is critical for the survival of the plants. This relationship underscores the interconnectedness of life and the importance of each species in maintaining the health of ecosystems.

To support bees in their role as pollinators, it is recommended to plant a variety of **native plants** that bloom at different times of the year. This ensures a continuous food supply for bees, supporting their health and the health of the hive. When selecting plants, consider those that provide high nectar and pollen yields, such as lavender, borage, and clover for temperate climates, and tropical milkweed and passionflower for warmer regions. Planting in clusters rather than single plants can also help attract more bees, as they prefer to visit larger, more concentrated food sources.

Additionally, the creation of **habitat features** such as hedgerows, wildflower meadows, and even small garden patches can significantly enhance local bee populations and biodiversity. These habitats not only provide food but also nesting and overwintering sites for bees, further supporting their survival and proliferation.

The use of **natural pest control methods** instead of chemical pesticides is another critical step in protecting bees and other pollinators. Chemical pesticides can be harmful or even lethal to bees; thus, adopting integrated pest management (IPM) strategies can help minimize these risks. IPM focuses on the use of biological control agents, mechanical controls, and cultural practices to manage pest populations with minimal impact on bees and other beneficial insects.

Water sources are also essential for bees, especially during hot and dry periods. Providing **clean, fresh water** in shallow dishes or birdbaths with landing platforms, such as stones or floating wood pieces, can help bees stay hydrated and cool. Ensuring these water sources are consistently available and safely accessible can significantly support bee health and vitality.

Decline of Bees: Impact on Ecosystems and Agriculture

The alarming decline of bee populations across the globe has raised significant concern among scientists, environmentalists, and agriculturists alike. This decline, often referred to as Colony Collapse Disorder (CCD), has been observed in numerous countries and poses a direct threat to the vast array of agricultural crops that rely on bees for pollination. To understand the gravity of this situation, it's essential to delve into the specific factors contributing to the decrease in bee numbers.

One of the primary causes of bee population decline is habitat destruction. As urbanization expands and natural landscapes are converted into cities or agricultural land, the diverse flora that bees depend on for nectar and pollen is significantly reduced. This loss of habitat not only diminishes food sources for bees but also eradicates nesting sites, further stressing bee populations. To mitigate this, efforts such as planting bee-friendly gardens and restoring natural habitats can provide crucial resources for bees. Selecting plants that bloom at different times of the year ensures a consistent food supply, aiding in the survival and health of bee colonies.

Pesticide use is another critical factor impacting bee health. Neonicotinoids, a class of insecticides, have been particularly scrutinized for their harmful effects on bees. These chemicals can disorient bees, making it difficult for them to find their way back to the hive, and can weaken their immune systems, making them more susceptible to diseases and parasites. Adopting integrated pest management (IPM) strategies that emphasize biological control methods over chemical pesticides can significantly reduce the risks to bee populations. For instance, encouraging the presence of natural bee predators or using mechanical barriers to protect crops can offer effective pest control without harming bees.

Climate change also plays a significant role in the decline of bee populations. The shift in weather patterns can disrupt the synchronicity between bee emergence from hibernation and the blooming of plants they rely on for food. This mismatch can lead to food shortages, weakening bee colonies and reducing their chances of survival. Moreover, extreme weather events, such as droughts and floods, can destroy habitats and food sources, further exacerbating the challenges bees face. Implementing conservation practices that enhance ecosystem resilience, such as creating green corridors and supporting organic agriculture, can help buffer some of the impacts of climate change on bees and other pollinators.

Diseases and parasites, notably the Varroa mite, have devastated bee colonies worldwide. These mites attach to bees and feed on their bodily fluids, weakening the bees and making them more vulnerable to viruses. Managing Varroa mite infestations through regular hive inspections and treatments is crucial for maintaining healthy colonies. Techniques such as drone comb removal, where frames designed to attract drone larvae (which are preferred by Varroa mites) are periodically removed and destroyed, can help control mite populations without the use of chemicals.

The multifaceted nature of the threats to bee populations underscores the need for a comprehensive approach to conservation.

Efforts to combat the decline of bee populations are as diverse as the challenges they face. One such initiative involves the breeding of bees that exhibit resistance to diseases and parasites, including the Varroa mite. By selecting for and propagating these traits, beekeepers can cultivate colonies that are more resilient to the threats that have decimated bee populations in the past. This approach not only helps in safeguarding the bees but also reduces the reliance on chemical treatments, which can have unintended consequences on bee health and the environment.

Public awareness and education play a pivotal role in bee conservation. Many people are unaware of the simple steps they can take to support bee populations, such as reducing pesticide use, planting native flowers, and providing water sources for bees. Community-based initiatives, such as local beekeeping clubs and school programs, can spread knowledge and foster a culture of conservation. These programs not only educate individuals about the importance of bees but also empower them to take action in their backyards and communities.

Legislation and policy changes are also critical in protecting bees. Governments around the world are beginning to recognize the importance of bees to agriculture and ecosystems, leading to regulations that limit the use of harmful pesticides and promote sustainable farming practices. For example, the European Union has imposed restrictions on the use of neonicotinoids, and other regions are considering similar measures. These policy changes, while often controversial, are vital steps in addressing some of the human-made challenges facing bees.

Research plays a fundamental role in understanding and mitigating the decline of bee populations. Scientists are investigating the complex interactions between bees and their environment, including the effects of pesticides, climate change, and habitat loss. This research is crucial for developing new strategies to protect bees, from improving habitat diversity to breeding disease-resistant bee varieties. Ongoing studies also explore the potential of technology in bee conservation, such as tracking bee movements with GPS to study their behavior and identifying areas critical for their survival.

Collaboration among farmers, beekeepers, scientists, and policymakers is essential for the successful conservation of bees. Farmers and beekeepers are on the front lines of this effort, implementing sustainable practices and managing hives in ways that prioritize bee health. Scientists contribute valuable insights and innovations, while policymakers can create the framework necessary for large-scale conservation efforts. Together, these groups can address the multifaceted challenges facing bees, ensuring their survival and the continuation of the vital services they provide.

In conclusion, the alarming decline of bee populations is a complex issue that requires a multifaceted approach to address. From habitat restoration and sustainable agriculture to research and public policy, each strategy plays a crucial role in conserving bees. By understanding the importance of bees to our ecosystem and taking action to protect them, we can ensure a healthy and productive future for bees and humans alike.

Global Initiatives for Bee Protection

Across the globe, numerous initiatives and efforts are underway to safeguard bees and promote sustainable practices within beekeeping and broader agricultural sectors. These global initiatives are spearheaded by a variety of stakeholders, including international organizations, government bodies, non-profit groups, and local communities, each contributing to a comprehensive approach aimed at addressing the multifaceted challenges bees face today.

One significant effort is the implementation of policies and regulations designed to protect bees from the harmful effects of pesticides. For instance, the European Union has taken a proactive stance by restricting the use of neonicotinoids, a class of pesticides linked to bee population declines. These restrictions represent a critical step towards reducing chemical exposure and ensuring safer foraging environments for bees. In the United States, the Environmental Protection Agency (EPA) has also initiated reviews of pesticide regulations with an eye towards better protecting pollinator health, demonstrating a growing recognition of the importance of regulatory action in bee conservation.

Another pivotal area of global initiative focuses on habitat restoration and the creation of bee-friendly environments. Projects such as the planting of wildflower corridors alongside agricultural fields aim to provide bees with essential resources, including nectar and pollen. These corridors not only offer nutrition but also serve as safe passages for bees to navigate across increasingly fragmented landscapes. Furthermore, urban beekeeping initiatives are transforming rooftops, balconies, and abandoned lots into productive spaces for hives, thereby contributing to biodiversity and pollination within city environments. These efforts underscore the importance of habitat in supporting healthy bee populations and the role that every individual and community can play in creating spaces where bees can thrive.

Research and development form another cornerstone of global efforts to protect bees. Scientific studies are crucial for understanding the complex dynamics of bee health, including disease management, genetic diversity, and the impact of environmental stressors. Collaborative research projects, often involving universities, government agencies, and beekeeping associations, are exploring innovative solutions such as breeding disease-resistant bees and developing non-toxic treatments for pests and pathogens. This research not only advances our knowledge but also provides practical tools and strategies that beekeepers can implement to enhance hive health and resilience.

International collaboration is evident in initiatives like the International Pollinators Initiative (IPI), coordinated by the Food and Agriculture Organization (FAO) of the United Nations. The IPI fosters global action to protect pollinating species by promoting sustainable agricultural practices, enhancing scientific research, and facilitating the exchange of information among countries. Through such collaborative frameworks, countries can share best practices, research findings, and policy successes, thereby amplifying the impact of individual efforts on a global scale.

In addition to these structured initiatives, grassroots movements and community-led projects play a vital role in bee conservation. From local beekeeping clubs that offer education and support to gardeners planting native species to attract pollinators, these community efforts embody a ground-up approach to conservation. They not only contribute to bee health directly but also raise awareness and engage the public in the importance of bees to our ecosystems and food systems.

As these initiatives illustrate, the global effort to protect bees and promote sustainable practices is multifaceted, involving a wide range of actions from policy reform and habitat creation to research and community engagement. Each of these efforts contributes to a larger strategy aimed at ensuring the health and longevity of bee populations worldwide. Through continued collaboration, innovation, and advocacy, there is hope for reversing the decline of bees and securing the future of pollination, a critical process upon which our food supply and natural ecosystems depend.

Building on the foundation of international cooperation and policy development, the role of technology and data collection in bee conservation has emerged as a critical tool. Advanced monitoring systems, such as remote hive monitoring, enable beekeepers and researchers to track hive health in real-time, providing valuable data on temperature, humidity, bee activity, and even sound patterns within the hive. This technology not only aids in early detection of issues such as disease or colony distress but also contributes to a broader understanding of bee health trends over time. By leveraging big data and analytics, scientists can identify patterns and predict potential threats to bee populations, informing both local and global conservation strategies.

Educational programs and public outreach are equally integral to the global initiative to protect bees. Schools, universities, and community organizations are increasingly incorporating pollinator conservation into their curricula and programs, fostering an early appreciation for the importance of bees and other pollinators. These educational efforts extend beyond formal settings, with many non-profits and conservation groups offering workshops, webinars, and resources online to reach a wider audience. By educating the public on simple actions that can support bee health, such as planting pollinator-friendly gardens or reducing pesticide use, these programs empower individuals to contribute to bee conservation in their daily lives.

The adoption of sustainable agricultural practices plays a pivotal role in supporting bee populations. Practices such as crop rotation, cover cropping, and reduced tillage not only improve soil health and reduce erosion but also create a more diverse and stable environment for bees. Farmers and agricultural producers are increasingly recognizing the benefits of these practices, not only for bee health but also for crop yield and quality. By integrating pollinator-friendly practices into agricultural management, such as minimizing pesticide use and preserving natural habitats, the agricultural sector can significantly contribute to the preservation of bee populations.

Corporate responsibility and private sector engagement have also become prominent in the bee conservation landscape. Many companies, from agriculture to retail, are investing in sustainability initiatives that include support for bee health. This includes funding research, supporting habitat restoration projects, and adopting bee-friendly practices in their operations. Corporate sponsorships and partnerships with conservation organizations are helping to amplify the impact of conservation efforts, demonstrating a growing recognition of the interconnectedness of business practices and environmental health.

As these diverse initiatives demonstrate, the global movement to protect bees and promote sustainable practices is a dynamic and evolving effort. It encompasses a wide range of strategies, from technological advancements and policy reforms to education and community action.

Chapter 2: A Brief History of Beekeeping

The advent of modern beekeeping marks a significant turning point in the history of bee management, primarily due to the invention of the **Langstroth hive** by Lorenzo Lorraine Langstroth in the 1850s. Langstroth's discovery of the "bee space," a specific distance between the frames and the walls of the hive that bees naturally leave free of comb, revolutionized beekeeping. This space, approximately 3/8 inch (or about 9.5 mm), allows bees to move freely around the hive but is too small for them to build comb and too large to fill with propolis. It is this precise measurement that enabled the creation of movable frames within the hive which could be easily removed and inspected without destroying the comb, thereby significantly reducing the stress on the bee colony and increasing the efficiency of managing hives.

The **Langstroth hive** design includes a series of rectangular frames within a box where the bees build their comb. The standardized size of these frames and boxes has allowed for a modular system where beekeepers can add or remove sections according to the needs of the hive or to facilitate honey harvesting. This design also incorporates the concept of a bottom board with an entrance for the bees, a series of boxes (brood chamber and supers) for the bees to live and store honey, and a cover to protect the colony from the elements. The introduction of the Langstroth hive enabled beekeepers to manage their colonies with greater precision, to increase the number of hives they could maintain, and to improve honey yields through more efficient extraction methods.

Following the invention of the Langstroth hive, another significant development in beekeeping was the creation of **standardized frames**. These frames, which hang inside the hive body, allow bees to build honeycomb within a confined space, making it easier for beekeepers to inspect the hive, manage bee health, and harvest honey without causing undue harm to the colony. The frames can be equipped with foundation, a thin sheet of beeswax or plastic embossed with the hexagonal pattern of honeycomb cells, to guide the bees in building straight, uniform comb. This innovation has facilitated the practice of extracting honey by centrifugal force using a honey extractor, a device that spins the frames, forcing honey out of the comb while leaving the structure intact for reuse by the bees.

The widespread adoption of the Langstroth hive and standardized frames has had a profound impact on beekeeping, transitioning it from a small-scale, subsistence activity to a more commercial and scientifically managed endeavor. These innovations have also paved the way for further advances in beekeeping technology, including the development of queen excluders (grids that allow worker bees to pass through but not the larger queens, thereby keeping the queen from laying eggs in the honey supers), ventilated bee suits for beekeeper protection, and various tools for hive management such as smokers (used to calm bees during hive inspections) and hive tools (used to pry apart frames and scrape wax and propolis).

Moreover, the evolution of modern beekeeping has been accompanied by a growing scientific interest in bee health and behavior, leading to better disease management practices, selective breeding programs for traits such as gentleness and high honey production, and research into the effects of pesticides and environmental changes on bee populations. The combination of practical innovations and scientific research has enabled beekeepers to face challenges such as pest and disease pressures, climate change, and habitat loss, contributing to the sustainability and resilience of beekeeping as a vital agricultural practice.

The transition to modern beekeeping practices, characterized by the use of the Langstroth hive and standardized frames, represents a pivotal moment in the history of beekeeping. These innovations have not only enhanced the efficiency of honey production and hive management but have also laid the foundation for ongoing advancements in beekeeping technology and bee conservation efforts. As beekeepers continue to adapt and innovate, the principles established by early modern beekeeping remain at the core of ensuring the health and productivity of bee colonies around the world.

Ancient Beekeeping Practices

Transitioning from the revolutionary advancements of the Langstroth hive, we delve deeper into the roots of beekeeping to uncover the practices of ancient civilizations, notably the Egyptians and Greeks, whose contributions to apiculture have shaped many aspects of modern beekeeping. The Egyptians, renowned for their agricultural prowess and reverence for bees as symbols of royalty and divine providence, were among the first to develop structured beekeeping. They crafted hives from mud and straw, materials readily available in the Nile Delta, which provided excellent insulation against the harsh Egyptian climate. These cylindrical hives, often stacked horizontally, facilitated the management of bee colonies and the harvest of honey without destroying the hive, a method that echoes in the sustainable practices advocated today.

The harvesting process was meticulously documented in tomb paintings, illustrating how beekeepers smoked out the bees to calmly extract honeycombs. This ancient technique of using smoke to pacify bees is a testament to the Egyptians' sophisticated understanding of bee behavior, a practice that remains a cornerstone of beekeeping. Moreover, the transportation of hives along the Nile to follow the blooming patterns of flowers showcases an early recognition of the importance of bees in pollination, a precursor to modern migratory beekeeping.

In contrast, the Greeks, who inherited and expanded upon Egyptian beekeeping knowledge, emphasized the philosophical and natural aspects of beekeeping, integrating it into their studies of nature and society. Aristotle, in his Historia Animalium, provided one of the earliest detailed accounts of bee biology and behavior, laying the groundwork for entomology. The Greeks favored the use of hollow logs, baskets, and pottery jars as hives, demonstrating an adaptation to their environment and materials at hand. Their approach to beekeeping underscored the importance of locality, a principle that resonates with today's emphasis on local ecotypes and bee genetics for healthy colonies.

Honey played a central role in Greek culture, not just as a food source but also as an offering to the gods, highlighting the spiritual and cultural significance of bees and beekeeping. The practice of offering honey to deities, combined with the medicinal applications documented by Hippocrates and other Greek physicians, illustrates the multifaceted value of bees and their products in ancient societies, a value that continues to expand with today's growing interest in hive products beyond honey, such as propolis and royal jelly for health and wellness.

Medieval Beekeeping Innovations

As we delve further into the Middle Ages, the innovation of skeps marks a significant advancement in beekeeping practices. Skeps, essentially baskets made from straw or wicker, were designed to house bee colonies. Unlike the earlier methods that often destroyed the hive to harvest honey, skeps allowed for a more sustainable approach, albeit with limitations in terms of hive inspection and disease management. The construction of skeps was a meticulous process, requiring the weaving of straw into a conical shape, often with a small entrance for the bees at the bottom. This design mimicked natural cavities, providing a suitable environment for the bees while also facilitating the collection of swarms.

Beekeepers of the time employed various techniques to harvest honey from skeps without completely destroying the colony. One common method was "driving," where bees were encouraged to move from an old skep to a new one, leaving behind honeycombs that could be harvested. This required careful timing and understanding of bee behavior to minimize the stress on the bees and ensure the survival of the colony. Another method involved the use of a "super," a smaller skep or box placed atop the main skep during the honey flow. Bees would fill this super with honey, which could then be removed with less disruption to the main colony.

The management of bees in skeps, however, presented challenges, particularly in terms of disease control and the inability to inspect the interior of the hive. The lack of removable frames meant that beekeepers had limited insight into the health and productivity of the colony. This often led to the loss of colonies to diseases or pests that could have been managed with more direct intervention.

Despite these challenges, skeps played a crucial role in the evolution of beekeeping. They represented a move towards more sustainable harvesting methods, laying the groundwork for future innovations. The use of skeps also highlighted the importance of understanding bee behavior and the environment, as beekeepers began to develop practices that aligned with the natural lifecycle of the bees. This period also saw the emergence of beekeeping as a more organized practice, with the establishment of beekeeping guilds and the sharing of knowledge among beekeepers. These guilds were instrumental in refining beekeeping techniques, including the construction and management of skeps, and played a key role in preserving beekeeping knowledge through the Middle Ages.

The transition from skeps to more modern hive designs, such as the movable-frame hives, can be seen as a direct response to the limitations encountered with skeps. The desire for better hive management, disease control, and honey harvesting drove the innovation of beekeeping equipment and practices. However, the legacy of skeps remains evident in modern beekeeping, with the principles of sustainable harvesting and the importance of understanding bee behavior continuing to influence beekeeping practices today.

The use of skeps also underscores the beekeeper's ingenuity in adapting to the challenges of their time, using available materials and knowledge to create environments that supported the health and productivity of bee colonies. This adaptability and respect for the natural world are at the heart of beekeeping, principles that continue to guide beekeepers in the pursuit of sustainable and ethical bee management. As we reflect on the history of beekeeping, the innovation of skeps during the Middle Ages stands out as a testament to the enduring relationship between humans and bees, a relationship built on mutual benefit, respect, and a deep understanding of the natural world.

Modern Beekeeping: Langstroth Hive Innovation

The Langstroth hive, named after Lorenzo Lorraine Langstroth who invented it in the mid-19th century, revolutionized beekeeping by introducing the concept of "bee space." Langstroth discovered that bees build their comb with a consistent space of approximately 3/8 inch between them, which allows bees to move freely around the hive. If the space is too small, bees will fill it with propolis, making the hive difficult to manage. If it's too large, they will fill it with comb, which also complicates hive management. Langstroth's design leveraged this insight to create a hive with removable frames that maintained this critical bee space, allowing beekeepers to inspect, manage, and harvest from the hive with minimal disruption to the bees.

The Langstroth hive consists of a series of rectangular frames that hang vertically in a box. The frames can be easily removed, which allows for the inspection of the bee colony, treatment of diseases, and extraction of honey without destroying the hive structure. This modularity also facilitates the expansion or reduction of the hive according to the colony's needs by adding or removing frames and boxes. The bottom of the hive features a removable board which helps in controlling pests and cleaning the hive. Above the frames, a cover board and a roof protect the hive from the elements.

The standardized frames are a critical component of the Langstroth hive. They not only maintain the necessary bee space but also guide the bees to build their comb within the frames, making the combs uniform and easily interchangeable among hives. This standardization was a significant advancement over previous hive designs, such as skeps or log hives, where honeycombs were fixed and harvesting often involved destructive methods that harmed the bees and reduced their productivity.

The adoption of the Langstroth hive design has several implications for modern beekeeping. Firstly, it allows for the efficient management of larger bee colonies, making commercial beekeeping viable. The ease of inspection and maintenance helps in early detection and treatment of diseases and pests, contributing to healthier bee populations. Additionally, the ability to add or remove frames makes it easier to manage honey production based on the colony's strength and the season's nectar flow.

For beekeepers, understanding the assembly and maintenance of a Langstroth hive is essential. The hive body or brood box is where the queen lays eggs, and it's typically placed at the bottom. Above this, honey supers, which are boxes of frames used exclusively for honey storage, are added as the colony grows and honey production increases. The frames within these boxes require foundation sheets made of wax or plastic, embossed with a honeycomb pattern to encourage the bees to build comb. When assembling a hive, ensuring all components fit snugly while maintaining the appropriate bee space is crucial to prevent bees from sealing gaps with propolis or building comb in undesired spaces.

In managing a Langstroth hive, beekeepers must regularly inspect the hive for signs of disease, check on the queen's health and productivity, and monitor the colony's size, adding or removing supers as necessary. During inspections, frames should be carefully lifted to minimize disturbance and checked for brood health, honey stores, and any signs of pests or diseases.

The Langstroth hive's design also facilitates honey harvesting with minimal disruption to the bees. Supers can be removed, and frames extracted without opening the brood chamber, reducing stress on the colony. Honey extraction involves uncapping the sealed honeycomb on the frames and using a centrifugal extractor to remove the honey, leaving the wax comb intact for the bees to clean and refill.

The invention of the Langstroth hive and standardized frames marked the advent of modern beekeeping by addressing the natural behaviors and needs of honeybees, thereby improving bee health, hive management, and honey production. This hive design remains the standard in beekeeping today, testament to its revolutionary impact on the practice.

CHAPTER 3: THE FASCINATION WITH BEES

Bees have long been subjects of fascination and reverence across various cultures, symbolizing hard work, community, and productivity. This deep-rooted interest is not only due to their essential role in pollination and the ecosystem but also because of their unique societal structure and the sheer complexity of their behaviors. The symbolism of bees extends into ancient cultures where they were often associated with wisdom, immortality, and diligence. These insects have been depicted in numerous artworks, literature, and scientific studies, showcasing their importance throughout human history.

The cultural significance of bees is evident in the way they have been portrayed in art and literature. For instance, bees have been featured in ancient Egyptian hieroglyphs, symbolizing royalty and the soul's immortality. In Greek mythology, bees were considered to be messengers of the gods, linking them directly to divine wisdom. Such depictions underscore the profound impact bees have had on human culture and thought, reflecting their importance beyond agricultural and ecological contributions.

In literature, bees have often been used as metaphors for human society, highlighting their industrious nature and complex social structures. The comparison between a bee colony's efficiency and human society has been a recurring theme, emphasizing cooperation and collective benefit. This analogy extends to the concept of the 'hive mind,' representing a collective consciousness and shared knowledge, which has found its way into discussions about human social and organizational structures.

Science, too, has been captivated by bees, with researchers dedicating extensive studies to understand their communication methods, navigation skills, and decision-making processes. The discovery of the waggle dance, a sophisticated method bees use to communicate the location of food sources to their hive mates, is just one example of their complex behavior. Such findings have not only expanded our understanding of animal communication but also provided insights into evolutionary biology and the development of social behaviors.

The appeal of bees goes beyond their biological and ecological roles, inspiring curiosity and respect. Their ability to work collectively for the good of the hive, their intricate communication methods, and their vital role in the ecosystem resonate with human ideals of community, cooperation, and sustainability. This enduring fascination with bees has led to their protection and conservation, recognizing their critical role in our world's biodiversity and the global food supply.

Bees continue to inspire through their resilience, adaptability, and the sheer complexity of their societies. As we learn more about these remarkable creatures, we are reminded of the interconnectedness of all life forms and the importance of preserving the natural world. The fascination with bees thus serves as a bridge between cultures, disciplines, and epochs, uniting us in our shared admiration and respect for these indispensable inhabitants of our planet.

Bee Myths and Cultural Stories

Delving into the realm of myths and cultural stories, bees have been central figures in narratives that span across continents and epochs, embodying a myriad of meanings and symbolisms. One of the most captivating myths originates from ancient India, where bees were revered in the Vedas, sacred texts that date back over 3,000 years. These texts depict bees as symbols of wisdom and diligence, drawing parallels between the bees' tireless work ethic and the spiritual pursuit of knowledge. The Rigveda, one of the four canonical sacred texts of Hinduism, contains hymns that praise the medicinal properties of honey, highlighting its significance not just as a food source but as a spiritual and healing substance.

In Norse mythology, bees were believed to ferry the souls of the dead to the afterlife, acting as messengers between the mortal world and the spiritual realm. This belief underscores the sacredness attributed to bees, viewing them as integral to the cycle of life and death. The Norse god Odin was said to have sought the mead of poetry, a drink made from honey that bestowed wisdom and poetic inspiration upon those who consumed it. This mead was guarded by a giant, and Odin's quest for it symbolizes the pursuit of knowledge and the high value placed on wisdom and eloquence.

Transitioning to the Celtic traditions, bees were considered bearers of secret knowledge, and honey was often used in rituals and to sweeten medicinal concoctions. The Druids, the priestly class in ancient Celtic societies, saw bees as symbols of the sun, celebrating their ability to create life-sustaining substances like honey. The connection between bees and sunlight was also reflected in their role in pollination, essential for the growth of plants and the continuation of life.

In Africa, the Yoruba people of Nigeria have a god named Oshun who is associated with water, purity, fertility, love, and sensuality. Bees are sacred to Oshun and are considered her messengers. Honey is used in many rituals to appeal to Oshun, often in matters of love and fertility. This illustrates the deep connection between bees, the divine, and the natural world, highlighting their importance in religious and cultural practices.

The ancient Egyptians held bees in high regard as well, associating them with royalty. Pharaohs were often referred to as "Beekeepers," a title that emphasized their role in maintaining order and ensuring the prosperity of their land, much like a queen bee does for her hive. Honey played a crucial role in Egyptian society, used not only as a sweetener but also in embalming practices, demonstrating the versatility and sacredness of honey in ancient cultures.

In the Americas, Mayan mythology tells of a bee god named Ah Muzen Cab, who was associated with honey, bees, and beekeeping. The Maya considered honey to be a gift from the gods, using it in various ceremonies and as an offering to deities. The significance of bees in Mayan culture is evident in their art and religious practices, which often depicted bees and included honey in ritualistic offerings.

These myths and stories from around the world highlight the universal reverence for bees and their byproducts, underscoring the deep connection humans have had with these remarkable creatures throughout history. The symbolic meanings attributed to bees—wisdom, immortality, diligence, and connection to the divine—reflect their importance in human culture and the natural world. As we continue to explore the fascinating world of bees, these cultural stories and myths remind us of the profound impact bees have had on human civilization, inspiring us to appreciate and protect these vital pollinators.

Symbolism of Bees in Ancient Cultures

The symbolism of bees in ancient cultures extends deeply into the realms of **industriousness** and **fertility**, two key aspects that have been revered across civilizations for their critical roles in societal success and continuity. These symbols are not mere representations but are imbued with significant cultural values and beliefs that highlight the bee's importance in human history.

In the context of **industriousness**, bees have been emblematic of hard work and diligence. This is vividly illustrated in the practices and lore of ancient Egypt, where bees were seen as models of efficiency and productivity. The Egyptians, known for their monumental architecture and agricultural advances, paralleled their societal achievements with the bee's ability to build complex hives and produce honey. They observed bees' tireless work ethic, drawing a parallel to their own societal efforts in building a civilization that stood the test of time. The bee's role in pollination, essential for the success of crops, further cemented its status as a symbol of industriousness.

Turning to **fertility**, bees have played a significant role in symbolizing fertility and abundance. This is particularly evident in ancient Greek culture, where bees were associated with Demeter, the goddess of agriculture and fertility. The Greeks recognized the bee's role in fertilizing flowers through pollination, a process that was essential for crop production and, by extension, the sustenance of life itself. In rituals and myths, bees were often invoked as symbols of abundance and prosperity, reflecting their integral role in ensuring the fertility of the land and the well-being of the people.

Moreover, the bee's ability to work collectively towards a common goal is a powerful symbol of community and cooperation. In ancient cultures, where communal harmony and joint efforts were pivotal for survival, bees served as a living example of how individual contributions, when harmonized towards a collective goal, can lead to the prosperity of the whole community. This symbolism is especially poignant in the context of bee societies, where each bee plays a specific role, whether as a worker, drone, or queen, contributing to the hive's overall productivity and survival.

In applying these ancient symbols to modern beekeeping practices, one can draw inspiration from the bee's industriousness by adopting diligent hive management practices. This includes regular inspections, disease management, and ensuring the bees have access to diverse floral resources. Emulating the bees' communal efforts, beekeepers can work together through local beekeeping associations, sharing knowledge and resources to support each other and the health of bee populations at large.

Furthermore, fostering the fertility of the land through sustainable beekeeping practices mirrors the ancient reverence for bees as symbols of abundance. Planting bee-friendly gardens, avoiding pesticides, and promoting biodiversity are modern ways to honor the bee's role in fertility and abundance. By creating environments where bees can thrive, beekeepers contribute to the health of the ecosystem, ensuring the continuation of the natural processes that bees have symbolized for millennia.

In essence, the ancient symbols of industriousness and fertility attributed to bees offer timeless lessons for contemporary beekeepers. By understanding and embracing these symbols, beekeepers can cultivate practices that not only benefit their hives but also contribute to the broader ecological and cultural landscapes, continuing a legacy that has been cherished by civilizations throughout history.

Bees in Art, Literature, and Science

The cultural significance of bees extends into the realm of art, where these creatures have been depicted in various forms, from ancient cave paintings to contemporary digital art. Artists have long been fascinated by bees' intricate social structures and their symbolic meanings, often using bees to represent concepts such as industriousness, community, and environmental awareness. In painting, bees have been rendered with meticulous attention to detail, capturing the delicate interplay of light on their wings and bodies. Techniques vary widely among artists, with some opting for hyper-realistic portrayals that emphasize the beauty and complexity of bees, while others adopt a more abstract approach, using bees to explore patterns, shapes, and colors that evoke the natural world. Materials used in bee art can range from traditional oils and acrylics to more unconventional media like beeswax itself, which is sometimes incorporated into mixed-media pieces for its texture and natural aroma. The use of beeswax, or encaustic painting, involves heating the wax and applying it to a surface, often with added pigments for color. This technique not only pays homage to the material origins within the hive but also creates a tactile connection between the artwork and its subject matter.

In literature, bees have buzzed through the pages of poetry, novels, and fables, serving as metaphors for human behavior, societal structures, and the natural cycles of life and death. Authors have drawn upon the rich behavioral patterns of bees, from their communication methods to their roles within the hive, to reflect on human society and individual relationships. In poetry, bees are often celebrated for their role in pollination, depicted as tireless workers that sustain the cycle of life through their interaction with flowers. Poets use a variety of literary devices, such as metaphor, simile, and personification, to draw parallels between the natural world of bees and human emotions, experiences, and values. For instance, a poem might compare a bee's search for nectar to a human's search for meaning or happiness, using vivid imagery and sensory details to bring this analogy to life. The rhythm and flow of the language in such poems often mimic the bees' movements, creating a dynamic reading experience that engages the senses.

In the realm of science, bees have been the subject of countless studies, with researchers delving into their behavior, genetics, and role in ecosystems. The scientific investigation into bees involves a range of methodologies, from field observations and experiments to advanced genetic analysis. Scientists study bees to understand their communication methods, such as the waggle dance, which involves complex movements to convey information about the location of food sources. This research often requires sophisticated technology, including high-speed cameras to capture the rapid movements of bees and software to analyze the data. Genetic studies on bees involve sequencing their DNA to understand the genetic basis of traits such as disease resistance, behavior, and adaptation to environmental changes. These studies typically require laboratory equipment for DNA extraction, amplification, and sequencing, as well as bioinformatics tools to analyze the genetic data.

The intersection of bees in art, literature, and science underscores their profound impact on human culture and knowledge. Through artistic depictions, literary metaphors, and scientific inquiry, bees continue to inspire and inform, reflecting the deep and enduring fascination humans have with these remarkable creatures. The study and celebration of bees in these diverse fields not only enrich our understanding of the natural world but also highlight the importance of bees in sustaining life on Earth. As we continue to explore and appreciate the world of bees, we are reminded of the intricate connections between all living beings and the critical role that bees play in maintaining the balance of ecosystems.

The Enduring Appeal of Bees

The enduring appeal of bees extends far beyond their ecological significance and into the very fabric of human curiosity and respect. This fascination is not merely a product of their vital role in pollination or their ability to produce honey but is deeply rooted in their complex behaviors and sophisticated societal structures. The hive operates as a finely tuned organism, with each bee playing a specific role that contributes to the survival and productivity of the colony. The queen, with her sole responsibility for laying eggs, the workers who tirelessly gather nectar and pollen, and the drones whose purpose is to mate with a new queen, all function within a framework of intricate cooperation and communication. This level of organization, where thousands of individuals work together for the common good, mirrors the ideals of human society and has long been a source of inspiration and admiration.

The communication methods employed by bees, particularly the waggle dance, are a testament to their intelligence and ability to convey complex information. A worker bee returning from a fruitful foraging expedition performs a dance that varies in duration and direction, indicating the distance and location of a food source relative to the sun's position. This remarkable form of communication demonstrates the bees' sophisticated navigational skills and their capacity to make decisions based on collective intelligence, further fueling human fascination with these creatures.

Moreover, the ability of bees to adapt to different environments and climates showcases their resilience and versatility. From the lush gardens of a suburban landscape to the flowering fields of rural farmlands, bees thrive in a variety of settings, pollinating plants and contributing to the diversity of life on Earth. Their role in sustaining natural and agricultural ecosystems underscores the interconnectedness of all species and the importance of preserving biodiversity for the health of the planet.

The practice of beekeeping itself, a tradition that dates back thousands of years, reflects the deep connection between humans and bees. Beekeepers not only reap the rewards of honey, wax, and other hive products but also develop a profound respect for the complexity and beauty of the bee society. The meticulous care involved in managing a hive, from monitoring the health of the bees to ensuring they have access to ample forage, requires knowledge, patience, and a sense of stewardship. It is this hands-on interaction with bees and the natural world that often leads to a lifelong passion for beekeeping and a commitment to conservation efforts.

In the realm of education and advocacy, bees serve as powerful ambassadors for environmental awareness. Teaching children and adults about the importance of bees and other pollinators encourages a greater appreciation for the natural world and the challenges it faces, such as habitat loss, climate change, and pesticide use. By understanding the critical role bees play in our food system and ecosystem health, individuals are more likely to support sustainable practices and policies that protect these vital insects.

The enduring appeal of bees lies in their ability to inspire wonder, curiosity, and respect. Their complex societies, sophisticated communication methods, and essential role in the ecosystem offer endless opportunities for discovery and learning. As we continue to study and appreciate bees, we are reminded of the delicate balance of nature and our responsibility to protect it. Through beekeeping, conservation, and education, we can ensure that bees continue to thrive and enchant future generations with their remarkable abilities and contributions to the Earth's vitality.

Part 2: Bee Biology and Behavior

CHAPTER 4: ANATOMY OF BEES

Bees possess specialized organs that play critical roles in their survival and the functioning of the hive. The wings and legs of a bee are not just for mobility but are equipped with unique features that aid in pollination and the collection of nectar and pollen. The wings of a bee, attached to the thorax, are vital for flight and are composed of two pairs: the larger forewings and the smaller hindwings. During flight, the wings lock together through a series of hooks called hamuli, allowing the bee to move through the air with precision and agility. The rapid movement of the wings, often beating over 200 times per second, not only propels the bee forward but also creates the characteristic buzzing sound. This buzzing is crucial during pollination, as it helps dislodge pollen from the anthers of flowers, a process known as buzz pollination.

The legs of a bee are multifunctional, serving purposes beyond locomotion. Each of the six legs, also attached to the thorax, is equipped with different tools and apparatuses suited for the tasks the bee must perform. The front legs have specialized structures for cleaning their antennae, which are essential for navigation and communication. The middle legs assist in walking and grooming, while the hind legs are designed for collecting and carrying pollen. The pollen is stored in a structure known as the pollen basket or corbicula, located on the outer side of the hind legs. The bees' ability to carry pollen is facilitated by the presence of stiff hairs on their legs, which trap pollen grains as the bee moves from flower to flower.

The stinger, or sting, is another specialized organ, found at the end of the abdomen. It is primarily a defensive tool used to protect the hive from predators. The bee's stinger is barbed, which means it can become lodged in the skin of its target. In honeybees, the act of stinging results in the bee's death, as the stinger and associated glands are torn from the bee's body when it tries to fly away after stinging. This sacrificial act is a testament to the bee's role in protecting the hive and underscores the communal nature of bee society.

Adaptations in bees extend to their ability to efficiently collect nectar and pollen, crucial for the hive's survival and the production of honey. The proboscis, a long, tube-like tongue, allows bees to access nectar deep within flowers. The proboscis is made up of two main parts: the labium and the glossa. The labium acts as a sheath, protecting the glossa, which is extended to lap up nectar. Once nectar is collected, it is stored in the bee's honey stomach, a separate compartment from its true stomach, to be transported back to the hive. There, it will be processed into honey by worker bees.

The efficiency with which bees collect nectar and pollen is not just a product of their physical adaptations but also of their behavioral strategies. Foraging bees communicate the location of food sources through the waggle dance, a sophisticated method of communication that involves a series of movements conveying information about the direction and distance of flowers from the hive. This remarkable behavior demonstrates the complexity of bee society and the advanced cognitive abilities of these insects.

Bee Anatomy: Head, Thorax, and Abdomen Functions

The head of a bee is a marvel of nature's engineering, housing critical sensory and processing organs. At the forefront, the bee's eyes are divided into two types: compound eyes and ocelli. Compound eyes, large and multifaceted, provide a wide field of vision, enabling the bee to detect movement and navigate through complex environments. These eyes are composed of thousands of tiny lenses, each contributing to the bee's ability to perceive patterns and colors, particularly in the ultraviolet spectrum, which is invisible to humans. This ability is crucial for identifying flowers and determining the presence of nectar. In contrast, the ocelli, three simple eyes located on the top of the head, are sensitive to light intensity and play a vital role in orienting the bee relative to the sun, aiding in navigation.

The bee's antennae, another critical feature on the head, serve as both olfactory and tactile sensors. These flexible appendages are covered in a multitude of sensory receptors that detect chemicals, humidity, and even air currents. The antennae are essential for communication within the hive, allowing bees to perceive pheromones released by other members of the colony, which can signal distress, mark trails, or indicate the health and status of the queen.

Beneath the antennae lies the mouthparts, a complex assembly designed for both consuming and manipulating food. The mandibles, strong and versatile, are used for grasping, biting, and cutting various materials, such as wax for building the hive or capping honey cells. The proboscis, a long, tubular structure, functions as a straw, enabling the bee to drink nectar from flowers. This mechanism is highly efficient, allowing bees to extract and transport nectar back to the hive, where it will be converted into honey.

Transitioning to the thorax, this segment serves as the powerhouse of the bee, housing the muscles that drive the wings and legs. The thorax is divided into three segments, each with a pair of legs and two of the segments supporting a pair of wings. The coordination between the wings and legs is exquisite, allowing for precise movements during flight, foraging, and communication. The bee's flight muscles are among the most powerful in relation to size within the animal kingdom, capable of rapid, agile maneuvers and sustained flight over considerable distances. This strength is vital for escaping predators, exploring new food sources, and maintaining body temperature through thermoregulation.

The abdomen, elongated and segmented, contains the bee's vital organs for digestion, reproduction, and venom production. The honey stomach, a specialized organ separate from the digestive stomach, is used for transporting nectar. The reproductive organs are located in the queen and drones, with the queen's capable of laying up to 2,000 eggs per day during peak season. Worker bees, although female, do not have fully developed reproductive organs, focusing their efforts on the colony's maintenance and growth. The venom sac and stinger, located at the end of the abdomen, are defense mechanisms used against predators. The process of stinging is often fatal to the bee, highlighting the sacrificial nature of their role in colony protection.

Each component of a bee's anatomy is a testament to the complexity and efficiency of their design, allowing them to fulfill their roles within the ecosystem and the hive. From the intricate workings of the head, facilitating sensory perception and food intake, to the thorax, powering flight and movement, and the abdomen, ensuring the continuation of the species and colony defense, bees are finely tuned organisms. Their anatomy not only supports their survival but also enables their critical role in pollination and biodiversity, underscoring the importance of their preservation and the sustainability of practices that support their health and habitat.

Pheromones and Hive Communication

Pheromones play a pivotal role in the complex social structure and functioning of a bee hive, acting as chemical messengers that influence behavior, development, and organization within the colony. These chemical signals are secreted by bees and detected by the olfactory receptors on their antennae, facilitating a wide range of communication needs, from alerting the colony to threats to directing them to food sources.

The queen bee produces a variety of pheromones, collectively known as the queen mandibular pheromone (QMP). This blend of chemicals serves multiple purposes, including suppressing the development of ovaries in worker bees, thus preventing them from laying eggs; attracting drones for mating; and helping the workers to recognize their queen. The presence of QMP throughout the hive plays a crucial role in maintaining social harmony and cohesion among the colony members. For instance, when a queen bee begins to age or her pheromone production decreases, workers may detect this change and initiate the rearing of a new queen to ensure the colony's survival.

Worker bees also produce pheromones, which are used for a variety of tasks essential to colony life. Alarm pheromones are released when a bee stings, serving as a warning to other bees of a potential threat. This pheromone can trigger a defensive response, rallying the colony to defend the hive against predators. Additionally, worker bees use pheromones to mark the locations of water, food sources, and the entrance to their hive, guiding foragers and helping to coordinate collective activities.

Brood pheromones are emitted by the larvae and play a critical role in the regulation of the hive's social structure and the distribution of labor. These pheromones signal the nutritional needs of the developing bees, prompting worker bees to adjust their foraging patterns and care behaviors accordingly. For example, if the brood pheromones indicate a need for more protein, worker bees may increase their collection of pollen to meet this demand.

The process of foraging is also heavily influenced by pheromones. Forager bees release a pheromone that suppresses the foraging behavior of younger bees, effectively regulating the number of foragers and ensuring that not all workers leave the hive at once. This balance is crucial for the colony's survival, as it maintains a sufficient workforce within the hive to care for the queen and brood, process incoming nectar, and perform other necessary tasks.

The intricate use of pheromones for communication within the hive exemplifies the remarkable adaptability and efficiency of bee colonies. By understanding the role of pheromones in bee behavior and hive management, beekeepers can better appreciate the complexities of bee society and the importance of maintaining a healthy, balanced colony. This knowledge can also inform practices such as requeening or managing hive disturbances, as beekeepers can take steps to minimize stress on the colony and support the natural communication and organization processes facilitated by pheromones.

Specialized Bee Organs for Survival and Hive Duties

The **wings** of a bee, beyond their primary function of flight, play a pivotal role in temperature regulation within the hive. Bees can vibrate their wings to generate heat, warming the hive during colder months. Conversely, by fanning their wings at the hive entrance or over water collected by foragers, they can induce evaporative cooling, lowering the hive's temperature on hot days. This ability to thermoregulate is critical for the survival of the colony, as maintaining an optimal temperature ensures the development of brood and the preservation of stored food.

Legs are equipped with a variety of tools that serve multiple purposes. The front legs, besides their role in cleaning the antennae, are used to manipulate wax for building the hive structure. The middle legs, supporting the bee's weight and aiding in grooming, also play a part in the packing of pollen into the pollen baskets on the hind legs. The hind legs, with their pollen baskets, are marvels of engineering designed for efficient pollen collection and transport. Pollen, essential for feeding the brood, is gathered using the stiff hairs on the legs as the bee moves among flowers. Upon returning to the hive, the bee uses its middle legs to offload the pollen into cells, a process that requires precision and dexterity.

The **stinger**, while a defense mechanism, has a nuanced role beyond simply deterring predators. The venom delivered by the stinger contains a complex cocktail of chemicals, including peptides and enzymes that have antimicrobial properties. This helps protect the hive from diseases. The act of stinging and releasing alarm pheromones also serves as a communication tool, signaling to other bees to join in defense of the hive. The sacrificial nature of the stinger in honeybees, where the bee dies following its use, underscores the communal prioritization of the hive's safety over individual survival.

The **proboscis** is another specialized organ critical for survival. This elongated, straw-like tongue is divided into two main parts: the labium, acting as a protective sheath, and the glossa, which is extended to collect nectar. The glossa's surface is covered in tiny hairs that increase its surface area, enhancing its ability to soak up liquid. The nectar collected is then stored in the honey stomach, a separate compartment from the digestive stomach, for transport back to the hive. Here, it is regurgitated and processed by worker bees into honey, a vital food source for the colony.

The **antennae**, though not directly involved in the collection of nectar or pollen, are indispensable for the bee's interaction with its environment. These sensory organs are loaded with receptors that allow bees to detect pheromones, sense air currents, and navigate using olfactory cues. The antennae's role in communication within the hive, particularly in the execution of the waggle dance, is crucial for the sharing of information about food source locations.

In summary, the specialized organs of bees—wings, legs, stinger, proboscis, and antennae—each play integral roles in the survival of the individual and the hive. From collecting nectar and pollen to defending the hive and regulating its temperature, these organs are examples of nature's ingenuity, enabling bees to perform a wide range of tasks essential for their survival and the functioning of the colony. Their complex behaviors and the anatomical structures that support them are a testament to the intricate world of bees, highlighting the importance of each organ in fulfilling the bee's ecological roles.

Bee Adaptations for Efficient Nectar and Pollen Collection

Bees have evolved a range of physical and behavioral adaptations that enable them to collect nectar and pollen with remarkable efficiency. One of the most notable adaptations is the **bee's proboscis**, a long, straw-like tongue that allows bees to access nectar deep within flowers. The proboscis is made up of two main parts: the **glossa**, which acts like a sponge to soak up liquid nectar, and the **labium**, which serves as a protective sheath. The glossa's surface is covered in tiny hairs that increase its surface area, making it an efficient tool for nectar collection. When not in use, the proboscis is retracted into the head.

For pollen collection, bees are equipped with **corbiculae**, or pollen baskets, located on their hind legs. These are concave structures surrounded by stiff hairs that trap and hold pollen grains. As a bee visits a flower, it uses its legs to brush pollen from the stamens onto its body. The bee then uses its front and middle legs to transfer the pollen grains to the pollen baskets. The **pollen is moistened with nectar** to ensure it sticks together and to the bee's legs, preventing loss during flight.

The **hairy body** of a bee also plays a crucial role in pollen collection. The hairs are electrostatically charged, attracting pollen grains as the bee moves among flowers. This not only aids in the direct collection of pollen for feeding but also in cross-pollination, as some of the pollen grains will fall off onto the next flower the bee visits, facilitating plant reproduction.

Bees exhibit a behavior known as **flower constancy**, which means they prefer to visit flowers of the same species during a single foraging trip. This behavior maximizes the efficiency of nectar and pollen collection because bees become more adept at handling the flowers of a specific species, reducing the time spent on each flower. Additionally, flower constancy enhances the chances of successful cross-pollination for the plants visited.

The **waggle dance** is another critical adaptation related to efficient foraging. After finding a rich source of nectar or pollen, a forager bee will return to the hive and perform a dance that communicates the distance, direction, and quality of the food source to other bees. This allows the foragers to efficiently find and exploit new food sources, reducing the energy and time required for individual search efforts.

To optimize their foraging efficiency, bees have developed a **time-compensated sun compass**. This internal mechanism allows them to navigate using the position of the sun, taking into account the time of day to adjust their flight paths accordingly. This sophisticated navigation system ensures that bees can find their way to and from food sources over long distances with remarkable accuracy.

These adaptations, from the structural features like the proboscis and pollen baskets to behaviors such as flower constancy and the waggle dance, illustrate the complex interplay between bees and their environment. They enable bees to effectively gather the resources necessary for the survival of the colony while also playing a pivotal role in the pollination of plants, showcasing the intricate connections within ecosystems.

CHAPTER 5: BEE SOCIETY

The division of labor among worker bees is a fascinating aspect of bee society, reflecting a highly organized and efficient system that changes as bees age, known as **age polyethism**. Initially, young worker bees, typically between 1 to 2 weeks old, are tasked with **brood care**, which involves feeding and caring for the larvae. This role is critical for the survival and future of the colony, as the health and wellbeing of the developing bees directly impact the hive's productivity and longevity. The brood care includes regulating the temperature of the brood area to ensure optimal growth conditions, a task that requires precise control and constant attention.

As worker bees grow older, their roles within the hive shift to tasks away from the brood, such as **wax production and comb building**. This transition occurs around the 2 to 3-week mark. Bees at this stage secrete wax from their abdominal glands, then chew the wax until it becomes malleable enough to mold into the hexagonal shapes that make up the honeycomb. This structure is not only used for storing honey and pollen but also for housing the next generation of eggs and larvae. The design of the honeycomb is incredibly efficient, providing maximum storage space with minimal material usage.

Following their stint as builders, worker bees then move on to the role of **guard bees**. This usually happens when they are around 3 weeks old. Guard bees are responsible for the security of the hive entrance, vetting each bee that seeks entry to ensure they belong to the colony and keeping out predators and parasites. They use their antennae to smell for specific pheromones that identify family members. This role is crucial for protecting the hive's resources and its inhabitants from threats.

The final stage in a worker bee's life, typically starting from 3 weeks until the end of their life span, is spent as a **forager**, venturing out of the hive to collect nectar, pollen, water, and propolis. Foraging is a demanding task that requires significant energy, as bees may travel up to 5 miles from their hive in search of resources. Foragers use a combination of visual landmarks, the sun's position, and magnetic fields to navigate to and from these sources. Upon returning to the hive, foragers perform the waggle dance to communicate the location, distance, and quality of the food sources to their hive mates.

The intricate cooperation and communication within the hive are what keep it functioning as a cohesive unit. This system ensures that all necessary tasks are covered, from feeding the young and building the hive structure to defending the colony and gathering food. It is a testament to the complexity of bee society and the evolutionary success of these social insects. The division of labor not only maximizes the efficiency of the hive but also ensures the survival and prosperity of the colony, showcasing the remarkable adaptability and organization of bees.

Hive Social Hierarchy: Queen, Workers, and Drones

At the core of the hive's social structure lies the queen bee, whose primary role is egg-laying and the production of pheromones that regulate the colony's activities. The queen is the only fully fertile female in the hive, and her ability to produce eggs is astonishing, with the potential to lay up to 2,000 eggs a day during peak season. This prolific capacity ensures the continuous growth and regeneration of the colony. The queen's pheromones serve as chemical signals that guide behavior and cohesion within the hive, influencing everything from the suppression of ovary development in worker bees to the maintenance of the hive's social order.

The worker bees, all of whom are female, form the backbone of the hive, performing a multitude of tasks vital for the survival of their community. Initially, as young adults, their duties are confined within the hive and include feeding the larvae (brood), cleaning, and regulating the temperature of the brood area to ensure optimal development conditions. As worker bees age, their responsibilities transition to tasks outside the hive, such as foraging for nectar, pollen, water, and propolis. This shift in roles is a seamless process, dictated by the needs of the colony and the age of the bees, showcasing a remarkable example of division of labor based on age, known as age polyethism.

Drones, the male bees of the colony, have a singular purpose: mating with a virgin queen. Unlike worker bees and the queen, drones do not have stingers and do not participate in nectar and pollen gathering or other colony maintenance tasks. Their presence is crucial, however, for the genetic diversity and continuation of bee populations. After mating, drones fulfill their life's purpose and soon die, as they are not equipped to survive independently without the support of worker bees.

The social hierarchy within a hive is a finely tuned system that ensures the survival and efficiency of the colony. The queen's role as the sole egg-layer and the producer of critical pheromones, the worker bees' execution of tasks necessary for the hive's maintenance and growth, and the drones' contribution to genetic diversity, each play a pivotal role in the complex society of bees. This hierarchy and division of labor are essential for the hive's productivity, health, and continuity, reflecting an evolutionary masterpiece of social organization and cooperation.

Selecting and Raising a New Queen Bee

When a colony decides it's time for a new queen, either due to the old queen's failing health, her absence, or a decision to swarm and form a new colony, the process of selecting and raising a new queen bee begins. This critical task is undertaken with remarkable precision and care by the worker bees, ensuring the future of the colony.

The initial step involves the worker bees selecting several young larvae to potentially become the next queen. These larvae must be less than three days old to be suitable candidates. The worker bees then proceed to feed these selected larvae with a nutrient-rich substance known as **royal jelly**. This special diet, which is richer and more abundant than what is fed to larvae destined to become worker bees, triggers the development of the larvae into queen bees. The composition of royal jelly includes vital proteins, vitamins, and fatty acids, making it a superfood that enables the larvae to develop fully functional ovaries necessary for egg-laying.

The cells housing these chosen larvae are transformed into **queen cups**, which are larger and differently shaped than the cells of worker bees or drones. These cups allow the developing queen larvae the space needed to grow. Worker bees diligently attend to these queen cups, continuously feeding and caring for the developing queens.

After about seven days of feeding on royal jelly, the larvae spin themselves into cocoons within their cells and begin the pupation stage. During this time, the future queens undergo a complete metamorphosis, emerging approximately seven to ten days later as fully developed queen bees. However, a colony only needs one queen, and if multiple queens emerge simultaneously, they will fight to the death until only one remains. Alternatively, a newly emerged queen may leave with a swarm of workers to establish a new colony if the old queen is still present and healthy, a natural method of colony propagation known as **swarming**.

Once a queen emerges, she begins her mating flight within a few days to a week after emergence. During this critical flight, she mates with multiple drones in mid-air, storing the sperm in her **spermatheca**. This single mating flight provides her with enough sperm to fertilize eggs for her entire life, which can be up to five years, though most commercial beekeepers replace queens every one to two years to ensure optimum colony productivity.

Upon returning to the hive after her mating flight, the new queen starts her egg-laying duties, marking the beginning of her reign over the colony. Her pheromones signal to the worker bees that a viable queen is now in charge, stabilizing the hive's social structure and ensuring the continuity of the colony's lifecycle.

Throughout this process, the role of the worker bees is paramount. They not only select and nurture the future queen but also regulate the temperature and humidity of the hive to create an optimal environment for the queen's development. Their ability to perform these tasks efficiently showcases the remarkable adaptability and social organization of bees, ensuring the survival and prosperity of the colony.

Age-Based Roles of Worker Bees

Understanding the division of labor among worker bees based on their age and stage of development reveals a meticulously organized system that ensures the hive's efficiency and survival. This age-based role assignment, known as **age polyethism**, is a dynamic process that adapts to the colony's immediate needs and the individual bee's capabilities.

Initially, worker bees serve as **nurse bees**. These bees, typically between 3 to 10 days old, are responsible for feeding the brood with a mixture of pollen and honey, known as **bee bread**, and with royal jelly for the first few days of the larvae's life. Nurse bees also maintain the brood's temperature at a constant 95°F (35°C) to ensure proper development. The nurse bees' hypopharyngeal glands are highly active during this period, producing royal jelly to feed both the larvae and the queen. The meticulous care provided by nurse bees is crucial for the health and future strength of the colony.

As worker bees age, transitioning into their second or third week of life, they take on the role of **house bees**. House bees are tasked with maintaining the cleanliness and structural integrity of the hive. This includes repairing the honeycomb, processing incoming nectar by reducing its water content to make honey, and ventilating the hive to control temperature and humidity levels. Ventilation is achieved through fanning their wings, a critical task during hot weather or when drying out nectar to create honey. House bees also pack pollen into cells for storage, using it as a resource for protein.

The transition from indoor to outdoor work marks a significant shift in a worker bee's life. As they approach the age of 3 weeks, worker bees begin to take on **guarding duties**, monitoring the hive entrance to protect against intruders and parasites. Guard bees challenge incoming bees to ensure they belong to the colony and are not attempting to rob the hive of its resources. This role is vital for the colony's security, as it prevents the spread of disease and the loss of valuable food stores.

Finally, the most mature worker bees, typically older than 3 weeks, become **foragers**. Foraging is a highly demanding task that requires complex navigational skills and significant energy expenditure. Foragers collect nectar, pollen, water, and propolis from the surrounding environment. The distance traveled and the weight of the collected resources put considerable strain on their bodies, leading to a relatively short lifespan once they begin foraging. Foragers play a critical role in the colony's survival by providing the necessary resources for food production and hive maintenance.

The division of labor in a bee colony is a fluid system, with bees moving between roles as needed. Factors such as colony size, resource availability, and environmental conditions can influence the timing and duration of each role. For example, a sudden need for more foragers due to a rich bloom of flowers can accelerate the transition of younger bees into the foraging role. Conversely, a loss of foragers due to predation or adverse weather may require older foragers to continue their role beyond the typical lifespan.

This age-based division of labor ensures that the colony operates at maximum efficiency, with each bee contributing to the hive's overall health and productivity according to its age and physical condition. It is a testament to the complex social structure and adaptability of honeybee colonies, allowing them to thrive in a wide range of environments and conditions.

Hive Cooperation and Communication

The intricate cooperation and communication within a bee hive are exemplified through the complex dance language and pheromone signaling systems that bees employ to sustain the hive's functionality. This section delves into the nuanced mechanisms of bee communication, highlighting how these methods contribute to the effective management of hive resources and the maintenance of social order among the colony's members.

The waggle dance, a sophisticated form of communication performed by forager bees, serves as a primary means of conveying detailed information about the location of food sources relative to the hive. When a forager bee discovers a rich source of nectar or pollen, it returns to the hive and performs a dance on the comb. This dance consists of a series of movements in a figure-eight pattern, with the direction of the straight run indicating the direction of the food source relative to the sun's position. The duration of the waggle phase of the dance correlates with the distance to the food source. Other foragers observe this dance and then fly out to collect the food, demonstrating an incredible level of social cooperation and information sharing that is vital for the colony's survival.

Pheromones play a crucial role in the regulation of hive activities and the maintenance of social harmony. These chemical signals are produced by various members of the colony, including the queen, workers, and drones, and serve multiple functions. For instance, the queen bee emits pheromones that regulate the reproductive functions of the hive, suppress the development of ovaries in worker bees, and attract drones for mating. Worker bees produce alarm pheromones that alert the colony to threats, enabling a coordinated defense response. They also use pheromones to mark the locations of water sources and to guide other workers to food sources.

The division of labor is another aspect of hive cooperation that ensures the efficient functioning of the colony. Worker bees perform tasks that are age-appropriate, with younger workers tending to the queen and brood, middle-aged workers building and repairing the hive, and older workers foraging for food. This age-based task allocation optimizes the use of the colony's workforce, ensuring that all necessary jobs are covered and that the hive operates smoothly.

Temperature regulation within the hive showcases yet another layer of cooperative behavior. Worker bees engage in thermoregulation to maintain the hive's temperature within a narrow range suitable for brood development. In cold conditions, workers cluster together to generate heat, while in hot conditions, they collect water and fan their wings to cool the hive through evaporation.

The seamless integration of these communication and cooperation strategies enables the bee colony to adapt to changing environmental conditions, optimize resource acquisition, and ensure the survival and growth of the hive. Through the waggle dance and pheromone signaling, bees coordinate their efforts in foraging, defense, and brood care, demonstrating a high degree of social organization and collective intelligence that is remarkable in the animal kingdom. This complex system of cooperation and communication is essential for the hive's productivity and longevity, reflecting the evolutionary success of the honeybee as a social insect.

CHAPTER 6: THE LIFE CYCLE OF BEES

The developmental stages of a bee, from egg to adult, are critical for the hive's functionality and overall health. Each stage is characterized by specific growth processes and physiological changes that contribute to the bee's role within the colony. The life cycle of a bee can be divided into four main stages: egg, larva, pupa, and adult.

Egg: The queen lays one egg per cell, which is oval and tiny, almost the size of a pinhead. The eggs are attached to the cell bottoms with a sticky substance the queen secretes. After three days, the egg hatches into a larva.

Larva: The newly hatched larva is fed by nurse bees with a diet of royal jelly for the first few days, followed by a mixture of pollen and honey, known as bee bread. This stage is crucial as the larva undergoes rapid growth, molting its skin several times. The feeding and care provided by the nurse bees are paramount during this stage, as the larva increases in size up to 1,500 times of its original size. The duration of the larval stage varies depending on the future role of the bee: worker bees remain in the larval stage for about six days, drones for about six and a half days, and queens for only five and a half days.

Pupa: After the larval stage, the cell is capped by the worker bees, and the larva spins a cocoon around itself, entering the pupal stage. This is a transformative phase where the larva metamorphoses into an adult bee. Inside the cocoon, the larva's organs and body structure undergo significant changes. The development of the wings, legs, eyes, and internal organs occurs during this stage. The pupal stage lasts for approximately 12 days for worker bees, 14.5 days for drones, and only 7.5 days for queens.

Adult: The final stage of development is the emergence of the fully formed adult bee from the cocoon. The new adult, known as an "emergent," spends a few hours drying its wings and acclimating to the hive environment. Initially, the emergent's exoskeleton is soft and pale, but it hardens and darkens within a few hours. Adult bees immediately begin to take on their roles within the colony, starting with cleaning their own cell and progressing through the various tasks as they age, from nursing to foraging.

The entire process from egg to adult bee takes approximately 21 days for worker bees, 24 days for drones, and only 16 days for queens. Environmental factors such as temperature and the colony's nutritional status can influence these development times. Optimal conditions within the hive, including stable temperatures between 93°F to 96°F (34°C to 35.5°C) and sufficient food supplies, are essential for the healthy development of bees.

Understanding the life cycle of bees provides beekeepers with insights into managing their hives more effectively. For instance, recognizing the signs of each developmental stage helps beekeepers in identifying the health and productivity of the colony. Monitoring the brood pattern can also indicate the queen's performance and whether she needs to be replaced. Additionally, knowledge of the life cycle is crucial for managing tasks such as splitting hives, preventing swarming, and ensuring the hive's growth and sustainability.

Bee Development: From Egg to Adult

The transformation from egg to adult in the life of a bee is not just a process of physical growth but a series of meticulously orchestrated changes that ensure the hive's survival and efficiency. Each stage of development plays a specific role in the communal life of the hive, contributing to its complex social structure. After emerging as adults, bees are immediately integrated into the hive's workflow, taking on tasks that are critical for the maintenance and productivity of their community.

The first role undertaken by the newly emerged adults is cleaning their own cell, a task that is crucial for maintaining the hygiene and order within the hive. This initial task is an introduction to the communal responsibilities that each bee will undertake throughout its life. It prepares the adult bee for a series of roles that it will assume as it ages, each with increasing levels of complexity and responsibility.

Following the cell cleaning, these adults, still soft and adjusting to their new environment, transition into the role of nurse bees. As nurse bees, they are responsible for feeding and caring for the brood, ensuring the future generation of workers, drones, and queens are well nurtured. This role is pivotal as it directly influences the health and strength of the colony's future members. The nurse bees feed the larvae with bee bread and royal jelly, substances that are rich in nutrients and essential for the larvae's development.

As bees grow older, their roles within the hive evolve. They progress from nurse bees to taking on tasks associated with hive maintenance, such as processing nectar into honey, building and repairing the honeycomb, and ventilating the hive. These tasks are vital for the survival of the colony, as they ensure the storage of food, the structural integrity of the hive, and the regulation of temperature and humidity levels within the hive.

Eventually, bees mature into foragers, a role that requires them to leave the safety of the hive and venture into the environment to collect nectar, pollen, water, and propolis. This stage of a bee's life is marked by significant physical exertion and exposure to external threats, making it the most perilous phase of their lifecycle. The foragers' contribution is critical, as the resources they gather are essential for the sustenance of the hive. They also play a key role in pollination, a process that is crucial for the reproduction of flowering plants and the production of a significant portion of the food consumed by humans.

The division of labor in the hive, dictated by the age and development stage of the bees, ensures that the colony operates efficiently. This system allows the hive to adapt to changing internal and external conditions, optimizing resource allocation and maximizing the productivity and health of the colony. The progression from egg to adult and through the various roles within the hive exemplifies the highly organized social structure of bee colonies and underscores the importance of each developmental stage in maintaining the balance and functionality of the hive.

The life cycle of bees, with its distinct stages and associated roles, highlights the complexity of the social organization within a hive. It reflects the interdependence of bees and the critical importance of each individual's contributions to the colony's success. By understanding the developmental stages of bees and their significance to the hive, beekeepers can better manage their hives, ensuring the health and productivity of their colonies. This knowledge is essential for effective hive management, enabling beekeepers to make informed decisions regarding brood care, hive expansion, and the timing of honey harvesting.

Lifespan of Worker Bees, Drones, and the Queen

Worker bees, drones, and the queen each have distinct lifespans that reflect their roles within the hive. Understanding these variations is crucial for effective hive management and ensuring the longevity and productivity of the colony.

Worker Bees: The lifespan of a worker bee varies significantly depending on the time of year they are born. Worker bees born in the spring and summer months typically live for 6 to 8 weeks. Their lives are labor-intensive, spending their initial weeks performing tasks within the hive such as feeding the brood and attending to the queen. As they age, their duties shift to foraging for nectar and pollen, which exposes them to greater risks and physical wear. Conversely, worker bees born in the late fall, known as "winter bees," can live up to 4 to 6 months. These bees are tasked with maintaining the hive through the winter, requiring less physical exertion as foraging ceases, and their primary role becomes insulating the queen and consuming stored food to generate heat.

Drones: Drones, the male bees, have the primary role of mating with a virgin queen. Their lifespan is markedly shorter, living up to 8 weeks during the mating season. Once a drone mates, it dies shortly after due to the physical consequences of the mating process. Drones that do not mate are often expelled from the hive in the autumn to conserve resources for the winter, as they do not contribute to foraging or maintenance of the hive. Therefore, the actual lifespan of a drone can be significantly impacted by the hive's needs and environmental conditions.

The Queen: The queen bee has the longest lifespan within the hive, living anywhere from 2 to 5 years. Her longevity is partly due to her protected status within the hive, where worker bees attend to all her needs, including feeding and grooming. The queen's primary role is to lay eggs, and her productivity in this regard determines her lifespan. Queens that remain highly productive in laying eggs can live several years; however, a decline in productivity often leads to the colony raising a new queen to replace her. The queen's lifespan is also influenced by the quality of care she receives from worker bees and the overall health of the hive.

For beekeepers, monitoring the lifespan and health of each group within the hive is essential. Replacing an aging or less productive queen before her decline can prevent disruptions in the hive's productivity. Similarly, understanding the seasonal variations in worker bee lifespans can aid in managing the hive's population and preparing for winter. Recognizing the limited utility and lifespan of drones can help in resource management, especially in preparing for the winter months when resources are scarce.

Each group within the hive plays a critical role in its survival and productivity, and their lifespans are a key aspect of the hive's overall health and longevity. By managing these variations effectively, beekeepers can ensure the sustainability and success of their hives.

Factors Influencing the Bee Life Cycle

Environmental conditions play a pivotal role in the life cycle of bees, influencing their development, behavior, and overall hive health. Temperature, humidity, and access to food sources are among the primary environmental factors that beekeepers must manage to ensure the well-being of their colonies.

Temperature is critical for bee development and hive activity. Bees maintain the brood area at a constant temperature of approximately 95°F (35°C) for the eggs and larvae to develop properly. In colder climates, bees cluster together to generate heat and maintain this optimal temperature, consuming stored honey as a source of energy. Beekeepers in these regions may need to provide additional insulation for hives during winter months to help minimize heat loss. Conversely, in hot climates, bees collect water and use evaporative cooling techniques to lower the hive's temperature. Here, ensuring adequate ventilation and shade can help prevent overheating, which can lead to brood development issues and reduced honey production.

Humidity within the hive also affects bee health, particularly in the development of brood. High humidity levels can encourage the growth of mold and fungi, which can harm or kill developing larvae. Beekeepers can manage humidity levels by ensuring proper ventilation in the hive, allowing excess moisture to escape, and by placing hives in locations with good air circulation.

Access to food sources is another environmental factor that significantly influences the bee life cycle. The availability of nectar and pollen affects the colony's ability to grow and sustain itself. During periods of scarcity, such as early spring or late fall, beekeepers may need to provide supplemental feeding to support the colony until natural food sources become available again. This can include sugar syrup or pollen substitutes, which help prevent starvation and support brood rearing.

Hive health is intrinsically linked to environmental conditions and is crucial for the successful development of bees through their life cycle. Diseases and pests, such as Varroa mites and Nosema, can devastate colonies, particularly if bees are already stressed by poor environmental conditions. Beekeepers must regularly monitor their hives for signs of disease and pest infestations, employing integrated pest management strategies that may include mechanical, chemical, or biological controls to manage threats without harming the bees.

Pesticide exposure is another factor affecting bee health and longevity. Bees exposed to harmful chemicals through foraging can experience impaired navigation, reduced foraging efficiency, and even death. Beekeepers can mitigate these risks by placing hives away from areas likely to be exposed to pesticides and by advocating for the use of bee-friendly pesticides and application methods among local farmers and landscapers.

In conclusion, the life cycle of bees is deeply influenced by environmental conditions and hive health. Beekeepers play a vital role in managing these factors, employing strategies to maintain optimal temperature and humidity levels, ensure access to food, and protect against diseases, pests, and pesticides. By understanding and addressing these environmental influences, beekeepers can support the health and productivity of their colonies, ensuring the successful development of bees from egg to adult.

CHAPTER 7: BEE COMMUNICATION

The **waggle dance** is a fascinating aspect of bee communication, serving as a means for foragers to inform their hive mates about the location of food sources. When a forager bee discovers a rich source of nectar or pollen, it returns to the hive and performs a dance on the comb. This dance is not merely a display of joy but a sophisticated means of conveying precise information. The direction of the dance in relation to the sun indicates the direction of the food source relative to the hive. The duration of the waggle phase of the dance correlates with the distance to the target. Bees can communicate distances of up to several miles with astonishing accuracy. For example, a short waggle run might indicate a food source close by, while a longer run suggests a more distant location. The intensity of the dance can also signal the quality of the food source, with more vigorous dances attracting more attention and indicating richer sources.

Pheromones play a critical role in the chemical communication within the hive. These chemical signals are produced by various glands in the bee's body and can influence a wide range of behaviors and physiological responses. The queen bee, for instance, produces a unique set of pheromones known as the queen mandibular pheromone (QMP). QMP has multiple functions, including attracting drones for mating, suppressing the development of ovaries in worker bees, and regulating many social behaviors within the hive. Worker bees also produce pheromones, such as the alarm pheromone, which is released when a bee feels threatened, signaling other bees to prepare for defense. Another example is the brood pheromone, secreted by larvae, which signals the need for care and feeding by nurse bees.

Sounds and vibrations are also integral to bee communication, with bees employing a variety of sounds to convey messages. The most well-known is the buzzing sound produced by their wings, which can vary in frequency and intensity. However, bees also use more subtle forms of sound communication. For instance, the "piping" sound, a series of vibrational signals produced by a queen bee, plays a role in swarming and hive establishment behaviors. Worker bees can produce a "stop signal," a brief vibration emitted when a bee bumps into another, which can serve to inhibit the waggle dance of a forager if the food source it advertises is considered dangerous.

Understanding these complex communication methods is crucial for beekeepers, as it allows for better insight into the health and needs of the colony. For example, a sudden increase in alarm pheromone levels could indicate a threat to the hive, such as a predator attack or human disturbance, necessitating a check and possible intervention. Similarly, observing changes in dance behavior can provide clues about the availability of food sources in the area, guiding beekeepers in decisions about supplementary feeding or the placement of hives.

In managing a hive, beekeepers can also manipulate these communication pathways to influence bee behavior. For instance, introducing synthetic pheromones can help in requeening a hive or calming bees during hive inspections. Monitoring the vibrational patterns within the hive can offer insights into the colony's reproductive status and alert beekeepers to the early stages of swarming preparations, allowing for timely interventions to manage hive population and prevent loss of bees.

Overall, the intricate communication systems of bees underscore the complexity of their social structure and the sophistication with which they navigate their environment. For beekeepers, a deeper understanding of these systems not only fosters a greater appreciation for the intricacies of bee behavior but also enhances their ability to care for and manage their hives effectively.

The Waggle Dance and Bee Communication

The waggle dance is a complex form of communication used by forager bees to inform their hive mates about the location of food sources, such as flowers rich in nectar and pollen. This dance is performed on the vertical comb within the dark hive, where visual cues are not visible, and relies on vibrations and movements to convey information.

When a forager bee returns to the hive after finding a productive food source, it begins the waggle dance by moving forward in a straight line, vibrating its body, then circling back to the starting point to repeat the process. The direction of the straight-line run relative to gravity inside the dark hive indicates the direction of the food source relative to the sun outside. For instance, a vertical run up the comb signals that the food source is directly towards the sun from the hive, while a run at a 60-degree angle to the right of the upward direction indicates the food source is 60 degrees to the right of the sun.

The duration of the waggle phase of the dance correlates with the distance from the hive to the food source. A longer waggle run means the food source is further away. The speed of the dance also plays a role; a slower dance indicates a farther distance. For example, a one-second waggle run might signify a food source about three-quarters of a mile away, whereas a half-second run could indicate a closer source, approximately a quarter of a mile away.

The intensity and enthusiasm of the dance can inform hive mates about the quality of the food source. A more vigorous and repeated dance suggests a richer source of nectar or pollen, prompting more bees to take notice and follow the directions provided.

Forager bees also incorporate a series of return phases, or loops, alternating between left and right after each straight run. The number of these loops can indicate the richness of the food source, with more loops signifying a more valuable find.

Beekeepers observing the waggle dance can gain insights into the foraging patterns of their bees, including the types of flowers they are visiting and the distances they are traveling to find food. This information can be valuable for understanding the health and productivity of the hive, as well as for making decisions about hive placement and management. For example, if dances indicate that bees are traveling long distances, a beekeeper might consider moving the hive closer to better foraging grounds or planting more bee-friendly flowers nearby.

Additionally, understanding the waggle dance can help beekeepers in selecting for traits such as foraging efficiency in their bee breeding programs. Bees that are more effective communicators and foragers can be more productive, leading to healthier colonies and potentially higher yields of honey and other hive products.

The waggle dance exemplifies the remarkable ways in which bees have evolved to communicate complex information. By decoding this dance, beekeepers can not only appreciate the sophisticated social behaviors of bees but also apply this knowledge to enhance hive management and conservation efforts.

Pheromones in Hive Communication

Pheromones, the chemical messengers used by bees, are pivotal in maintaining the complex social structure and efficiency of the hive. These chemical signals are not just simple messages but carry detailed information that can trigger a wide array of responses from the colony's members. The queen mandibular pheromone (QMP), for instance, is a sophisticated blend of chemicals that manages to maintain the social order within the hive. It not only attracts drones for mating but also plays a crucial role in suppressing the development of ovaries in worker bees, ensuring that the queen remains the sole egg-layer in the colony. This pheromone blend also influences the workers to feed and take care of the queen, showcasing the depth of control exerted by chemical communication.

The alarm pheromone is another critical chemical signal, predominantly released by guard bees or bees that are threatened. This pheromone composition can change the behavior of the entire colony within moments. When released, the alarm pheromone causes an immediate increase in aggression levels among the hive's inhabitants, preparing them for collective defense. It's a potent tool that transforms a normally peaceful community into a formidable force against intruders. The specificity of the alarm pheromone's composition means that beekeepers must handle hives with care, avoiding actions that might trigger its release, such as rough handling or excessive smoke.

For hive coordination, the brood pheromone plays a significant role. Secreted by the larvae, this pheromone signals the need for care and feeding. It's fascinating how this chemical signal can dictate the behavior of nurse bees, directing them to where their attention is needed most. The brood pheromone is vital for the survival of the next generation of bees, ensuring that the larvae receive adequate nutrition and care for their development.

In the context of marking locations, bees utilize pheromones to communicate about resources or new home sites. Foragers use the Nasonov pheromone to mark flowers that have been visited for nectar and pollen, which helps in optimizing foraging efficiency by preventing the duplication of effort among the foragers. When swarming, scout bees use the same pheromone to signal the location of potential new nesting sites to their hive mates. This use of pheromones for navigation and resource management underscores the adaptability and efficiency of bees, enabling them to thrive in various environments.

Understanding the role of pheromones in bee communication offers invaluable insights for beekeepers. By recognizing the signs of pheromone release and understanding their implications, beekeepers can better manage their hives. For example, during hive inspections, the detection of alarm pheromone can serve as a warning to proceed with caution or to reassess the handling techniques to minimize stress on the colony. Similarly, awareness of the brood pheromone's importance can guide beekeepers in maintaining optimal brood chamber conditions, ensuring the colony's future.

Moreover, beekeepers can use synthetic pheromones to manipulate hive behaviors for various purposes, such as introducing a new queen or calming the bees during hive management activities. However, the use of synthetic pheromones requires precision and understanding of the natural pheromone signals to avoid unintended consequences.

In essence, the complex system of pheromone communication in bees is a testament to the sophistication of these creatures. For beekeepers, a deep understanding of pheromone signals and their functions within the hive can enhance hive management practices, contributing to healthier, more productive colonies. This knowledge not only helps in the practical aspects of beekeeping but also fosters a greater appreciation for the intricate social dynamics that underpin the survival of bee colonies.

Bee Communication: Sounds and Vibrations

Beyond the waggle dance and pheromonal messages, **sounds and vibrations** serve as another layer of sophisticated communication among bees. These auditory signals play a crucial role in the coordination and survival of the hive. Bees produce sounds through a variety of methods, including wing beats, body vibrations, and by scraping body parts against the hive structure. Each sound and vibration pattern carries specific information, from alerting the colony about threats to indicating the need for food or water.

One notable sound-based communication is the **"tooting and quacking"** noises made by queen bees. These sounds are particularly significant during the swarming season or when a new queen is emerging. The mated queen inside the hive produces a tooting sound, which is a series of pulses followed by a buzz. This sound can be heard by beekeepers during hive inspections. In response, virgin queens still in their cells reply with a quacking sound. The tooting and quacking exchange helps prevent the virgin queens from emerging simultaneously, which could lead to fatal conflicts. It also signals worker bees about the presence of multiple queens, prompting them to take necessary actions, such as preparing for swarming or supporting the emergence of a new queen.

Another critical sound signal is the **stop signal**. This vibration-based communication is produced when a forager bee vibrates its body and wings after returning to the hive. The stop signal is often directed at receiver bees and serves to inhibit the waggle dance of a forager bee. This can occur if the foraging bee encountered danger, such as a predator or a competitor, at the food source. The stop signal effectively warns other bees of the potential threat, preventing them from heading to the same location. This form of communication is vital for the safety and efficiency of foraging activities.

Vibrational signals also play a role in the **regulation of hive temperature**. Worker bees can generate heat by vibrating their flight muscles without moving their wings. This behavior is essential during colder months or when incubating brood. Conversely, to cool the hive, bees fan their wings to create airflow, reducing the hive's temperature. These temperature regulation behaviors are communicated through vibrations felt throughout the hive structure, prompting other bees to join in the effort as needed.

Understanding these auditory and vibrational communications requires beekeepers to be attuned to the sounds and movements within the hive. Monitoring for changes in sound patterns can alert beekeepers to shifts in the colony's health, the emergence of new queens, or the presence of stressors. For instance, an increase in quacking without subsequent tooting may indicate that a new queen has been unable to emerge, necessitating beekeeper intervention.

Furthermore, beekeepers can use this knowledge to **manage their hives more effectively**. For example, recognizing the stop signal can inform beekeepers of potential dangers to their foraging bees, prompting a review of the hive's location or the need for supplementary feeding. Similarly, understanding the temperature regulation signals can guide beekeepers in providing insulation or ventilation to support the hive's needs.

Incorporating an awareness of bee sound and vibration communication into hive management practices not only enhances the beekeeper's ability to care for their bees but also deepens the connection between the beekeeper and the hive. By paying close attention to these signals, beekeepers can better anticipate and respond to the needs of their colonies, ensuring their health and productivity.

Part 3: Types of Bees

Chapter 8: Honeybees

Honeybees, known scientifically as Apis mellifera, are among the most significant pollinators in our ecosystems, playing a pivotal role in the pollination of fruit trees, vegetables, and wildflowers. Their ability to produce honey, a natural sweetener, and other valuable hive products like beeswax, propolis, and royal jelly, further underscores their importance to human agriculture and economies. Understanding the key characteristics of honeybees is essential for effective beekeeping and conservation efforts. Honeybees are distinguished by their complex social structure, which is organized around a single queen, numerous workers, and, during certain seasons, drones. The queen bee is the only fertile female in the colony and is responsible for laying all the eggs. Worker bees, which are female but not fertile, perform various tasks necessary for colony survival, including foraging for nectar and pollen, caring for the queen and brood, and maintaining and defending the hive. Drones are male bees whose sole purpose is to mate with a virgin queen, after which they die.

Honeybees are also remarkable for their sophisticated methods of communication. The waggle dance, a unique behavior performed by returning foragers, informs other bees in the hive about the direction and distance of food sources. This dance is a critical component of their foraging efficiency and the success of the colony. Additionally, honeybees use a variety of pheromones to communicate within the hive. These chemical signals regulate everything from the development of the hive to emergency responses.

Beekeepers must pay close attention to the health and productivity of their honeybee colonies. Monitoring for signs of disease, such as American foulbrood or Varroa mite infestation, is crucial. Effective management practices, including regular hive inspections and the provision of additional food sources during scarce times, can help maintain a strong and healthy colony. Furthermore, understanding the seasonal behaviors of honeybees, such as swarming in spring to form new colonies, is vital for successful beekeeping. Swarming is a natural part of the honeybee lifecycle and indicates a healthy, thriving colony, but it also requires beekeepers to take steps to manage their hives to prevent loss of bees or potential issues with neighbors in urban and suburban settings.

In terms of habitat, honeybees thrive in a variety of environments, from rural farmlands to urban gardens. They require access to a diverse range of flowers for nectar and pollen throughout the growing season. Beekeepers can support their colonies by planting bee-friendly gardens or by placing their hives in locations that offer ample foraging opportunities. This not only benefits the bees but also enhances the pollination of plants, contributing to the health of local ecosystems.

The adaptability of honeybees to different climates and settings is another aspect that makes them fascinating. They can regulate the temperature inside the hive to ensure the survival of the brood, using their wings to fan air in or out depending on the need to cool or warm the hive. This thermoregulation is critical during extreme weather conditions and is a testament to the complex social behavior and cooperation among hive members.

For those interested in beekeeping, starting with honeybees offers a rewarding experience that not only yields hive products like honey and beeswax but also provides essential pollination services that benefit agricultural crops and wild plants. However, it's important to approach beekeeping with a commitment to learning and respecting the natural behaviors and needs of honeybees. This includes choosing the right type of hive, understanding the basics of bee biology and behavior, and adopting sustainable practices that support the health and productivity of the colony.

Given the intricate nature of honeybee societies, beekeepers are encouraged to engage in practices that promote the well-being of their colonies. One such practice is the selective breeding of bees for desirable traits such as gentleness, disease resistance, and high productivity. By carefully selecting for these traits, beekeepers can enhance the resilience of their colonies against pests and diseases, reduce the need for chemical treatments, and improve their overall yield of honey and other hive products.

Another critical aspect of beekeeping is the management of the hive's environment to prevent stress on the colony. Stressors for honeybees can include overcrowding, poor nutrition, exposure to pesticides, and extreme weather conditions. Beekeepers can mitigate these stressors by ensuring that hives have enough space to grow, providing supplemental feeding during times of nectar scarcity, using organic farming practices to minimize pesticide exposure, and situating hives in locations that offer protection from harsh weather.

The practice of rotational cropping and planting a diversity of flowering plants can also significantly benefit honeybee colonies. By offering a continuous and varied source of nectar and pollen, beekeepers can support the nutritional needs of their bees throughout the year. This practice not only aids in the health of the honeybee colony but also supports the broader ecosystem by promoting biodiversity.

In addition to these management practices, beekeepers play a crucial role in the conservation of honeybees by participating in community science projects and advocacy. By monitoring their hives for signs of disease and pest infestations, beekeepers can contribute valuable data that help scientists track the health of bee populations on a larger scale. Furthermore, engaging in advocacy efforts to promote bee-friendly policies and practices within local communities can help ensure the sustainability of beekeeping and the preservation of bee populations for future generations.

Finally, the education of new beekeepers and the public about the importance of honeybees in our ecosystems is vital. Through workshops, beekeeping clubs, and educational outreach, experienced beekeepers can share their knowledge and passion for beekeeping, inspiring others to take part in the stewardship of honeybees. By fostering a community of informed and responsible beekeepers, we can collectively contribute to the health and longevity of honeybee colonies, ensuring their continued role in pollinating our crops and wild plants.

In conclusion, successful beekeeping with honeybees requires a comprehensive understanding of their biology, behavior, and needs. By adopting sustainable and informed beekeeping practices, beekeepers can support the health and productivity of their colonies, contributing to the conservation of honeybees and the ecosystems they help sustain.

Key Characteristics of Honeybees

Honeybees, scientifically known as Apis mellifera, possess the remarkable ability to produce honey and wax, two of the most valued products in beekeeping. The process of honey production begins when forager bees collect nectar from flowers using their long, tube-shaped tongues and store it in their special stomach, known as the honey stomach. Upon returning to the hive, the nectar is transferred to the worker bees inside the hive through regurgitation. These workers then chew the nectar, breaking down complex sugars into simpler forms, and spread it into the honeycombs. The unique design of the honeycomb, coupled with the fanning of bees' wings, facilitates rapid evaporation of water from the nectar, thickening it into honey. Once the moisture content is reduced to about 18%, the honey is considered ripe. The bees then seal the honeycomb with a thin layer of beeswax, preserving the honey for future use.

Beeswax production, on the other hand, is an equally fascinating process that showcases the industrious nature of honeybees. Worker bees possess special glands on the underside of their abdomen that secrete beeswax in the form of tiny flakes. Consuming honey activates these glands, providing the energy required for wax production. The bees chew these wax flakes, mixing them with saliva to make the wax malleable. This softened wax is then used to construct the honeycomb structure within the hive. The hexagonal shape of each cell in the honeycomb is a marvel of natural engineering, providing maximum storage space with minimal material.

The honeycomb serves multiple purposes; it is the nursery for new bees, the pantry for honey and pollen, and the heart of the hive's social structure. The construction of the honeycomb is a communal effort, with multiple bees working in tandem to build and maintain this critical infrastructure. The precision and cooperation required for this task are testaments to the complex social behavior and communication skills of honeybees.

The ability to produce honey and wax is not just a remarkable aspect of honeybee biology but also a critical component of their survival strategy. Honey serves as a vital food source for the colony, especially during winter or periods of scarcity when foraging is not possible. Beeswax, apart from constructing the honeycomb, plays a crucial role in regulating the internal temperature of the hive and protecting the colony from diseases and parasites.

For beekeepers, understanding the intricacies of honey and wax production is crucial. Effective hive management techniques, such as regular harvesting to stimulate productivity and prevent overcrowding in the hive, can enhance honey and wax yield. Moreover, adopting practices that support the health and well-being of the honeybee colony, such as providing a diverse range of foraging options and minimizing the use of chemicals, can improve the quality of these hive products.

The production of honey and wax by honeybees is a testament to their importance not only in natural ecosystems as pollinators but also in human economies. These products have been used by humans for centuries, not just as food and material for candle making but also in cosmetics, pharmaceuticals, and other industries. The sustainable harvesting and utilization of honey and wax offer a way for beekeepers to support their livelihoods while contributing to the conservation of honeybee populations. By fostering a deeper understanding of these processes and the key characteristics of honeybees, beekeepers can ensure the health of their colonies, the quality of the products, and the sustainability of beekeeping as a practice.

European vs. Africanized Honeybee Differences

European and Africanized honeybees, while both belonging to the species Apis mellifera, exhibit significant differences in their behavior and physical characteristics, which are crucial for beekeepers to understand for effective hive management.

Behavioral Differences:

1. **Aggression Level**: Africanized honeybees are notably more aggressive than their European counterparts. This aggression manifests in their defensive behavior; they are more likely to sting in defense of their hive, often pursuing perceived threats over longer distances and in larger numbers. This trait requires beekeepers to adopt additional safety measures, such as wearing full protective gear even for routine hive inspections and ensuring hives are located far from public areas.

2. **Swarming Frequency**: Africanized bees have a higher swarming frequency, which is their natural method of colony reproduction. They are more likely to abandon their hive in search of a

new home, especially if they perceive the current environment as threatening or if the colony becomes too crowded. Beekeepers managing Africanized bees need to monitor their hives closely for signs of swarming intent and may need to provide additional space or perform splits more frequently to manage this behavior.

3. **Foraging Behavior**: Africanized honeybees are more vigorous foragers than European bees. They start foraging at younger ages and will forage further and under less favorable weather conditions. This trait can lead to higher productivity in terms of pollination and potentially honey production but also requires beekeepers to ensure a consistent and ample forage supply to support the colony's needs.

Physical Differences:

1. **Size**: Africanized honeybees are generally smaller than European honeybees. This size difference affects the amount of honey they consume as food; smaller bees require less food, making Africanized bees more adept at surviving in environments where food sources are scarce or seasonal.

2. **Wing Shape**: Though subtle, differences in wing shape between Africanized and European honeybees can be detected by experts using morphometric analysis. This characteristic is primarily used for research and identification purposes rather than practical beekeeping.

3. **Wax Production**: Due to their smaller size and different foraging habits, Africanized bees may produce wax differently from European bees. The implications of this difference can affect how beekeepers manage hive resources and extract hive products.

Management Implications:

Beekeepers working with Africanized bees need to adopt specific strategies to safely and effectively manage their hives. This includes enhanced safety protocols to protect against stings, frequent hive inspections to monitor for swarming behavior, and ensuring ample foraging opportunities to support the colony's vigorous foraging behavior. Additionally, understanding the physical differences can help in tailoring hive management practices, such as adjusting the harvesting schedule to accommodate the bees' smaller size and potentially different wax production patterns.

Given these differences, beekeepers should carefully consider their environment, resources, and personal safety when deciding whether to keep European or Africanized honeybees. Those in regions where Africanized bees are prevalent may require additional training and preparation to manage these bees successfully and safely.

Popular Honeybee Breeds in Beekeeping

Selecting the right breed of honeybee for your beekeeping endeavors is a decision that can significantly influence the success and productivity of your hive. Among the most popular breeds, **Carnica** (Apis mellifera carnica), **Buckfast** (a hybrid bee), and **Ligustica** (Apis mellifera ligustica) stand out for their unique traits and adaptability to various beekeeping challenges.

Carnica bees are renowned for their gentle nature, making them an excellent choice for beginner beekeepers. They are less prone to swarming compared to other breeds, which simplifies hive management. Carnicas are also hardy bees, capable of withstanding colder climates due to their ability to form tight winter clusters and efficiently use their stored honey. When selecting a location for Carnica hives, ensure it offers ample foraging opportunities within a short flight distance, as Carnicas are efficient foragers who prefer not to travel far from their hive.

Buckfast bees are a testament to selective breeding, developed by Brother Adam at Buckfast Abbey in England. They are known for their exceptional resistance to diseases and mites, a trait that reduces the beekeeper's need for chemical treatments. Buckfast bees exhibit a balanced temperament, being both docile and industrious. They are versatile foragers that can adapt to a variety of climatic conditions, making them suitable for diverse environments. When managing Buckfast hives, pay close attention to their space requirements; these bees are prolific honey producers and may require additional supers during peak foraging seasons to prevent swarming.

Ligustica bees, or Italian bees, are favored for their prolific brood-rearing capabilities and extended foraging seasons, attributed to their ability to adapt to warmer climates. Their gentle disposition and reluctance to sting make hive inspections more manageable. However, their propensity for brood rearing necessitates diligent management to ensure adequate food stores for the colony. Providing Ligustica hives with a mix of floral sources can enhance their foraging efficiency and honey production.

For beekeepers, understanding the characteristics of these popular breeds—Carnica, Buckfast, and Ligustica—is crucial for selecting the breed that best fits their climate, management style, and beekeeping goals. Each breed offers distinct advantages, from disease resistance and temperament to foraging behavior and climate adaptability, guiding beekeepers towards a rewarding beekeeping experience.

Adaptability of Honeybees to Climates and Environments

The adaptability of honeybees to various climates and environments is a testament to their resilience and a crucial factor for beekeepers to consider when managing their hives. Honeybees, Apis mellifera, have thrived in diverse geographical locations, from the temperate regions of Europe to the arid landscapes of Africa, showcasing their remarkable ability to adjust to different environmental conditions. This adaptability is largely attributed to their sophisticated social structure, efficient foraging behavior, and the physiological mechanisms they have developed to cope with changes in temperature and humidity.

To ensure the health and productivity of a bee colony, beekeepers must understand how honeybees adapt to their surroundings and how to support these adaptations. For instance, in colder climates, honeybees form tight clusters to conserve heat and maintain the temperature of the hive's core where the queen resides. This behavior underscores the importance of providing adequate insulation for hives in winter to reduce heat loss. Materials such as polystyrene or wrapped tar paper can be used to insulate the hive, while ensuring proper ventilation to prevent moisture accumulation that can lead to mold growth and threaten the colony's health.

In contrast, in warmer climates, honeybees employ a different set of strategies to cool the hive. They collect water and distribute it throughout the hive, then fan their wings to evaporate the water and reduce the temperature. This behavior highlights the necessity of ensuring a reliable water source near the hive in hot environments. A shallow water dish with stones or floating debris allows bees to land and drink without drowning, providing them with the resources they need to regulate hive temperature effectively.

Honeybees also adapt to their environment through their foraging behavior. They are capable of visiting a wide variety of flowers, which allows them to thrive in areas with diverse plant life. However, the availability of forage throughout the year can vary significantly depending on the location. Beekeepers can support their colonies by planting a selection of native flowers, shrubs, and trees that bloom at different times of the year, ensuring a continuous supply of nectar and pollen.

This not only aids in the survival and productivity of the hive but also enhances the local ecosystem's biodiversity.

Furthermore, the adaptability of honeybees extends to their ability to defend against pests and diseases, which can vary by region. For example, the Varroa mite, a significant threat to honeybee colonies worldwide, requires beekeepers to monitor their hives regularly and adopt integrated pest management strategies. These might include mechanical methods like drone comb removal, chemical treatments with organic acids such as oxalic or formic acid, and breeding for mite-resistant bee strains. Each method's effectiveness can depend on the local climate and the specific challenges faced by the bee colony.

Understanding the adaptability of honeybees to different climates and environments is critical for successful beekeeping. It involves a comprehensive approach that includes selecting the appropriate hive type that suits the local weather conditions, providing resources for temperature regulation and foraging, and implementing pest and disease management practices tailored to the environment. By supporting the natural adaptations of honeybees, beekeepers can enhance the resilience of their colonies, promote sustainable beekeeping practices, and contribute to the preservation of these vital pollinators for future generations.

Chapter 9: Solitary Bees

Solitary bees, unlike their honeybee counterparts, do not live in colonies but rather lead independent lives. Each female solitary bee lays her eggs in individual nests, provisioning them with pollen and nectar for the larvae to consume upon hatching. This nesting behavior significantly impacts the selection of nesting materials and locations, which can range from hollow stems or wood tunnels to burrows in the ground. For beekeepers and enthusiasts looking to support solitary bees, understanding and facilitating their unique nesting requirements is key.

Ground-nesting bees, which make up a large percentage of solitary bee species, prefer well-drained, bare soil exposed to full sun. Avoiding mulch or dense vegetation in certain areas of the garden can encourage ground-nesting bees to settle there. For those interested in promoting these beneficial pollinators, maintaining patches of undisturbed, uncultivated land can provide ideal nesting sites. Additionally, providing a south-facing slope can enhance the site's attractiveness by ensuring more warmth and sunlight.

Wood-nesting bees, such as mason bees, seek out pre-existing cavities to lay their eggs. Beekeepers can support these species by setting up bee hotels or nesting blocks made from untreated wood with holes of varying diameters drilled into them. The holes should be smooth to prevent injury to the bees and deep enough to accommodate multiple larvae, typically between 5/16-inch and 3/8-inch in diameter and 3 to 6 inches deep. Placing these nesting aids in a sheltered location, protected from rain and facing the morning sun, can increase occupancy rates.

For both ground and wood-nesting solitary bees, protection from predators and pesticides is crucial. Using natural pest control methods in the garden and avoiding chemical treatments near nesting sites can help ensure the safety and health of these pollinators. Additionally, providing a diverse array of native flowering plants that bloom at different times of the year can offer a continuous food source for adult bees.

Water sources are also essential for solitary bees, though their requirements differ from those of honeybees. A shallow water dish with pebbles or twigs for bees to land on can prevent drowning while allowing them to hydrate. Ensuring these water sources are clean and replenished regularly will support the health and longevity of the bee population.

Finally, engaging in practices that promote the conservation of solitary bees can have a broader impact on the ecosystem. By fostering a diverse and healthy bee population, gardeners and beekeepers contribute to the pollination of plants, many of which are vital for producing fruits, vegetables, and seeds. This not only benefits the solitary bees themselves but also supports local wildlife and promotes biodiversity. Encouraging the presence of solitary bees through targeted actions can thus play a significant role in sustaining healthy ecosystems and enhancing agricultural productivity.

Solitary Bees vs. Honeybees: Lifestyle and Behavior

Solitary bees exhibit a lifestyle and behavior markedly different from that of honeybees, primarily due to their non-social nature. Unlike honeybees, which thrive in large colonies with a complex social structure, solitary bees work alone and do not produce honey or beeswax. Each female solitary bee is responsible for her own nesting and provisioning, laying her eggs in individual cells filled with a mix of nectar and pollen. This solitary behavior influences their interaction with the environment and their role in pollination.

One of the most striking differences is in their nesting habits. Solitary bees often nest in the ground or in natural cavities, such as hollow stems, wood, or even snail shells. For those interested in supporting solitary bees, creating a bee-friendly habitat involves providing nesting materials and sites. This can include leaving patches of bare, undisturbed soil for ground-nesting species or installing bee hotels with tubes of different diameters to cater to various species. These bee hotels should be placed in a location that receives morning sunlight and is protected from the elements, ideally with a mud source nearby for certain species like mason bees that use mud to construct their nests.

Pollination is another area where solitary bees stand out. Due to their foraging behavior, solitary bees are highly efficient pollinators. They exhibit flower constancy, meaning they tend to visit the same type of flower repeatedly during a single foraging trip, which significantly enhances the pollination process. This behavior contrasts with honeybees, which may visit various flower types, leading to less efficient pollination for certain crops. Gardeners and farmers can support solitary bees by planting a diversity of native flowering plants that bloom at different times throughout the growing season, providing a continuous food source for these pollinators.

Solitary bees also have unique adaptations that make them resilient to different environmental conditions. For example, some species have developed mechanisms to control the sex of their offspring based on the amount of food available, ensuring their population can adapt to changing conditions. Additionally, solitary bees often have specific relationships with the plant species they pollinate, making them critical for the reproduction of those plants and the overall health of ecosystems.

Protecting solitary bees from pesticides is crucial, as they can be more susceptible to chemical exposure than honeybees. Using natural pest control methods and avoiding the use of pesticides when bees are active can help safeguard these important pollinators. Furthermore, educating the community about the importance of solitary bees and how to protect them can lead to broader conservation efforts, enhancing biodiversity and ecosystem health.

In summary, supporting solitary bees involves understanding their unique needs and behaviors. By providing suitable nesting sites, ensuring a diversity of flowering plants, protecting them from pesticides, and raising public awareness, individuals can contribute to the conservation of solitary bees and the vital pollination services they provide.

Solitary Bees and Their Ecological Roles

Bumblebees, a familiar sight in gardens and wildlands alike, are among the most effective pollinators due to their size, flight capabilities, and behavior. Unlike honeybees, bumblebees can perform buzz pollination, a technique where the bee grabs onto a flower and vibrates its muscles without flapping its wings, causing pollen to dislodge and collect on its body. This method is particularly effective for pollinating plants with tubular flowers or those that require a significant vibration to release pollen, such as tomatoes, blueberries, and cranberries. To support bumblebees, planting native flowering plants that bloom at different times throughout the growing season is crucial. This ensures a consistent food source for bumblebees, aiding their survival and the pollination of various plant species. Furthermore, creating undisturbed areas in gardens or landscapes, such as piles of leaves or unmowed sections of grass, provides essential nesting sites for bumblebees. These habitats can help increase their populations and enhance their pollination activities in the surrounding area.

Mason bees, named for their use of mud or clay to build partitions within their nesting cavities, are solitary bees known for their incredible pollination efficiency. One mason bee can pollinate what would require about 100 honeybees to accomplish, making them invaluable in the pollination of spring-blooming fruits and vegetables, such as apples, cherries, and almonds. To attract and support mason bees, one can set up mason bee houses, which consist of tubes or drilled holes of specific diameters, usually between 5/16-inch and 3/8-inch, in untreated wood or bamboo. These bee houses should be placed facing east or southeast to catch the morning sun, at least 3 feet off the ground, and secured to prevent swaying in the wind, which can deter mason bees from nesting. Additionally, providing a source of mud or clay near the bee house can significantly increase the likelihood of mason bees taking up residence, as they need it to construct their nests. By fostering a welcoming environment for mason bees, gardeners and beekeepers can significantly boost local pollination rates, benefiting a wide range of plant species.

Both bumblebees and mason bees, like all pollinators, face threats from habitat loss, pesticide use, and climate change. Minimizing pesticide application, especially during the blooming period when bees are most active, can mitigate some of these risks. Opting for organic or natural pest control methods not only protects bee populations but also supports the broader ecosystem's health. Additionally, conserving natural habitats and incorporating a variety of plant species into gardens and agricultural landscapes can help maintain and even increase the diversity and abundance of solitary bees. These efforts, combined with community engagement and education about the importance of solitary bees, can lead to more sustainable practices that benefit both the bees and the environment they help to sustain.

By understanding the unique characteristics and ecological roles of bumblebees and mason bees, individuals can take targeted actions to support these vital pollinators. Through the creation of bee-friendly habitats, the reduction of pesticide use, and the planting of native flowering plants, we can contribute to the health and longevity of solitary bee populations. These efforts not only aid in the conservation of solitary bees but also enhance the pollination of crops and wild plants, underscoring the interconnectedness of all species within an ecosystem.

Solitary Bees: Key Pollinators of Crops and Wild Plants

Solitary bees, encompassing a vast array of species such as leafcutter bees, sweat bees, and carpenter bees, play a pivotal role in the pollination of crops and wild plants, a process that is crucial for the production of food and the preservation of biodiversity. Each solitary bee species has evolved unique foraging behaviors and plant preferences, making them specialized pollinators for certain types of crops and wildflowers. For instance, the blue orchard bee, a type of mason bee, is particularly effective at pollinating fruit trees, including almonds, apples, and cherries. Their ability to carry pollen more efficiently compared to honeybees is due to their hairy bodies, which trap pollen and facilitate its transfer between flowers.

The significance of solitary bees in agriculture cannot be overstated. They contribute to the increase in yield and quality of many crops, enhancing food security and agricultural sustainability. To optimize the pollination services provided by solitary bees, it is essential to understand their life cycle and habitat needs. For example, providing nesting resources such as hollow stems, untreated blocks of wood with drilled holes, or even simple piles of brush and leaves can attract solitary bees to gardens and farms. These nesting sites are critical for the next generation of bees, as females typically lay their eggs in these secluded spaces, provisioning them with pollen and nectar for the larvae to consume upon hatching.

Moreover, the conservation of natural habitats around agricultural lands supports the diversity and abundance of solitary bees. Maintaining hedgerows, wildflower meadows, and uncultivated areas within or adjacent to farmlands offers foraging and nesting sites, promoting the health and resilience of solitary bee populations. It is also beneficial to implement farming practices that minimize the use of pesticides or opt for targeted applications that reduce the risk to pollinators. Selecting plants for agricultural and landscaping purposes that bloom at different times throughout the growing season can ensure continuous food sources for a variety of solitary bee species, supporting their survival and pollination activities.

In addition to their agricultural value, solitary bees are vital for the pollination of wild plants, contributing to the maintenance of ecological balance and biodiversity. Many wild plant species rely exclusively on specific bee species for pollination, and without these pollinators, the reproduction and survival of these plants would be jeopardized, potentially leading to declines in plant diversity and the animals that depend on those plants for food and habitat. Therefore, the presence of solitary bees in natural ecosystems is a key factor in sustaining the health and functionality of these environments.

By understanding the ecological roles and requirements of solitary bees, individuals and communities can take action to support these indispensable pollinators. Whether through creating pollinator-friendly gardens, adopting bee-friendly farming practices, or participating in conservation efforts, each action contributes to the well-being of solitary bees and the essential ecosystem services they provide. Through such efforts, the vital importance of solitary bees in pollinating crops and wild plants is recognized and preserved, ensuring the sustainability of our natural and agricultural landscapes for future generations.

CHAPTER 10: OTHER BEE SPECIES

Stingless bees, often overlooked in discussions about pollinators, play a crucial role in tropical ecosystems. Unlike their more familiar counterparts, honeybees, stingless bees are adapted to a wide range of environmental conditions found in tropical and subtropical regions. These bees are smaller in size and, as their name suggests, lack the ability to sting, relying instead on biting to defend their hives. Their hives are often constructed in hollow trees or underground cavities, utilizing propolis and wax to create intricate structures that can house thousands of individuals. The management of stingless bee colonies for pollination and honey production requires a unique set of practices due to their different social structures and nesting behaviors.

To support stingless bees, it is essential to preserve their natural habitats, which are increasingly threatened by deforestation and urbanization. Planting a variety of native flowering plants that provide nectar and pollen year-round can help sustain these bees. Additionally, understanding the specific nesting preferences of stingless bees is crucial for their conservation. For those interested in keeping stingless bees, it is advisable to research the particular species native to their region, as the requirements for successful hive management can vary significantly.

Wild bees, including solitary and social species that are not typically kept for commercial pollination or honey production, are vital for the pollination of wild plants and many crops. These bees often have specific foraging preferences and play key roles in the pollination of particular plant species, contributing to the genetic diversity and resilience of ecosystems. To support wild bees, creating habitats free from pesticides and providing a diversity of plants that flower at different times throughout the year are effective strategies. Additionally, leaving natural areas undisturbed can help preserve the nesting sites of ground-nesting bees and provide refuge for all types of wild bees.

The conservation of bee diversity is critical in the face of challenges such as habitat loss, climate change, and the spread of diseases and pests. Initiatives aimed at protecting bees should focus on preserving natural habitats, promoting sustainable agricultural practices that benefit pollinators, and raising awareness about the importance of all bee species. Engaging in citizen science projects that monitor bee populations can also contribute valuable data to conservation efforts, helping scientists and policymakers to develop strategies that support bee health and biodiversity.

In the context of beekeeping, understanding the ecological contributions of different bee species can inform more sustainable practices. For example, integrating flower strips or hedgerows into agricultural landscapes can provide forage for both managed honeybees and wild pollinators, enhancing crop yields and ecological health. Beekeepers can also play a role in conservation by choosing to use organic methods to control pests and diseases, reducing the reliance on chemical treatments that can harm bees and other beneficial insects.

The preservation of bee diversity worldwide is not only a matter of ecological health but also of cultural and economic importance. Many communities depend on the pollination services provided by a variety of bee species to sustain their crops and livelihoods. By valuing and protecting all bee species, from the familiar honeybee to the lesser-known solitary and stingless bees, we can ensure the resilience of our food systems and the health of our planet.

Stingless Bees: Characteristics and Habitats

Stingless bees, a key group within the meliponini tribe, exhibit fascinating characteristics that set them apart from their more well-known relatives, the honeybees. Unlike honeybees, stingless bees have a highly social structure within their colonies, which can range in size from a few hundred to over 80,000 individuals. These bees are smaller, typically measuring between 3 to 5 mm in length, and possess a sting that has become atrophied, rendering them unable to sting as a defense mechanism. Instead, they may bite or use their resinous propolis to entrap or deter invaders.

The **nests** of stingless bees are marvels of natural architecture, often located in hollow trees, rock crevices, or even underground. The entrance to their nests is typically guarded and uniquely designed, sometimes featuring elaborate tunnels or propolis structures that serve to protect the colony from predators and environmental hazards. Inside, the bees construct vertical brood combs made of cerumen, a mixture of beeswax and plant resin, where the queen lays her eggs. Surrounding these brood combs, pots filled with honey and pollen are stored, which are spherical in shape and made from the same cerumen material.

Habitats of stingless bees are predominantly found in tropical and subtropical regions of the world, including Central and South America, Africa, Southeast Asia, and Australia. These bees thrive in diverse environments ranging from rainforests to urban areas, provided there is an abundance of flowering plants from which they can collect nectar and pollen. Stingless bees are essential pollinators in their ecosystems, often specializing in the pollination of native flora that other bee species may overlook.

To support stingless bees, it is crucial to **preserve their natural habitats** by protecting forests and other natural landscapes from deforestation and degradation. In areas where natural habitats are diminishing, creating **bee-friendly environments** by planting a variety of native flowering plants can provide alternative sources of nectar and pollen for these bees. Additionally, avoiding or minimizing the use of pesticides and chemicals in gardening and farming practices helps protect stingless bees and other pollinators from harmful exposure.

For those interested in **beekeeping** with stingless bees, it's important to research and understand the specific needs and behaviors of the species native to their region. Beekeeping practices for stingless bees differ significantly from those of honeybees, particularly in terms of hive design, colony management, and honey harvesting. Hive designs must accommodate the unique nesting preferences of stingless bees, including the need for smaller entrance holes and internal structures that mimic their natural nesting environments.

In conclusion, stingless bees play a vital role in pollination and biodiversity, contributing to the health of ecosystems and the production of crops. By understanding their characteristics, habitats, and needs, individuals and communities can take steps to protect and support these invaluable pollinators.

Wild Bees: Biodiversity and Plant Reproduction

Wild bees, encompassing a vast array of species beyond the commonly known honeybees and bumblebees, play an indispensable role in the ecological processes that sustain biodiversity and plant reproduction. Their activities contribute significantly to the pollination of a wide range of plant species, including many that are crucial for human food production and natural ecosystems. Understanding the ecological contributions of wild bees involves examining their interactions with plants and the specific mechanisms through which they aid in pollination and plant diversity.

Pollination is a fundamental ecological service provided by wild bees. They transfer pollen from the male structures of flowers (anthers) to the female structures (stigmas) of the same or different flowers, facilitating the fertilization process. This service is not only vital for the production of fruits, seeds, and vegetables but also for the genetic diversity within plant populations. Wild bees exhibit a variety of foraging behaviors and body sizes, enabling them to pollinate a diverse array of plants. For instance, some species are specialized pollinators that have co-evolved with specific plants, which means they are uniquely adapted to pollinate certain flowers, thereby ensuring the survival of both the bee species and the plant species.

Diversity in bee species is crucial for effective pollination because different bees are active at different times of the year and day, and they have preferences for different flowers. This diversity ensures a more continuous and comprehensive pollination service across a wide variety of plants and conditions. For example, while some bees are generalists, visiting a wide range of flowers, others are specialists, relying on a limited number of plant species. This specialization can lead to a mutual dependence where both the plant and the bee species benefit from this relationship, enhancing the resilience of their ecosystems.

Habitat creation and maintenance are other vital ecological contributions of wild bees. Through their nesting and foraging activities, wild bees help in the aeration of the soil and the dispersal of seeds, contributing to healthy and diverse habitats. Many wild bees are ground-nesters, digging tunnels in well-drained soil, while others nest in cavities, such as hollow stems or dead wood, creating microhabitats for a variety of other organisms.

To support the ecological contributions of wild bees, several conservation strategies can be implemented:

- **Preserving natural habitats** and creating new ones is essential to provide nesting sites and diverse floral resources throughout the year.

- **Planting native plants** in gardens, parks, and agricultural landscapes offers crucial forage for many wild bee species. Choosing a variety of plants that bloom at different times of the year ensures a continuous food supply.

- **Reducing or eliminating pesticide use**, especially those harmful to bees, can protect wild bee populations from decline. If pest control is necessary, opting for targeted, less harmful methods can mitigate negative impacts on bees.

- **Educating communities** about the importance of wild bees and how to protect them can foster greater appreciation and conservation efforts.

By understanding and supporting the ecological roles of wild bees, individuals and communities can contribute to the health and sustainability of ecosystems, ensuring the continued provision of the vital services that these pollinators offer.

Threats to Lesser-Known Bee Species

Habitat loss poses a significant threat to lesser-known bee species, fundamentally altering the landscapes these pollinators rely on for nesting and foraging. As urban expansion and agricultural development continue to encroach on natural areas, the diverse ecosystems that support a wide range of bee species are increasingly fragmented and diminished. This reduction in habitat not only decreases the availability of nesting sites but also limits the variety and abundance of flowering plants that bees depend on for nectar and pollen throughout the year. To mitigate these effects, it is crucial to prioritize the conservation and restoration of natural habitats. Strategies include setting aside protected areas that are managed for biodiversity, implementing sustainable land use practices that incorporate ecological principles, and rehabilitating degraded lands with native plant species that provide critical resources for bees.

Pesticides, particularly neonicotinoids, have been identified as a major hazard to bee health, contributing to declines in bee populations worldwide. These chemicals, widely used in agriculture and landscaping, can be lethal to bees or impair their ability to forage, navigate, and reproduce. The impact of pesticides is not limited to direct toxicity. Sublethal exposure can also weaken bees' immune systems, making them more susceptible to diseases and parasites. To protect lesser-known bee species from pesticide exposure, it is essential to advocate for and adopt integrated pest management (IPM) practices. IPM emphasizes the use of non-chemical pest control methods and the selective application of safer pesticide alternatives when necessary. Encouraging the adoption of organic farming practices can further reduce the reliance on synthetic pesticides, benefiting both pollinators and the environment.

Creating pollinator-friendly landscapes by planting native, flowering plants in urban, suburban, and agricultural areas can provide essential forage. Maintaining undisturbed areas with natural vegetation helps preserve the nesting sites that many solitary and ground-nesting bees need. Encouraging community involvement and education about the importance of all bee species can foster greater appreciation and conservation actions. Citizen science projects offer opportunities for individuals to contribute to monitoring bee populations and health, providing valuable data that can inform conservation strategies.

Furthermore, policymakers and stakeholders must collaborate to develop and implement regulations that protect bee habitats and regulate pesticide use. Financial incentives for landowners and farmers who adopt bee-friendly practices can encourage wider participation in conservation efforts. By integrating these efforts, it is possible to create a more sustainable and resilient environment that supports the full spectrum of bee diversity, ensuring their vital role in pollination and ecosystem health is preserved for future generations.

Bee Conservation Initiatives Worldwide

Conservation initiatives aimed at preserving bee diversity worldwide have become increasingly important as the threats to bee populations, including habitat loss, pesticide use, and climate change, continue to escalate. These initiatives are multifaceted, involving a range of strategies designed to protect and enhance the natural environments that bees depend on for survival. One of the key approaches is the establishment of protected areas specifically managed for pollinator health. These areas serve as safe havens for bees and other pollinators, providing them with the necessary resources for nesting and foraging without the risk of pesticide exposure or habitat destruction.

Another significant conservation strategy is the promotion of sustainable agriculture practices that support bee health. This includes the adoption of integrated pest management (IPM) techniques, which prioritize biological and mechanical control methods over chemical interventions. Farmers are encouraged to plant cover crops and maintain hedgerows and wildflower margins around agricultural fields. These practices not only offer bees a rich source of nectar and pollen but also help to control pests naturally and enhance soil health, creating a more resilient agricultural ecosystem.

The restoration of native plant species in urban and rural landscapes is also crucial for bee conservation. Native plants are better adapted to local soil, climate conditions, and native bee species, providing optimal nutrition and support for a wide range of pollinators. Community-driven initiatives, such as the creation of pollinator gardens in public parks, schoolyards, and private gardens, contribute significantly to the preservation of bee diversity. These gardens are designed to include a variety of native flowering plants that bloom at different times throughout the year, ensuring a continuous food supply for bees.

Education and outreach programs play a vital role in bee conservation efforts. By raising public awareness about the importance of bees to ecosystems and human agriculture, these programs encourage individuals and communities to take action in support of bee health. Workshops, seminars, and school programs offer practical advice on how to create bee-friendly environments and advocate for policies that protect pollinators.

Research and monitoring are foundational to the success of bee conservation initiatives. Scientists and researchers study bee populations, behavior, and health to identify the most effective conservation strategies and monitor the impact of existing measures. Citizen science projects, where members of the public participate in data collection and monitoring, have become an invaluable resource for researchers. These projects not only contribute to our understanding of bee populations and their challenges but also engage and educate the public, fostering a community of advocates for bee conservation.

Collaboration among stakeholders, including governments, non-profit organizations, farmers, beekeepers, and the public, is essential to the success of bee conservation efforts. Policies and regulations that support habitat conservation, sustainable agriculture, and the reduction of pesticide use are critical for creating a more sustainable future for bees. Financial incentives for landowners and farmers who implement bee-friendly practices can accelerate the adoption of conservation measures.

Through a combination of protected areas, sustainable agriculture, habitat restoration, education, research, and collaborative policies, conservation initiatives worldwide aim to preserve bee diversity and ensure the health of ecosystems for future generations.

CHAPTER 11: INTERACTIONS AMONG SPECIES

Supporting local pollinators goes beyond just planting a variety of flowering plants; it involves creating a comprehensive environment that caters to their diverse needs. For instance, providing water sources such as birdbaths or shallow dishes with stones for bees to land on ensures they have access to hydration, especially during hot weather. Additionally, leaving areas of your garden undisturbed with leaf litter, dead wood, and bare soil can offer essential nesting sites for solitary bees and other beneficial insects. When selecting plants, aim for native species that bloom at different times throughout the year to provide a consistent food source. Incorporating plants like lavender, sunflowers, and bee balm can attract a wide range of pollinators, while herbs such as rosemary and thyme are not only beneficial for bees but can also enhance your garden's aroma and culinary offerings.

Partnering with local farmers and gardeners to create larger pollinator-friendly areas can significantly impact bee populations. Collaboration can lead to the development of green corridors that connect isolated patches of habitat, allowing bees and other pollinators to move freely across landscapes, increasing their chances for survival and reproduction. These corridors are particularly crucial in urban and suburban areas where natural habitats are scarce. By working together, communities can transform vacant lots, roadside verges, and even rooftops into thriving spaces for pollinators.

Education plays a pivotal role in supporting local pollinators. Hosting workshops or creating informational materials on the importance of bees in the ecosystem can raise awareness and inspire action within your community. Schools can incorporate pollinator conservation into their curriculum, encouraging children to become stewards of their environment from a young age. Community gardens can serve as living classrooms, providing hands-on learning experiences about gardening and beekeeping.

Monitoring and studying bee populations in your area can offer insights into their health and the challenges they face. Participating in citizen science projects such as bee counts or habitat assessments can contribute valuable data to researchers and help guide conservation efforts. These activities can also deepen your understanding of bee behavior and ecology, enriching your beekeeping practice or gardening endeavors.

Finally, advocating for policies that protect pollinators is essential. Engaging with local government officials to support ordinances that limit pesticide use, preserve green spaces, and promote sustainable landscaping practices can create a more favorable environment for bees and other pollinators. By taking these steps, individuals and communities can play a crucial role in supporting local pollinator populations, ensuring their survival and the continued health of ecosystems they support.

Bee and Pollinator Resource Competition

In ecosystems where bees and other pollinators coexist, competition for resources such as nectar and pollen can be intense, especially during periods of scarcity. However, the diversity of pollinator species actually contributes to the resilience and productivity of plant communities. Different pollinators have evolved unique foraging behaviors and physical adaptations that allow them to access different parts of flowers or forage at different times of the day, reducing direct competition and facilitating the coexistence of multiple species within the same habitat.

For example, **honeybees** are generalists that can forage on a wide variety of flowers, but they may prefer certain types of flowers over others when given a choice. **Bumblebees**, with their larger bodies and longer tongues, can access nectar from flowers with deeper corollas that are inaccessible to honeybees. This differentiation in flower preference and foraging capability helps to minimize competition between these two pollinator groups. Similarly, **solitary bees**, which often specialize in pollinating specific plant species, play a crucial role in the pollination of plants that may not be as attractive to more generalist pollinators like honeybees and bumblebees.

Temporal separation is another strategy that reduces competition among pollinators. Some flowers release their nectar at specific times of the day to attract particular pollinators. For instance, flowers that open at night cater to nocturnal pollinators such as moths and bats, while those that open early in the morning are more accessible to bees and other diurnal pollinators. This temporal segregation ensures that different pollinators can access the resources they need without directly competing with each other.

Moreover, the spatial arrangement of plants within an ecosystem can influence pollinator interactions. Diverse plant communities that offer a variety of flowering plants can support a higher diversity of pollinators by providing a range of foraging opportunities. Planting native flowering plants in gardens, parks, and agricultural landscapes creates a mosaic of habitats that cater to different pollinator species. For beekeepers and gardeners, understanding the flowering periods of different plants and ensuring a succession of blooms throughout the growing season can help support a wide range of pollinators.

To further mitigate competition and promote coexistence among pollinators, conservation efforts focus on preserving and restoring habitats that provide abundant foraging resources and nesting sites. Protecting natural areas, establishing pollinator-friendly gardens, and implementing agricultural practices that enhance floral diversity are key strategies. These efforts not only support pollinators but also enhance ecosystem services such as crop pollination and biodiversity conservation.

In summary, the coexistence of bees and other pollinators in shared ecosystems is facilitated by a combination of behavioral, morphological, and ecological adaptations that reduce competition for resources. By understanding and supporting the complex interactions among pollinator species, we can contribute to the health and sustainability of pollinator communities and the ecosystems they support.

Bee Species and Ecological Balance

Understanding the role of different bee species in maintaining ecological balance requires a deep dive into the specific contributions each species makes to their ecosystems. **Honeybees**, for example, are prolific pollinators of both wild and cultivated plants. Their ability to travel significant distances from their hive allows for the pollination of plants spread over a wide area. This not only aids in plant reproduction but also in the production of fruits, nuts, and seeds that form the basis of the diet for a variety of animals.

Bumblebees are particularly effective at pollinating certain types of flowers, thanks to their size and strength. They perform a unique process known as "buzz pollination," where they grab onto a flower and vibrate their bodies to dislodge pollen that is otherwise inaccessible. This method is crucial for the pollination of crops like tomatoes, peppers, and cranberries. By ensuring the pollination of these plants, bumblebees support not only human agriculture but also the habitats of numerous other species that rely on these plants for food and shelter.

Solitary bees, which encompass a wide range of species, including mason bees and leafcutter bees, often specialize in the pollination of specific plant species. This specialization makes them extremely efficient pollinators for those plants, often more so than honeybees or bumblebees. Their role is vital in supporting the biodiversity of plant species, some of which might not survive without their specific pollinator. By maintaining plant diversity, solitary bees also support a wide array of wildlife that depends on various plants for nutrition and habitat.

To support these different bee species, and by extension the ecological balance they help maintain, it is crucial to adopt practices that protect their habitats and food sources. This includes planting native plants that provide nectar and pollen throughout the growing season. When selecting plants, consider a range of species that bloom at different times to ensure a consistent food supply. For example, early bloomers like crocus and snowdrop can support bees emerging from hibernation in early spring, while late bloomers like goldenrod and aster provide essential resources before winter.

Creating habitats for nesting and overwintering is also essential. Leaving patches of bare soil or undisturbed grass can provide nesting sites for ground-nesting solitary bees. Similarly, leaving dead wood in place can offer habitat for cavity-nesting species. For honeybees and bumblebees, maintaining healthy, disease-free hives and nesting sites is crucial. This includes regular monitoring for signs of disease or pests and adopting integrated pest management strategies that minimize the use of chemicals.

Water sources are another critical component of a bee-friendly environment. Providing shallow water sources with landing spots, such as stones or floating wood, can help bees hydrate safely without drowning. This is especially important during hot weather when bees require more water to cool their hives and digest their food.

In summary, each bee species plays a unique and critical role in maintaining ecological balance through pollination, plant diversity support, and as part of the food web. By understanding and supporting the needs of different bee species, we can contribute to the health of ecosystems and the survival of countless plant and animal species dependent on these intricate relationships.

Effects of Beekeeping on Wild Pollinators

Beekeeping practices have a profound impact on wild pollinator populations and biodiversity, influencing the ecological dynamics in various ways. When beekeepers introduce domesticated honeybees into an area, these managed bees compete with native pollinators for the same nectar and pollen resources. This competition can be particularly intense in regions where floral resources are limited, either seasonally or due to environmental degradation. To mitigate these effects, beekeepers can adopt specific strategies to ensure their practices support, rather than harm, wild pollinator populations and the broader ecosystem.

Firstly, selecting the **location** of apiaries is critical. Placing hives in areas rich in floral diversity can reduce the competition for resources between domesticated honeybees and wild pollinators. Beekeepers should aim to position their hives near natural habitats that offer a wide range of flowering plants blooming at different times throughout the year. This not only supports the health and productivity of the managed hives but also contributes to the nourishment of wild pollinators.

Floral diversity in the surrounding landscape is another key factor. Beekeepers can play a role in enhancing this diversity by planting native flowers, shrubs, and trees that provide nectar and pollen sources throughout the growing season. By creating or restoring habitats with a rich variety of flowering plants, beekeepers help support a wide range of pollinators. This approach benefits not only the honeybees but also solitary bees, bumblebees, butterflies, and other pollinating insects that contribute to the ecological balance.

Pesticide use is a significant concern for all pollinators. Beekeepers should adopt integrated pest management (IPM) practices that minimize or eliminate the use of chemical pesticides. Natural pest control methods, such as encouraging the presence of beneficial insects that prey on honeybee pests, can help maintain hive health without adversely affecting wild pollinators. When chemical treatments are necessary, selecting products and application times that pose the least risk to non-target species is crucial.

Water management practices also play a role. Providing water sources within or near apiaries can reduce the need for bees, both domesticated and wild, to compete for limited water supplies, especially during dry periods. Shallow water dishes with stones or floating debris allow bees to drink without drowning and can be a vital resource during hot weather.

Beekeepers should also be aware of the **carrying capacity** of their local environment. Overstocking an area with hives can lead to resource depletion, negatively affecting both managed honeybees and wild pollinators. Monitoring foraging patterns and adjusting the number of hives based on the abundance of floral resources can help prevent overcompetition.

Finally, engaging in **community and conservation efforts** to protect and restore pollinator habitats is essential. Beekeepers can work with local conservation groups, agricultural organizations, and governmental agencies to support initiatives aimed at enhancing pollinator health and biodiversity. This collaborative approach can lead to more effective conservation strategies that benefit all pollinators.

By adopting these practices, beekeepers can ensure their activities contribute positively to the health of wild pollinator populations and the overall biodiversity of the ecosystems in which they operate. Through careful management and a commitment to sustainable practices, beekeeping can coexist harmoniously with the natural world, supporting the vital role pollinators play in our environment.

Bee Species Interactions and Conservation

Understanding the complex interactions among bee species is crucial for informing conservation efforts and ensuring the sustainability of both wild and managed bee populations. Monitoring these interactions requires a multifaceted approach that includes field observations, data collection, and the analysis of environmental factors affecting bee behavior and habitat. To effectively study these interactions, one must first establish a baseline of the types of bees present in a given area and their respective roles within the ecosystem. This involves identifying not only the various bee species but also their foraging patterns, nesting preferences, and seasonal behaviors.

Field observations are a primary method for gathering data on bee interactions. This involves visually monitoring bees as they forage for nectar and pollen, observing their preferences for certain plants over others, and noting any competitive behaviors exhibited towards other pollinators. Such observations can be supplemented with photographic documentation and GPS mapping of foraging routes and floral resources. This detailed record-keeping allows for the analysis of competition and cooperation among different bee species, as well as between bees and other pollinators.

Another critical aspect of studying bee interactions is the use of tracking technologies. Radio frequency identification (RFID) tags and harmonic radar can track the movements of individual bees, providing insights into their foraging ranges and how they might overlap with those of other species. This technology can reveal the extent to which different bee species compete for the same resources or how they might partition resources temporally or spatially to minimize competition.

Environmental DNA (eDNA) sampling is another innovative tool that can be used to monitor bee populations and their interactions. By collecting soil, water, or air samples from a particular habitat, researchers can identify the DNA of various bee species present, even those that are elusive or difficult to observe directly. This method offers a non-invasive way to assess biodiversity and can be particularly useful in tracking changes in bee populations over time, indicating shifts in species dominance or the introduction of non-native species.

To support the health of bee populations, it is also essential to understand the impact of environmental changes on their interactions. Climate change, habitat destruction, and the use of pesticides can alter the availability of floral resources, forcing bees to adapt their foraging behaviors or face increased competition. Long-term monitoring of bee populations and their interactions under changing environmental conditions can help identify species at risk and inform conservation strategies.

Collaboration among researchers, beekeepers, and citizen scientists is vital for the success of these monitoring efforts. Citizen science projects, such as those that encourage the public to report sightings of bees or participate in habitat restoration projects, can greatly expand the scope of data collection. Meanwhile, beekeepers can provide valuable insights into the health and behavior of managed bee populations, contributing to a broader understanding of how domesticated and wild bees interact.

Incorporating these monitoring and study methods into conservation efforts requires a coordinated approach that balances the needs of bee populations with those of agricultural and urban development. By fostering habitats that support a diversity of bee species and implementing practices that minimize harmful impacts, it is possible to promote healthy interactions among bee species. This, in turn, enhances pollination services and supports the resilience of ecosystems. Through diligent monitoring and research, we can gain a deeper understanding of the intricate relationships among bee species and take informed actions to protect these vital pollinators for future generations.

Part 4: The Importance of Biodiversity

CHAPTER 12: BEES AND PLANTS

To foster a thriving environment for bees and ensure the success of beekeeping endeavors, selecting the right plants is crucial. This involves understanding which plants are most beneficial to bees and how to incorporate them into the landscape effectively. The goal is to create a garden or habitat that provides a continuous bloom throughout the growing seasons, offering bees a consistent source of nectar and pollen.

Early Spring Blooms: Start with early bloomers like crocus, snowdrop, and winter aconite. These plants are among the first to flower, providing essential food sources for bees emerging from hibernation. Ensure these are planted in an area that receives early spring sunlight to encourage blooming as soon as the temperature starts to rise.

Mid-Spring to Early Summer Plants: As the season progresses, focus on planting a variety of flowering trees and shrubs like crabapple, cherry, and lilac, along with herbaceous plants such as lupine, foxglove, and penstemon. These species offer a rich source of nectar and pollen during this critical period of hive expansion and brood rearing.

Summer Bloomers: To support bees during the height of summer, incorporate plants like lavender, echinacea, and bee balm, which thrive in full sun and provide abundant nectar. Additionally, consider adding sunflowers and zinnias to the mix, as they are not only excellent sources of food for bees but also add vibrant color to the landscape.

Late Summer to Fall Flowers: As many plants begin to fade, it's important to have late bloomers like goldenrod, aster, and sedum in place. These plants will ensure that bees have access to the resources they need to prepare for the coming winter. Plant these in areas that continue to receive full sunlight as the angle of the sun shifts in late summer.

Plant Diversity and Placement: A diverse selection of plants will attract a wider variety of bees and other pollinators. When planting, group the same species or colors together in clusters to make them more attractive to bees. This also makes foraging more efficient, conserving the bees' energy.

Native Plants: Whenever possible, choose native plants. These species are well adapted to the local climate and soil conditions, requiring less water and maintenance than non-natives. They also tend to attract native bee species, which are often more efficient pollinators for the local flora.

Avoiding Pesticides: It's critical to avoid the use of pesticides on plants that attract bees. Even pesticides labeled as "safe for bees" can be harmful if not used correctly. Opt for organic pest control methods and introduce beneficial insects that naturally keep pest populations in check.

Water Sources: In addition to food, bees need water. Incorporate shallow water features with landing spots, such as a birdbath with stones or a shallow dish filled with pebbles and water. This will provide bees with a safe place to drink and cool off, especially during hot weather.

Habitat Features: Finally, consider adding features that encourage bees to nest and take refuge in your garden. Leave areas of bare soil for ground-nesting bees and install bee hotels for solitary bees. For honeybees, ensure your hives are placed in a location that mimics their natural nesting preferences – sunny, dry, and protected from strong winds.

By carefully selecting and planting a variety of bee-friendly plants and providing a safe habitat, beekeepers and gardeners can play a crucial role in supporting the health and productivity of bee populations. This not only benefits the bees but also enhances the biodiversity and beauty of the environment, creating a vibrant ecosystem where plants and pollinators thrive together.

Bee and Flower Co-Evolution

The intricate dance between bees and flowers is a marvel of nature, showcasing a phenomenon known as co-evolution. This process involves two or more species influencing each other's evolutionary path over millions of years. In the case of bees and flowers, the relationship is symbiotic, meaning both parties derive significant benefits that contribute to their survival and reproductive success.

Flowers have evolved to become more attractive to bees for pollination purposes. This attractiveness is not just limited to vibrant colors visible to the human eye but extends to ultraviolet patterns, which bees can see and are drawn to. These patterns often guide bees to the nectar, acting as landing strips. The nectar itself has evolved to cater to the bees' dietary needs, offering them a rich source of energy. In terms of scent, flowers have developed specific fragrances to lure bees from considerable distances. The shape of flowers also plays a critical role in this co-evolutionary relationship. Certain flowers have evolved shapes that perfectly accommodate the body of a bee, ensuring that when a bee lands to collect nectar, its body brushes against the stamens and pistils, facilitating effective pollination.

Bees, on their part, have evolved in ways that enhance their efficiency as pollinators. The development of hairy bodies in bees is a prime example. These hairs trap pollen grains as bees move from flower to flower, making them highly effective at cross-pollination. Bees' legs have adapted to form pollen baskets, enabling them to collect and carry large amounts of pollen back to their hives. Even the bees' foraging behavior has evolved in response to the flowering patterns of plants they pollinate, optimizing their routes to conserve energy and maximize pollen collection.

The co-evolution of bees and flowers is not just a matter of physical adaptations but also involves intricate behavioral patterns. For example, some flowers have evolved to release their scent at specific times of the day when their primary pollinators are most active. In response, bees have developed a remarkable ability to remember the location and timing of flowering in various plant species, allowing them to forage more efficiently.

This co-evolutionary relationship has profound implications for biodiversity. The diversity of bee species is mirrored by the diversity of flowering plants, each pair having co-evolved with specific adaptations that enhance their mutual survival. This biodiversity is crucial for ecosystem health, as it supports a wide range of other organisms and contributes to the resilience of natural systems.

However, this delicate balance is threatened by human activities. Habitat destruction, pesticide use, and climate change pose significant risks to both bees and flowering plants, disrupting their co-evolutionary relationship. Protecting this relationship requires conscious efforts to preserve habitats, reduce chemical use, and mitigate climate change impacts. Encouraging the planting of native flowers, supporting organic farming practices, and creating pollinator-friendly spaces in urban areas are steps in the right direction.

The co-evolution of bees and flowers exemplifies the interconnectedness of life on Earth, highlighting the importance of conserving natural habitats and biodiversity. By understanding and supporting the conditions that have allowed these relationships to flourish, we contribute to the health of our planet and the survival of the species that call it home.

Flowers Attractive and Beneficial to Bees

Selecting the right flowers for your garden can significantly impact the health and productivity of your local bee population. When choosing flowers, aim for a mix that provides nectar and pollen sources throughout the entire growing season. Here, we delve into specific flowers that are not only attractive but also highly beneficial to bees, ensuring they have access to the resources they need from early spring to late fall.

Spring: Begin with **Almond (Prunus dulcis)**, which blooms early and offers an excellent source of nectar and pollen for bees as they emerge from winter. Planting **Willow (Salix spp.)** is also advisable, as its early blooming catkins provide crucial pollen for bees when few other plants are flowering. Both should be planted in well-drained soil and in areas that receive full sun to partial shade.

Early to Mid-Spring:Fruit trees such as **Apple (Malus domestica)**, **Pear (Pyrus communis)**, and **Cherry (Prunus avium)** are invaluable. Their blossoms not only support a wide range of bee species but also contribute to the pollination of your fruit crops, enhancing their yield and quality. These trees prefer sunny locations and loamy, well-drained soil.

Late Spring to Early Summer:Borage (Borago officinalis) is a must-have for its high nectar production. Plant in full sun and well-drained soil. **Foxglove (Digitalis purpurea)**, while toxic to humans and animals, is a favorite of bees. Ensure it's planted in slightly acidic to neutral soil, in partial shade to full sun. **Chives (Allium schoenoprasum)**, with their purple flowers, not only attract bees but also add flavor to your dishes. They thrive in full sun and well-drained soil.

Summer:Lavender (Lavandula spp.) is renowned for its fragrance and nectar-rich flowers. Plant in full sun and well-drained, slightly alkaline soil. **Coneflower (Echinacea purpurea)** and **Black-eyed Susan (Rudbeckia hirta)** are also summer favorites, providing bees with nectar and pollen. Both prefer full sun and can tolerate a variety of soil types, though well-drained soil is best.

Late Summer to Fall:Goldenrod (Solidago spp.) and **Aster (Aster spp.)** are critical for bees preparing for winter, offering abundant nectar and pollen. Plant in full sun to partial shade in well-drained soil. **Joe-Pye weed (Eutrochium purpureum)**, with its tall, pinkish-purple flowers, is another excellent late-season plant for bees. It prefers full sun to partial shade and moist, fertile soil.

Year-Round Planting Considerations:

- **Soil Preparation:** Before planting, enrich your garden soil with compost to improve its fertility and water-holding capacity. This will benefit both your plants and the bees by encouraging strong, healthy growth and abundant flowering.

- **Watering:** While establishing plants, ensure regular watering to help root development. Once established, most of these plants are relatively drought-tolerant, but during prolonged dry spells, additional watering will help maintain flower production.

- **Mulching:** Apply a layer of organic mulch around your plants to conserve moisture, suppress weeds, and provide a steady supply of nutrients to the soil as it decomposes.

By incorporating these plants into your garden, you not only create a beautiful and vibrant landscape but also contribute significantly to the health and sustainability of bee populations. Remember, a garden that's good for bees is good for the environment and, ultimately, for us all.

Bees' Impact on Agricultural Productivity

The critical role bees play in enhancing agricultural productivity cannot be overstated. Bees are instrumental in the pollination process, a vital step in the reproduction of many plant species, including those crucial for agriculture. Pollination by bees leads to the growth of fruits, vegetables, and nuts, directly influencing crop yields and, consequently, the global food supply. Understanding the mechanics of bee pollination and its impact on agriculture provides a clear insight into the indispensable role these insects play in our food ecosystem.

Pollination occurs when bees collect nectar and pollen from flowers. The pollen collected on their bodies is then transferred from one flower to another, facilitating the fertilization process. This cross-pollination is essential for the genetic diversity of plants, leading to stronger, more resilient crops. For instance, almond trees are almost entirely dependent on honeybees for pollination, and without their assistance, the almond industry would face significant challenges. Similarly, crops like blueberries and cherries are 90% dependent on bee pollination, highlighting the critical role bees play in the production of fruits that are staples in the American diet.

To maximize the benefits of bee pollination in agriculture, it is crucial to support a healthy bee population. This involves the implementation of bee-friendly farming practices, such as reducing pesticide use, planting cover crops that provide bees with nutritional diversity, and maintaining hedgerows and natural habitats to support wild bee populations. Moreover, the strategic placement of bee hives in or near agricultural fields can enhance pollination efficiency and crop yields. Farmers and beekeepers often collaborate to ensure hives are placed at optimal locations, taking into account factors like sun exposure, wind patterns, and proximity to water sources, to create an ideal environment for bees to thrive and work.

Crop rotation and polyculture farming practices also play a significant role in supporting bee populations by providing them with a continuous source of food. By planting a variety of crops that bloom at different times, farmers can ensure that bees have access to a diverse range of nectar and pollen sources throughout the growing season. This not only supports the health and productivity of bee colonies but also contributes to soil health and reduces the need for chemical fertilizers and pesticides, creating a more sustainable agricultural ecosystem.

In addition to these practices, the use of technology and research in beekeeping and agriculture has led to innovative solutions to enhance bee health and pollination services. For example, the development of bee-friendly pesticides and the use of precision agriculture techniques to minimize the impact of farming on bee populations. Research into bee genetics and breeding programs aimed at enhancing bees' resilience to diseases and environmental stresses further supports the sustainability of bee populations and their critical role in agriculture.

The symbiotic relationship between bees and agriculture underscores the importance of conservation efforts and sustainable farming practices in ensuring food security. By protecting and supporting bee populations, we safeguard our ability to produce a diverse and abundant food supply, highlighting the interconnectedness of ecosystems and the human food chain. As such, the role of bees in enhancing agricultural productivity serves as a compelling argument for the conservation of these essential pollinators and the adoption of practices that support their health and habitat.

CHAPTER 13: THREATS TO BIODIVERSITY

The rapid advancement of **technological innovations** in beekeeping presents a promising horizon for addressing the **challenges** bees face today, particularly in the realms of **habitat loss**, **pesticide exposure**, and **climate change**. These technologies, ranging from **precision agriculture** to **hive monitoring systems**, offer beekeepers and researchers unprecedented tools to enhance the health and productivity of bee populations, thereby supporting biodiversity.

Precision agriculture employs **satellite imagery** and **GPS technology** to optimize farming practices, reducing the need for pesticides by targeting only areas that require treatment. This approach minimizes the exposure of bees to harmful chemicals, safeguarding their health and the ecosystems they inhabit. For beekeepers, adopting **integrated pest management (IPM)** strategies that rely on biological control agents and **mechanical pest controls** can further reduce reliance on chemical treatments, promoting a healthier environment for bees.

Hive monitoring technologies, including **temperature sensors**, **humidity sensors**, and **acoustic monitoring devices**, enable beekeepers to track the health and behavior of their colonies in real time. These tools can alert beekeepers to potential issues such as **disease outbreaks**, **pest infestations**, or **queen failure**, allowing for timely interventions that can prevent colony losses. By maintaining stronger, healthier colonies, beekeepers contribute to the **resilience of bee populations** and the ecosystems they support.

Genetic research and **breeding programs** are also at the forefront of technological innovations in beekeeping. Scientists are working to develop bee strains that are more resistant to diseases, pests, and environmental stresses. By enhancing the genetic diversity and resilience of bee populations, these efforts aim to mitigate the impacts of **climate change** and other threats to bee health. Beekeepers can support these initiatives by participating in **breeding programs** and adopting **best practices** for maintaining the genetic health of their colonies.

In the realm of **public engagement** and **education**, technology plays a crucial role in raising awareness about the importance of bees and the challenges they face. **Social media platforms**, **online forums**, and **educational websites** offer platforms for sharing information, fostering community involvement, and mobilizing support for bee conservation efforts. Beekeepers and enthusiasts can leverage these tools to advocate for policies and practices that protect bees and their habitats.

Global trends in beekeeping reflect a growing recognition of the importance of sustainable practices and the role of technology in achieving them. From **urban beekeeping initiatives** that create pollinator-friendly spaces in cities to **international collaborations** focused on bee health research, there is a concerted effort worldwide to address the challenges facing bees. By embracing technological innovations and sustainable beekeeping practices, beekeepers around the world can contribute to the health and vitality of bee populations, ensuring their ability to support biodiversity and agricultural productivity for generations to come.

The integration of technology in beekeeping and the adoption of sustainable practices are essential for the future of beekeeping. These innovations not only enhance the efficiency and effectiveness of beekeeping but also play a critical role in conserving bee populations and the biodiversity they support. As beekeepers, researchers, and the public continue to collaborate and innovate, the prospects for bees and the ecosystems they inhabit become increasingly hopeful. By prioritizing the health and well-being of bees, we invest in the future of our planet, ensuring a rich diversity of life and a sustainable environment for all species.

Impact of Habitat Destruction on Bees

Habitat destruction poses a significant threat to bee populations and, by extension, the ecosystems they support. When natural habitats are altered or eliminated, bees lose the resources they rely on for survival, including food sources and nesting sites. This loss not only affects bee populations but also the broader ecosystem services they provide, such as pollination, which is crucial for the reproduction of many wild and agricultural plants.

To mitigate the impact of habitat destruction, it is essential to focus on habitat restoration and preservation. This involves identifying and protecting critical habitats that are vital for bee populations. Restoration efforts might include planting native flora that provides bees with nectar and pollen throughout the year. Selecting plants that bloom at different times creates a continuous food supply, supporting a diverse range of bee species. For instance, incorporating **native wildflowers**, **shrubs**, and **trees** such as **milkweed**, **goldenrod**, **willow**, and **maple** can offer vital resources to both solitary and social bees.

Creating protected areas free from development and agricultural expansion is another crucial strategy. These areas serve as refuges for bee populations, allowing them to thrive and maintain the ecological balance. In urban and suburban settings, developing green spaces, such as parks and community gardens, with bee-friendly plants can provide essential habitats in environments otherwise dominated by concrete and asphalt.

In addition to habitat creation, the reduction of pesticide use is vital. Pesticides can have lethal and sub-lethal effects on bees, further exacerbating the challenges posed by habitat loss. Adopting **integrated pest management (IPM)** strategies can help minimize pesticide reliance. IPM emphasizes the use of biological control agents, mechanical controls, and cultural practices to manage pests with minimal chemical intervention. For example, encouraging the presence of natural predators or using barriers to protect plants from pests can reduce the need for chemical pesticides.

Furthermore, public education and community involvement are key components in addressing habitat destruction. Educating the public about the importance of bees and the threats they face can lead to more support for conservation initiatives. Community involvement in planting bee-friendly gardens, participating in citizen science projects, and advocating for policies that protect natural habitats can collectively make a significant difference in conserving bee populations.

Lastly, supporting local and national conservation organizations that work towards habitat preservation and restoration can amplify efforts to protect bees. These organizations often have the expertise and resources to implement large-scale conservation projects, influence policy, and conduct research that contributes to the understanding and protection of bee populations.

By implementing these strategies, it is possible to mitigate the impact of habitat destruction on bees and ensure the preservation of biodiversity and ecosystem services they support. Through concerted efforts in habitat restoration, pesticide reduction, public education, and support for conservation initiatives, the resilience of bee populations and their habitats can be enhanced, securing their vital role in ecosystems for future generations.

Pesticides' Impact on Bee Health and Behavior

Pesticides, chemicals designed to control pests that threaten agricultural production and public health, have a profound impact on bee behavior and health. While these substances are targeted at pests, their broad application can inadvertently affect non-target species, including bees, which are crucial pollinators for many crops and wild plants. Understanding the mechanisms through which pesticides affect bees is vital for developing strategies to mitigate these impacts.

One of the primary ways pesticides affect bees is through direct toxicity. Certain pesticides, particularly those belonging to the class of neonicotinoids, have been found to be highly toxic to bees. Neonicotinoids act on the central nervous system of insects, causing paralysis and death. When bees are exposed to these chemicals, either through direct contact during spraying operations or by consuming contaminated nectar and pollen, they can suffer immediate health effects leading to increased mortality rates. The lethal dose can vary, with some pesticides causing death shortly after exposure, while others may have a delayed effect.

Sub-lethal exposure to pesticides can also disrupt bee behavior and physiological functions in ways that are not immediately fatal but can significantly impact colony health and survival. For example, exposure to sub-lethal levels of pesticides can impair bees' ability to navigate, forage, and communicate. Bees rely on complex behaviors to locate food sources and return to their hives; however, pesticides can interfere with their homing ability, leading to disorientation and an inability to find their way back to the hive. This not only reduces the efficiency of food collection but can also result in the loss of foragers, further straining the colony's resources.

Moreover, pesticides can compromise bees' immune systems, making them more susceptible to diseases and parasites. A healthy bee colony can usually fend off pathogens and parasites to some extent, but pesticide exposure can weaken their defenses. For instance, bees exposed to certain fungicides have shown reduced ability to metabolize and detoxify other chemicals, including varroa mite treatments. This can lead to an increased prevalence of diseases within the hive, such as Nosema, a microsporidian infection that affects bees' digestive systems, and can contribute to colony collapse disorder (CCD).

The impact of pesticides on bee reproduction is another area of concern. Exposure to pesticides can affect the queen's fertility, egg-laying patterns, and the viability of her eggs. A healthy queen is crucial for the maintenance and growth of a bee colony, as she is the sole egg-layer. Any impairment in her ability to reproduce can have severe consequences for the colony's future. Additionally, pesticides can affect the development of brood, leading to a reduction in the number of workers and foragers available to support the colony.

The cumulative effects of pesticide exposure on bee colonies are complex and multifaceted, involving direct and indirect impacts on bee health, behavior, and colony dynamics. As bees play a critical role in pollination, their decline due to pesticide exposure poses a significant threat to biodiversity, ecosystem stability, and agricultural productivity. Recognizing the intricate relationships between pesticides, bee health, and the environment is essential for developing integrated pest management strategies that minimize harm to bees and other beneficial organisms.

To address the challenges posed by pesticide exposure, it is crucial to adopt and promote practices that reduce the risk to bee populations. Integrated Pest Management (IPM) is a sustainable approach that combines biological, cultural, physical, and chemical tools in a way that minimizes economic, health, and environmental risks. IPM strategies include monitoring pest populations to determine whether action is necessary, using biological control agents such as predators and parasites of pests, and applying chemical controls as a last resort and in a targeted manner to minimize non-target exposure.

Crop diversification is another practice that can enhance the resilience of bee populations. By planting a variety of crops that bloom at different times, farmers can provide bees with a continuous source of nectar and pollen. This not only supports the nutritional needs of bees but also reduces the reliance on any single crop that may be treated with pesticides. Additionally, maintaining buffer zones of untreated plants can offer safe foraging areas for bees, further reducing their exposure to harmful chemicals.

The selection of pesticides, when their use is deemed necessary, is also critical. Products that are less toxic to bees, applied at times when bees are less active, or using methods that limit drift can significantly reduce the impact on bee populations. For instance, applying pesticides in the evening, when bees are less likely to be foraging, can minimize direct exposure. Furthermore, the development and use of precision application technologies can ensure that pesticides are applied only where needed, reducing the amount used and the potential for non-target exposure.

Beekeepers play a vital role in mitigating the effects of pesticides on bees. By engaging in open communication with farmers and land managers, beekeepers can advocate for bee-friendly practices and coordinate the timing and methods of pesticide application. Monitoring hive health and implementing management practices that strengthen colony resilience, such as supplemental feeding and disease management, can also help mitigate the adverse effects of pesticide exposure.

Public awareness and education are essential components of protecting bee populations from pesticides. Consumers can support bee health by choosing products from farms that use bee-friendly practices, including organic produce. Advocacy for policies that regulate pesticide use, promote research on bee-friendly pesticides, and support habitat conservation can also contribute to a healthier environment for bees.

Climate Change Impact on Bee Habitats

Climate change significantly impacts bee habitats and food availability, altering ecosystems at a rapid pace. Rising temperatures, shifting weather patterns, and extreme weather events contribute to habitat loss, changes in plant phenology, and the availability of nectar and pollen sources crucial for bee nutrition and survival. These environmental shifts necessitate a detailed understanding and strategic response to mitigate adverse effects on bee populations and ensure their role in pollination and biodiversity is maintained.

As temperatures increase, the flowering times of many plants advance, leading to a mismatch between the availability of floral resources and the biological cycles of bees that rely on them. This temporal mismatch can result in bees emerging before or after their primary food sources are available, leading to nutritional stress and reduced survival rates. To counteract this, planting a diverse range of plant species that flower at different times can provide bees with a continuous food source throughout the season. Selection should focus on native plants adapted to local climate conditions, as they are more likely to thrive and support bee populations effectively.

Extreme weather events, such as droughts and floods, further exacerbate the challenge by directly destroying bee habitats and reducing the abundance of flowering plants. Drought conditions can lead to a scarcity of water sources for bees, essential for cooling hives and diluting honey. Implementing water conservation practices, such as setting up bee waterers or maintaining natural water bodies near bee habitats, can provide bees with the necessary hydration. Additionally, constructing rain gardens and using mulch can help manage water runoff and retain moisture in the soil, supporting plant growth and ensuring bees have access to floral resources even during dry periods.

The alteration of landscapes due to climate change also encourages the spread of invasive species, which can outcompete native flora and reduce the diversity of nectar and pollen sources available to bees. Managing invasive species through manual removal, careful selection of plants for gardens and landscapes, and supporting natural predators can help preserve native plant diversity. Encouraging the growth of native plants not only supports bee populations but also enhances the overall resilience of ecosystems to climate change.

Furthermore, climate change affects the distribution of bees and plants, leading to shifts in their geographical ranges. Some bee species may move to higher altitudes or latitudes in search of suitable habitats, while others may face population declines if unable to adapt or migrate. Monitoring bee populations and their movements can inform conservation strategies, such as creating wildlife corridors or protecting key habitats that facilitate bee migration and adaptation.

Adapting beekeeping practices to the changing climate is also crucial. Beekeepers can monitor hive temperatures and humidity levels more closely, using insulation and ventilation to maintain optimal conditions within hives. Shifting the timing of beekeeping activities, such as colony splitting or supplemental feeding, to align with changing plant phenology and weather patterns can help support bee health and productivity. Engaging in community science projects that track flowering times, bee activity, and climate conditions can provide valuable data to guide these adjustments.

Collaboration among beekeepers, researchers, policymakers, and the public is essential to address the multifaceted challenges posed by climate change to bee habitats and food availability. Supporting research into bee adaptation mechanisms, advocating for climate-resilient agricultural and land-use policies, and raising awareness about the importance of bees in ecosystems are critical steps toward ensuring the sustainability of bee populations. Through concerted efforts, it is possible to mitigate the impacts of climate change on bees and preserve the essential ecosystem services they provide.

Diseases and Parasites Affecting Bees

The prevalence of diseases and parasites, such as Varroa mites, presents a significant threat to bee populations worldwide, affecting both wild and managed colonies. These challenges are not only a concern for the bees' health but also pose a risk to biodiversity, ecosystem stability, and agricultural productivity due to the crucial role bees play in pollination. Understanding the impact of these threats is essential for developing effective management strategies to protect bee populations and, by extension, the environments they support.

Varroa mites, specifically Varroa destructor, are external parasitic mites that attack and feed on honey bees. They are considered one of the most severe pests of honeybees globally and have been implicated in the spread of various bee viruses. Varroa mites attach themselves to the body of the bee, weakening the bee by sucking its hemolymph (a fluid equivalent to blood in insects), which can lead to a host of problems including reduced lifespan, impaired development, and increased susceptibility to diseases. The mite's life cycle is closely intertwined with that of the honeybee, allowing it to spread rapidly within and between colonies.

The management of Varroa mites is a complex issue that requires careful consideration of the mite's biology and the colony's health. Traditional control methods include the use of chemical acaricides, which, while effective, can lead to residue issues in hive products and potential mite resistance over time. Therefore, beekeepers are increasingly turning to integrated pest management (IPM) strategies to control Varroa populations. These strategies may include mechanical methods, such as drone brood removal (since Varroa mites preferentially reproduce in drone cells) and screened bottom boards to physically remove mites from the colony. Additionally, selective breeding for Varroa-resistant bee strains offers a promising avenue for long-term control, though it requires significant investment in research and development.

Beyond Varroa mites, bees face threats from a variety of other diseases and parasites. Bacterial diseases like American Foulbrood and European Foulbrood can devastate colonies, leading to the destruction of infected hives to prevent spread. Viral diseases, such as Deformed Wing Virus, further compound the challenges faced by bee populations, often in conjunction with Varroa mite infestations. The complexity of these health issues underscores the need for comprehensive monitoring and management practices to ensure the vitality of bee colonies.

Effective management of bee health necessitates a multifaceted approach, incorporating regular hive inspections, disease and mite monitoring, and the application of treatment strategies as needed. Beekeepers play a crucial role in this process, as early detection and intervention can significantly reduce the impact of diseases and parasites on bee colonies. Additionally, the adoption of bee-friendly agricultural practices, such as minimizing pesticide use and planting pollinator-friendly crops, can support the health of bee populations at a broader ecological level.

As we delve further into the intricacies of bee health management, it becomes clear that collaboration among beekeepers, researchers, and policymakers is essential for developing and implementing strategies that address the multifaceted challenges posed by diseases and parasites. Through continued research, education, and advocacy, it is possible to enhance the resilience of bee populations, ensuring their ability to fulfill their critical roles in ecosystems and agriculture.

To tackle the pervasive issue of Varroa mites and other bee diseases, beekeepers and researchers are exploring innovative and less invasive control methods. One such approach involves the use of organic acids like formic and oxalic acid, which have shown effectiveness in reducing mite populations without leaving harmful residues in hive products. These acids mimic the natural defense mechanisms found in some bee populations, offering a more sustainable alternative to synthetic chemicals. However, the application of these treatments requires precision in timing and dosage to avoid harming the bee colony, underscoring the importance of skilled beekeeping practices.

Biological control methods, employing natural predators of Varroa mites or pathogens that specifically target bee pests without impacting the bees, are also under investigation. While this area of research is still in its early stages, it holds promise for developing long-term, sustainable solutions to bee health challenges. Additionally, enhancing the genetic diversity of bee populations through selective breeding programs can increase their resilience to diseases and parasites. By identifying and propagating traits that confer resistance to specific threats, beekeepers can gradually build more robust colonies capable of withstanding the pressures of pests and pathogens.

The role of environmental management in supporting bee health cannot be overstated. Creating habitats that provide a rich diversity of floral resources can strengthen bee populations by improving nutrition, which in turn enhances their disease resistance. Conservation efforts that focus on restoring native plant species and establishing pesticide-free zones contribute to healthier bee ecosystems. Moreover, practices such as crop rotation and the maintenance of uncultivated areas within agricultural landscapes can promote a more balanced environment, reducing the stressors that exacerbate bee health issues.

Public engagement and education play critical roles in addressing the threats to bee populations. By raising awareness of the importance of bees to our food systems and ecosystems, communities can mobilize to support conservation efforts and adopt more bee-friendly practices in gardening and land management. Citizen science initiatives that involve the public in monitoring bee health and reporting on bee activity can provide valuable data to researchers and policymakers, helping to inform more effective conservation strategies.

In the fight against diseases and parasites threatening bees, ongoing research and innovation are vital. Advances in technology, such as precision beekeeping tools that allow for real-time monitoring of hive conditions, offer new avenues for early detection and intervention in bee health issues. These technologies can help beekeepers make informed decisions about when and how to intervene, minimizing the impact of treatments on the bees and maximizing the effectiveness of disease and pest control measures.

Collaborative efforts among the beekeeping community, scientists, and government agencies are essential to develop regulations and guidelines that support bee health while also ensuring the sustainability of beekeeping practices. By fostering an integrated approach that combines scientific research, practical beekeeping knowledge, and public support, it is possible to address the complex challenges facing bee populations. Through such collaboration, we can work towards a future where bees continue to thrive, supporting biodiversity and agriculture for generations to come.

CHAPTER 14: THE BEEKEEPER'S ROLE IN CONSERVATION

Beekeepers are instrumental in advancing conservation efforts by implementing sustainable beekeeping practices that prioritize the health of bee populations and their environments. One effective strategy involves the establishment of **bee sanctuaries**. These sanctuaries provide safe havens for bees, offering an abundance of native flowering plants that supply bees with essential nutrients throughout the year. To create a bee sanctuary, select a variety of native plants that bloom at different times, ensuring a continuous food source for bees. Incorporate plants like lavender, borage, and echinacea, which are known for their high nectar and pollen content. Additionally, avoid the use of pesticides and herbicides, which can be harmful to bees, and opt for natural pest control methods instead.

Another crucial aspect of bee conservation is **water source management**. Bees need access to clean water for drinking and hive cooling. Set up shallow water sources with landing platforms, such as stones or floating debris, to prevent bees from drowning. Regularly clean and refill these water sources to maintain their usability and prevent the spread of diseases.

Habitat creation and restoration play a significant role in supporting bee populations. Beyond establishing bee sanctuaries, beekeepers can engage in habitat restoration projects that focus on replanting native vegetation and restoring natural landscapes. Participating in local conservation efforts to rehabilitate public lands and natural reserves can amplify the impact on bee populations. Collaborate with local conservation groups to identify areas in need of restoration and choose native plant species that will thrive in the local climate and soil conditions.

Educational outreach is a powerful tool for bee conservation. Beekeepers can share their knowledge and passion for bees with the community through workshops, school programs, and public demonstrations. Educating the public about the importance of bees to ecosystems and agriculture, as well as how to protect them, can lead to broader community support for conservation initiatives. Partner with local schools, libraries, and community centers to organize educational events. Develop informative materials that highlight simple actions individuals can take to support bees, such as planting bee-friendly gardens or avoiding pesticide use.

Advocacy for bee-friendly policies is essential for creating a larger impact. Beekeepers can advocate for local, state, and national policies that protect bees and their habitats. This includes supporting legislation that limits the use of harmful pesticides, promotes habitat conservation, and encourages sustainable farming practices. Engage with policymakers by attending public meetings, writing letters, and joining forces with environmental organizations to lobby for bee-friendly policies. Stay informed about current issues affecting bees and participate in public comment periods for new regulations or policies.

In summary, beekeepers have a unique opportunity to lead by example and inspire others to take action in support of bee conservation. Through the creation of bee sanctaries, careful water source management, active participation in habitat creation and restoration, educational outreach, and advocacy for bee-friendly policies, beekeepers can make a significant contribution to the preservation of bee populations and the broader ecosystem. By adopting these practices, beekeepers not only enhance the health and productivity of their own hives but also contribute to the global effort to protect these vital pollinators for future generations.

Responsible Beekeeping and Biodiversity

Responsible beekeeping practices extend beyond the boundaries of the hive, contributing significantly to the health of ecosystems and promoting biodiversity. One such practice is **selective planting**, where beekeepers can plant a variety of native flowers, shrubs, and trees that bloom at different times throughout the year. This ensures that bees have a consistent source of nectar and pollen, which is crucial for their nutrition and the production of honey. When selecting plants, it's important to choose species that are native to the area, as these are most likely to thrive and provide the best support for local bee populations. For example, planting wildflowers like purple coneflower and foxglove, alongside trees such as willow and maple, can create a diverse and rich environment for bees.

Another practice is the **creation of wildflower meadows** instead of manicured lawns. Wildflower meadows provide a habitat for a wide range of pollinators, not just honeybees. To establish a wildflower meadow, one should start by selecting a mix of perennial and annual seeds that are suited to the local climate and soil conditions. Preparing the soil by removing existing grass and weeds, and ensuring good soil contact for the seeds, will increase the chances of a successful meadow. Regular maintenance in the first few years will help establish the meadow, after which it will require minimal intervention.

Habitat boxes for solitary bees and other pollinators are also a key aspect of promoting biodiversity. Many species of bees are solitary and do not live in hives but instead, require individual nesting sites. Providing habitat boxes, which can be made from untreated wood with holes of various sizes drilled into it, offers these bees a place to lay their eggs. Placing these boxes in a sheltered location, facing south to catch the morning sun, can encourage solitary bees to take up residence.

Chemical-free beekeeping is crucial for protecting bee populations and the wider environment. Avoiding the use of synthetic pesticides and herbicides in and around the beekeeping area helps maintain a healthy ecosystem for bees and other pollinators. Instead, natural pest control methods, such as planting marigolds to deter pests or using essential oils like thyme and lemongrass oil in the hive, can be effective alternatives that do not harm bees or the biodiversity of the area.

Water conservation practices play a role in responsible beekeeping as well. Bees need water for cooling the hive and diluting honey, but water sources can become scarce in dry conditions. Collecting rainwater in barrels and creating shallow water stations with stones or marbles for bees to land on can provide a sustainable water source for bees throughout the year.

Implementing these responsible beekeeping practices requires a commitment to learning and adapting to the needs of the local ecosystem. By focusing on creating a diverse and chemical-free environment, providing resources for a wide range of pollinators, and conserving water, beekeepers can play a vital role in supporting and enhancing biodiversity. Engaging with local conservation groups and sharing knowledge with other beekeepers can further amplify the positive impact on local ecosystems.

Creating Bee-Friendly Environments

Creating bee-friendly environments extends beyond the immediate vicinity of the hive and encompasses a broader commitment to enhancing the local ecosystem for the benefit of bees and other pollinators. One effective initiative is the establishment of **pollinator pathways**. These pathways are corridors of native plants that provide continuous forage for bees and bridge the gap between isolated green spaces. To implement a pollinator pathway, collaborate with neighbors and local community groups to plant native flowering plants along a designated route. This can include residential gardens, public parks, and even roadside verges. Focus on selecting plant species that bloom at various times throughout the year to ensure a steady supply of nectar and pollen. For instance, early bloomers like crocus and snowdrop can provide critical food sources in late winter and early spring, while asters and goldenrod can offer sustenance in the fall.

Another initiative involves **reducing lawn areas** in favor of more diverse landscapes that support bee populations. Traditional lawns offer little nutritional value to bees and require high inputs of water, pesticides, and fertilizers, which can be harmful to pollinators. Transitioning even a small portion of a lawn to a wildflower meadow or a native plant garden can significantly increase the biodiversity of the area and provide essential resources for bees. Start by selecting a sunny spot and remove existing grass. Choose a mix of native wildflowers and grasses that are suited to your soil type and climate. Sow the seeds or plant seedlings in the fall or spring, and provide regular water until the plants are established. Over time, this low-maintenance area will attract a variety of pollinators and contribute to the health of the local ecosystem.

Supporting local agriculture that practices sustainable farming methods is another critical step in creating bee-friendly environments. Sustainable farms often use crop rotation, organic pest control, and other practices that enhance soil health and reduce the need for harmful chemicals. By purchasing products from these farms, beekeepers can help support agricultural practices that are beneficial to bees. Additionally, participating in community-supported agriculture (CSA) programs or visiting local farmers' markets can provide opportunities to engage with farmers about the importance of pollinator-friendly practices and encourage the planting of bee-friendly crops and cover crops that provide additional forage for bees.

Urban beekeeping initiatives can also play a significant role in supporting bee populations in cities and towns. Urban environments offer a unique opportunity to create pockets of biodiversity that can serve as refuges for bees and other wildlife. Establishing rooftop gardens, balcony planters, and community gardens with a variety of flowering plants can provide essential forage in urban areas. Additionally, advocating for city policies that support green spaces, reduce pesticide use, and encourage the planting of native species in public landscaping projects can help create more sustainable urban environments for bees.

Implementing these initiatives requires a commitment to stewardship of the local environment and a willingness to engage with the community to promote practices that support bee health. By taking these steps, beekeepers can contribute to the creation of landscapes that not only support the needs of their hives but also enhance the overall biodiversity and resilience of the ecosystem. Through collaboration, education, and advocacy, beekeepers can help ensure that bees, along with other pollinators, continue to thrive in diverse environments.

Engaging Communities in Bee Conservation

Engaging local communities in bee conservation requires a multifaceted approach that leverages education, awareness, and hands-on involvement. One effective method is to **organize community beekeeping workshops**. These workshops should cover topics such as the basics of bee biology, the importance of bees to our ecosystem, and practical steps for creating bee-friendly gardens. Utilize visual aids, such as diagrams of bee anatomy and lifecycle, to enhance understanding. Ensure that workshops are led by experienced beekeepers or entomologists who can share insights and answer questions in an accessible manner.

School programs are another vital avenue for fostering bee conservation awareness. Collaborate with local schools to integrate bee conservation topics into the science curriculum. Activities could include building bee hotels, planting native flowers in school gardens, and interactive presentations by beekeepers. Provide teachers with resources like lesson plans and activity guides focused on pollinators and their role in food production. For example, a lesson plan might involve students mapping out a garden with plants that bloom at different times of the year, ensuring a continuous food source for bees.

Public demonstrations at community events can also raise awareness and interest. Set up informational booths at farmers' markets, fairs, and environmental events. Display live bee colonies in secure observation hives, if possible, to give people a close-up view of bee behavior. Offer brochures and flyers with information on how to support bee populations, such as lists of bee-friendly plants suitable for the local climate, tips for reducing pesticide use, and instructions for creating simple bee waterers from household items.

Citizen science projects engage the community in data collection and research, fostering a deeper connection to bee conservation efforts. Projects might involve tracking bee sightings, photographing flowers visited by bees, or monitoring the health of local bee populations. Use platforms like iNaturalist or Bumble Bee Watch to facilitate data submission and sharing. Provide clear instructions on how to observe and report bee activity safely and respectfully, emphasizing the importance of not disturbing bees or their habitats.

To support these initiatives, create a **community resource center** either online or in a local library. This center should offer access to materials on beekeeping, bee biology, and conservation strategies. Include books, documentaries, and links to reputable websites. Organize a monthly newsletter or an online forum where community members can share experiences, ask questions, and learn from each other. Highlight success stories from the community, such as successful garden transformations or schools that have incorporated bee conservation into their curriculum.

Finally, **partnerships with local businesses** can amplify conservation efforts. Encourage businesses to sponsor bee conservation projects or to adopt bee-friendly practices, such as maintaining flower beds with native plants or reducing pesticide use. Recognize and promote businesses that support bee conservation, creating a positive feedback loop that encourages others to participate.

By implementing these strategies, beekeepers and conservationists can build a strong, informed community that is actively engaged in supporting bee populations. Through education, hands-on involvement, and collaboration, it's possible to foster a culture of conservation that benefits both bees and humans.

Urban Beekeeping: Boosting Biodiversity and Pollination

Urban beekeeping serves as a vital link in enhancing **biodiversity** and **pollination** within city environments, where green spaces are often limited. By introducing hives into urban settings, beekeepers can significantly contribute to local ecosystems, supporting a wide range of plant and animal species. One of the first steps in urban beekeeping is selecting the right location for your hives. Rooftops, balconies, and community gardens are ideal places, offering bees access to a variety of urban flora. When choosing a site, ensure it receives ample sunlight throughout the day, is shielded from strong winds, and has some form of rain protection to keep the hives dry.

The choice of **bee species** is crucial in urban environments. Opt for species known for their gentleness and adaptability to confined spaces, such as the Italian honeybee (*Apis mellifera ligustica*). These bees are less likely to become stressed in urban settings, making them ideal for close proximity to human populations.

Plant selection around the hive area is another key consideration. Prioritize native plants, herbs, and flowering fruits and vegetables that bloom at different times of the year to provide continuous forage for the bees. Lavender, thyme, rosemary, and fruit trees like apple and cherry are excellent choices, offering both nectar and pollen to support hive health and productivity.

Water source management is equally important. Set up a **bee water station** by filling a shallow container with water and adding stones or floating platforms for bees to land on. This prevents drowning and ensures bees have easy access to water, especially during hot summer months.

Urban beekeepers should also be mindful of **hive management practices** that minimize stress on the bees. Regular inspections should be conducted to check for signs of disease or pests, but with minimal disturbance to the bees. Utilize smoke gently to calm the bees before inspections. Additionally, during peak flowering seasons, be prepared to add extra supers to the hive to accommodate honey production, ensuring bees have ample space to work and store their honey.

Community engagement is a powerful aspect of urban beekeeping. Educating neighbors and local communities about the benefits of bees can foster a supportive environment and dispel common misconceptions about bees being a nuisance. Hosting workshops, open hive days, and school presentations can increase awareness and appreciation for bees and their role in urban ecosystems.

Lastly, urban beekeepers can participate in **local conservation efforts** by collaborating with city planners and parks departments to create more green spaces and pollinator-friendly areas. Advocating for policies that limit pesticide use and support the planting of native species in public spaces can further enhance urban biodiversity and provide safer environments for both bees and humans.

By integrating these practices, urban beekeepers not only contribute to the health and productivity of their hives but also play a crucial role in sustaining urban biodiversity and pollination networks. Through careful planning, community involvement, and sustainable management, urban beekeeping can thrive, offering a beacon of hope for pollinators in city landscapes.

CHAPTER 15: PROMOTING BIODIVERSITY

Creating **green corridors** is a critical strategy in promoting biodiversity, especially in urban and suburban areas where habitat fragmentation can severely limit the movement of bees and other pollinators. These corridors are essentially continuous areas of natural or semi-natural habitats, planted with a diversity of native flora, that connect larger green spaces such as parks, nature reserves, and undeveloped land. By establishing these corridors, beekeepers and conservationists can facilitate the safe passage of bees, allowing them to access a wider range of habitats and foraging resources, which is crucial for their survival and the pollination of plants across different areas.

To implement a green corridor effectively, one must first identify potential routes that connect existing green spaces. These routes can be along railway lines, riverbanks, or even less obvious paths like the verges of roads or utility rights-of-way. Once a route is identified, the next step involves selecting the right mix of plants to support a diverse range of pollinators throughout the year. This selection should include a variety of flowering plants that bloom at different times, providing continuous nectar and pollen sources. For example, early bloomers like willow and crocus can provide critical food sources in late winter and early spring, while plants like goldenrod and aster will support bees into the late fall.

The planting process in green corridors should prioritize species that are native to the area, as these plants are best suited to the local climate and soil conditions, and they typically require less maintenance once established. Additionally, native plants have co-evolved with local bee populations and other pollinators, making them an ideal choice for supporting these species. It's also beneficial to include plants of varying heights and structures to cater to different types of bees and wildlife, from ground-nesting bees to birds and small mammals.

Maintenance of green corridors is another important aspect, involving the management of invasive species, which can outcompete native plants and reduce biodiversity. Regular monitoring and removal of such species help to maintain the health and diversity of the corridor. Furthermore, minimizing the use of pesticides and herbicides in these areas is crucial, as these chemicals can harm bees and other beneficial insects. Instead, natural pest control methods should be employed, such as encouraging the presence of pest predators and using organic mulches to suppress weeds.

Engaging the community is essential for the success of green corridors. This can be achieved by organizing planting days, educational walks, and other activities that raise awareness about the importance of these corridors and the role of bees in our ecosystems. Involving schools, local businesses, and other stakeholders not only helps in the physical creation of these corridors but also fosters a sense of ownership and responsibility towards their maintenance and success.

Finally, monitoring the effectiveness of green corridors in supporting bee populations and biodiversity is key. This can involve tracking bee diversity and abundance through regular surveys, which can provide valuable data on the health of the corridor and its impact on local ecosystems. Such data can inform future conservation efforts and help to demonstrate the value of green corridors to policymakers and the wider community, potentially leading to increased support and funding for these and similar projects.

By carefully planning, planting, and maintaining green corridors, beekeepers and conservationists can make significant contributions to the preservation of bee populations and the enhancement of biodiversity. These corridors not only serve as lifelines for bees and other pollinators but also enrich our landscapes, making them more resilient and vibrant for all forms of life.

Bee-Friendly Gardens and Landscapes

Creating bee-friendly gardens and landscapes involves selecting plants that provide bees with nectar and pollen, crucial for their nutrition and the colony's survival. The process starts with understanding the local climate and soil conditions, as these factors greatly influence which plants will thrive. For instance, in areas with dry, sandy soil, consider planting lavender or rosemary, which are drought-resistant and provide excellent sources of nectar throughout the year. Conversely, in wetter climates, bee balm and asters are suitable as they can tolerate more moisture while still offering rich nectar sources.

When designing a garden, aim for a succession of blooming plants to ensure that bees have a continuous food supply from early spring through late fall. Early bloomers like crocus, snowdrop, and willow are vital for bees emerging from hibernation, seeking their first nectar sources. Mid-season flowers, such as echinacea and lavender, sustain bees during the summer. Late bloomers like goldenrod and aster support the colony's preparation for winter. Incorporating a variety of flower shapes and colors can attract a wider range of bee species, as different bees are adapted to different types of flowers.

In addition to flowering plants, consider adding features that mimic natural bee habitats. Bare patches of soil, for example, can provide nesting sites for ground-nesting bees, while piles of twigs and dead wood can offer refuge for cavity-nesting species. Installing a water feature, such as a shallow birdbath with stones for bees to land on, can provide a much-needed water source, especially in urban or suburban settings where natural sources may be scarce.

To maximize the impact of a bee-friendly garden, avoid the use of pesticides and herbicides, which can be harmful to bees and other beneficial insects. Opt for organic gardening practices, such as using compost for fertilization and manually removing pests or diseased plants. Engaging neighbors and community members in creating bee-friendly spaces can amplify the benefits, creating larger, interconnected habitats for bees to thrive.

By carefully selecting plants and creating conducive habitats, gardeners and landscapers can play a pivotal role in supporting bee populations, contributing to the health of ecosystems and the success of agricultural endeavors.

Green Corridors for Pollinator Habitats

Establishing green corridors requires a strategic approach to selecting and connecting areas that can support pollinators like bees. The first step in this process involves mapping out existing green spaces within the targeted urban or suburban landscape. This could include public parks, nature reserves, community gardens, and even less formal spaces like roadside verges or unused land. Utilizing geographic information system (GIS) technology can provide precise data on land use, helping to identify potential corridor paths that link these green spaces effectively.

Once potential corridors are identified, the selection of plant species becomes paramount. Opt for a mix of native flowering plants that offer a succession of blooms from early spring through late fall. This ensures that pollinators have a consistent source of food throughout their active seasons. For early spring, consider planting **red maple** (*Acer rubrum*) and **pussy willow** (*Salix discolor*), which are among the first to provide nectar and pollen. Summer bloomers should include species like **purple coneflower** (*Echinacea purpurea*) and **lavender** (*Lavandula*), known for their attractiveness to bees. For late-season forage, **New England aster** (*Symphyotrichum novae-angliae*) and **goldenrod** (*Solidago*) are excellent choices, offering vital resources as bees prepare for winter.

The physical planting of these corridors should be done with consideration to the pollinators' flight patterns and needs. Plants should be grouped in clusters to create rich feeding sites, rather than spaced far apart, which could make foraging inefficient and energy-consuming for bees. Incorporating a variety of plant heights and types, including trees, shrubs, and herbaceous plants, will cater to a wide range of bee species, including those that nest above ground, in cavities, or in the soil.

In urban environments, where soil and space may be limited, utilizing container gardens and green roofs can also contribute to green corridors. Containers can be filled with a mix of annual and perennial plants that are known to attract bees, while green roofs can offer refuge and foraging opportunities in densely built-up areas. When selecting containers, ensure they have adequate drainage and are large enough to support the root systems of the chosen plants. For green roofs, consider the weight capacity of the building and choose soil mediums and plants that are suited to this unique environment.

Water access within these corridors is another critical component. Installing simple water features, such as birdbaths or shallow dishes with pebbles for landing spots, can provide bees with hydration. These water sources should be regularly cleaned and refilled to prevent the spread of diseases among pollinator populations.

Community involvement is key to the success of green corridors. Engage local residents, schools, and businesses in the planning and planting process to foster a sense of ownership and stewardship over these areas. Educational signage along the corridors can inform the public about the importance of pollinators and how to protect them. This not only aids in the conservation effort but also enriches the community's connection to their local environment.

Monitoring and maintenance are ongoing requirements for green corridors. Regular assessments can help identify issues such as plant disease, invasive species encroachment, or habitat disruption. Adaptive management, including replacing non-performing plants with more suitable species, controlling invasive species, and adjusting water features or nesting sites as needed, ensures the corridors continue to serve their intended purpose effectively.

By meticulously planning, planting, and maintaining green corridors, communities can significantly enhance urban and suburban landscapes for pollinators. These corridors not only provide crucial resources and habitat connectivity for bees but also contribute to the overall ecological health of the area, supporting a diverse range of wildlife and increasing the aesthetic and recreational value of green spaces for human residents.

Importance of Pollinators

Raising public awareness about the importance of pollinators is a critical step in promoting biodiversity and ensuring the sustainability of our ecosystems. This endeavor begins with education, spreading knowledge about how pollinators, especially bees, play a vital role in the pollination of many crops and wild plants, which is essential for food production and environmental health. To effectively communicate this message, it is essential to utilize a variety of platforms and strategies to reach a broad audience.

One effective approach is to host informational workshops and seminars in local communities, schools, and libraries. These events can be tailored to different age groups, ensuring that content is accessible and engaging for everyone from young children to adults. For younger audiences, interactive activities such as building bee hotels or planting pollinator-friendly gardens can make the learning experience fun and memorable. For adults, detailed presentations on the science of pollination, the challenges faced by pollinators, and how individuals can contribute to their protection can inspire action.

Social media platforms offer another powerful tool for raising awareness. Creating engaging content such as infographics, short videos, and interactive posts that highlight fascinating facts about bees and other pollinators can capture the public's interest. Utilizing hashtags can help in spreading the message further, encouraging users to share information and participate in global conversations about pollinator conservation.

Local media outlets, including newspapers, radio stations, and television networks, can also play a significant role in educating the public. Writing articles, participating in interviews, or sponsoring community segments dedicated to pollinators and their importance to biodiversity and food security can reach a wide audience. Collaborating with local journalists to cover stories on community-led conservation efforts, beekeeping workshops, or the establishment of green corridors can highlight positive actions being taken and inspire others to contribute.

Incorporating pollinator education into school curricula is another crucial step. Working with educators to develop lesson plans and educational materials that cover the role of pollinators, the threats they face, and how students can help can instill a sense of responsibility and interest in conservation from a young age. School projects such as creating pollinator gardens or citizen science projects tracking local bee populations can provide hands-on learning experiences.

Finally, partnering with local businesses and organizations to support pollinator-friendly practices can amplify the impact of awareness-raising efforts. Encouraging businesses to adopt pollinator-friendly landscaping, support local beekeepers by selling their honey, or sponsor educational events can help in creating a community-wide approach to pollinator conservation. Restaurants and cafes can also contribute by highlighting pollinator-dependent ingredients in their dishes and providing information about the importance of bees and other pollinators to their customers.

Through these multifaceted strategies, raising public awareness about the importance of pollinators can lead to increased community engagement, support for conservation initiatives, and positive changes in individual and collective actions towards creating a more pollinator-friendly world. Engaging the public in understanding and protecting our vital pollinator populations is not just about saving bees; it's about securing the future of our food supply and preserving the health of our planet for generations to come.

Part 5: Preparing for Beekeeping

Chapter 16: Essential Beekeeping Tools for Beginners

Moving on to the next set of essential tools, every beekeeper needs to have a **smoker** and understand its proper use. A smoker is a device used to generate smoke, which calms bees and makes them less likely to sting during hive inspections or when you're harvesting honey. The smoker should be filled with materials that produce cool, white smoke, such as pine needles, untreated burlap, or cotton rags. Lighting the smoker involves placing a lit piece of newspaper or cardboard at the bottom of the smoker, then gently adding your chosen fuel on top. Pump the bellows gently to get the smoke going before you approach the hive. It's crucial to use the smoke sparingly, directing a few puffs at the hive entrance before opening the hive and then under the lid to calm the bees inside.

Another indispensable tool is the **hive tool**, a versatile instrument used for various tasks around the hive. The hive tool is primarily used to pry apart hive bodies and frames that have been glued together by the bees' propolis. It can also be used to scrape off excess propolis or wax from hive components. The best hive tools are made of stainless steel, which is durable and easy to clean. Look for a hive tool with a flat end for prying and a hooked end that can be used to lift frames.

Bee brushes are gentle tools used to move bees off of combs or out of the way when you're working in the hive. A bee brush should have soft bristles to avoid harming the bees. Use light, sweeping motions to encourage bees to move, being careful not to crush or agitate them, which could provoke stinging.

For beekeepers, **personal protective equipment (PPE)** is non-negotiable. A bee suit or jacket, veil, and gloves can protect you from stings while allowing you to work comfortably around the bees. Bee suits and jackets are available in various materials, but those made from a lightweight, breathable fabric will keep you cool and protected in the heat. Veils come in different styles, but ensure yours provides clear visibility and is securely attached to your hat or helmet to keep bees away from your face. Gloves should be thick enough to prevent stings but flexible enough to allow easy manipulation of hive components. Leather gloves are a popular choice, offering a good balance between protection and dexterity.

Finally, **feeding equipment** is essential, especially for new colonies or during times of scarce nectar flow. Feeders come in several types, including entrance feeders, top feeders, and frame feeders. Entrance feeders are placed at the hive entrance and are easy to monitor and refill without opening the hive. Top feeders sit on top of the uppermost hive body, allowing bees to feed without leaving the warmth of the hive. Frame feeders replace one of the frames inside the hive body, providing bees direct access to syrup without exposure to the outside. When choosing a feeder, consider the size of your colony, the time of year, and how frequently you can check and refill the feeders.

Each of these tools plays a crucial role in effective hive management, ensuring the health and productivity of your bee colonies. By selecting the right tools and learning to use them properly, you can enjoy a rewarding beekeeping experience while contributing to the health of the bee population.

Types of Hives and Their Unique Features

When selecting the right type of hive, it's essential to consider the specific needs of your beekeeping practice as well as the local climate and the bees' health. Beyond the traditional Langstroth hive, two other popular types are the Top-bar and Warré hives, each offering unique benefits and challenges.

Top-bar hives are characterized by a horizontal layout where bees build their comb hanging from removable bars, mimicking the natural structure of a tree limb. This design allows beekeepers to manage the hive with minimal disturbance to the bees, as individual bars can be lifted out for inspection or honey harvesting. Top-bar hives are particularly suited to small-scale beekeeping or educational projects because they offer a more natural living environment for the bees and require less physical strength to manage. However, they may produce less honey compared to Langstroth hives and require more frequent inspections to ensure the comb is being built straight.

Warré hives, often referred to as "the People's Hive," emphasize vertical expansion and mimic the bees' natural upward growth pattern. They consist of stacked boxes without frames, where bees are free to build their comb. This design is intended to interfere as little as possible with the natural behaviors of the bee colony, promoting a healthy and stress-free environment. Warré hives are equipped with a quilt box on top for insulation and moisture regulation, making them particularly well-suited for colder climates. Beekeepers may find these hives to be low-maintenance since they are designed to be opened only once or twice a year for honey harvesting. However, the lack of frames makes it more challenging to inspect individual combs without causing damage.

Each hive type has its **material requirements**. For instance, Top-bar hives benefit from lightweight, insulating materials that make the hive easy to move and suitable for different climates. Cedar or pine provides excellent durability and resistance to weather. For Warré hives, thicker wood can help with insulation, making hardwoods or dense softwoods like fir a good choice.

In conclusion, selecting the right hive involves weighing factors such as ease of management, honey production, and the well-being of the bee colony. Whether opting for a Langstroth, Top-bar, or Warré hive, the key is to ensure that the design aligns with your beekeeping goals and local environmental conditions.

Personal Protective Equipment for Beekeeping Safety

When delving into the realm of beekeeping, prioritizing safety through personal protective equipment (PPE) is paramount. This equipment, designed to shield beekeepers from bee stings and other potential hazards, includes veils, suits, and gloves, each serving a unique purpose in the beekeeper's arsenal. The selection of these items should be made with careful consideration of material, fit, and comfort to ensure they provide the necessary protection without hindering the beekeeping activities.

Veils are a critical component of beekeeping PPE. They are designed to protect the face and neck, areas particularly sensitive to bee stings. A good veil should offer clear visibility and be made from a mesh material that is fine enough to prevent bees from getting through but still allows for ample airflow. It should fit comfortably over the head, with some models designed to integrate with hats or helmets for added stability. The veil should extend down to the shoulders and be able to be securely fastened to prevent bees from entering underneath. For those wearing glasses, it's essential to choose a veil that accommodates this without compromising visibility or comfort.

Beekeeping suits and jackets provide body protection and are available in various materials and designs. Lightweight, breathable fabrics are preferable, especially in warmer climates, to keep the beekeeper cool while offering protection. Suits come in full-length or jacket styles, with the full-length offering the most comprehensive coverage. They should feature elastic cuffs at the wrists and ankles or be designed to securely fasten over boots and gloves to prevent bees from crawling inside. Zippers should be durable and covered with protective flaps to ensure no gaps for bees to enter. Pockets are useful but should have secure closures to prevent bees from becoming trapped inside. When choosing a suit or jacket, the fit is crucial; it should be loose enough to allow for comfortable movement but not so baggy as to catch on equipment or hive components.

Gloves are the final essential piece of PPE, designed to protect the hands while allowing for the dexterity needed to handle hive components and tools. Leather gloves are a popular choice, offering a balance between protection and flexibility. They should extend up the forearm and fit snugly at the end to prevent bees from getting inside. Some beekeepers prefer nitrile or rubber gloves for increased dexterity, though these may not provide as much protection from stings. The choice of gloves often comes down to personal preference and the specific tasks being performed, with some beekeepers using different gloves for different activities.

In addition to these primary pieces of PPE, beekeepers should consider footwear that covers the ankles and is easy to clean, as well as hats or helmets that integrate with veils for added protection. When selecting PPE, durability, ease of cleaning, and the ability to perform necessary tasks comfortably and safely should all be considered. Proper care and maintenance of PPE, including regular cleaning and inspection for wear or damage, are essential to ensure it continues to provide adequate protection.

By carefully selecting and properly using personal protective equipment, beekeepers can enjoy the fascinating world of beekeeping while minimizing the risk of stings and other injuries. This protective gear, combined with a calm, measured approach to beekeeping tasks, forms the foundation of safe and successful beekeeping practices.

Hive Management Tools

Moving beyond the foundational elements of personal protective equipment, the beekeeper's toolkit is further enriched by specialized instruments designed for hive management. Among these, smokers, hive tools, and brushes stand out as indispensable for maintaining healthy colonies and facilitating hive inspections.

A smoker, a cylindrical device equipped with a bellows, plays a crucial role in calming bees during hive inspections or when performing any activity that might disturb the colony. The smoke generated by the smoker masks alarm pheromones released by guard bees, thus reducing the likelihood of defensive behavior. For effective use, the smoker should be filled with materials that produce a cool, thick smoke. Pine needles, untreated burlap, or cotton rags are recommended due to their slow-burning properties and minimal risk of overheating the smoker. Lighting the smoker involves placing a small, easily ignitable item such as a piece of cardboard at the bottom, adding the chosen fuel on top, and then gently pumping the bellows until the fuel smolders, producing smoke. It's essential to direct the smoke towards the hive entrance and under the lid gently, using just enough to calm the bees without overdoing it, as excessive smoke can agitate them or mask important olfactory cues within the hive.

The hive tool, a flat, metal implement, is another cornerstone of the beekeeper's arsenal. Its primary function is to pry apart hive components that bees have sealed with propolis, a resinous mixture used as a construction material within the hive. The hive tool's design, typically featuring a flat end for scraping and a curved end for lifting, allows for versatility in hive management tasks. When selecting a hive tool, durability and comfort are paramount. Stainless steel construction ensures longevity and resistance to corrosion, while a handle with a comfortable grip reduces hand fatigue during prolonged use. The hive tool's effectiveness lies in its simplicity; however, mastering its use requires practice. Techniques such as leveraging the tool to gently break propolis seals without damaging the hive components or injuring bees are developed over time.

Lastly, the bee brush, a soft-bristled brush, is used to gently remove bees from surfaces where they are not wanted, such as frames during honey extraction or the beekeeper's clothing. The bristles should be soft enough to avoid harming the bees yet firm enough to effectively brush them away. Synthetic fibers are often preferred for their durability and ease of cleaning. Using the brush requires a light touch and patience, as aggressive brushing can agitate bees, increasing the risk of stinging. Instead, gentle strokes in the direction of the bees' movement encourage them to relocate without causing distress.

Together, these tools form the core of effective hive management practices. Their use not only facilitates routine inspections and maintenance but also underscores the beekeeper's role as a steward of the colony's health and productivity. Mastery of these tools, combined with a deep understanding of bee behavior and colony dynamics, enables beekeepers to nurture their hives through the seasons, ensuring the well-being of the bees and the sustainability of beekeeping endeavors.

CHAPTER 17: CHOOSING THE IDEAL HIVE LOCATION

Ensuring access to abundant nectar and water sources near the hive is paramount for the health and productivity of your bee colony. Bees require a diverse range of flowering plants that bloom at different times throughout the year to provide a continuous source of nectar and pollen. Ideal locations are those near wildflower meadows, gardens rich in bee-friendly plants, or agricultural areas that do not use harmful pesticides. It's beneficial to research and plant native flora that thrives in your region to support your bees. Additionally, a steady water source is crucial, especially during hot weather. A shallow water dish with stones or floating wood pieces can prevent drowning and serve as an excellent water source for bees.

Avoiding risky zones is another critical consideration. Areas with heavy pesticide use, such as certain agricultural or industrial zones, pose significant risks to bee health and can lead to colony collapse. Similarly, places with frequent disturbances from humans, pets, or machinery can stress the bees and make them more prone to disease or abandonment of the hive. It's essential to choose a location that minimizes these risks, perhaps by setting up hives in secluded parts of a property or in areas shielded by natural barriers like hedges or trees.

Integrating hives seamlessly into urban or rural environments requires thoughtful placement to ensure bees and humans can coexist peacefully. In urban settings, rooftops, balconies, or backyards with adequate safety measures can be suitable locations. It's important to communicate with neighbors and educate them about beekeeping to alleviate concerns. In rural areas, placing hives at the edges of fields or near natural forage can provide bees with ample resources while keeping them at a safe distance from residential and livestock areas. Regardless of the setting, ensuring that hives are not directly facing paths or entrances can minimize the chance of bee-human interactions.

When considering the ideal hive location, it's also essential to account for the microclimate of the chosen spot. Areas that receive morning sunlight can encourage bees to start their foraging early, while providing afternoon shade can help protect the hive from overheating in summer. Wind protection is crucial to prevent hives from being toppled over during storms or strong winds; a natural windbreak such as a hedge, wall, or row of trees can be very effective. Additionally, the hive entrance should ideally face away from prevailing winds to prevent cold air from entering the hive during winter months.

The ground on which the hive stands should be stable and well-drained to avoid moisture accumulation, which can lead to mold growth and weaken the bee colony. Elevating the hive slightly off the ground using a hive stand or pallets can aid in ventilation and keep the bees dry. It's also beneficial to ensure the hive location is accessible year-round for maintenance and honey harvesting activities, considering factors like snow accumulation or flooding in certain areas.

By meticulously evaluating each of these factors—access to forage, avoidance of risky zones, integration into the environment, microclimate considerations, and ground stability—beekeepers can select an optimal location that supports the health, safety, and productivity of their bee colonies. This careful selection process underscores the beekeeper's role in fostering a harmonious relationship between bees and their surroundings, ultimately contributing to the success of beekeeping endeavors and the well-being of the local ecosystem.

Environmental Factors for Hive Placement

In the careful selection of a hive location, understanding and optimizing environmental factors are crucial to the success and health of the bee colony. Sunlight plays a pivotal role in the daily activities of bees, including foraging and temperature regulation within the hive. A site that benefits from direct sunlight, especially during the morning hours, can stimulate an early start for the bees' foraging activities, which is beneficial for the productivity of the hive. An ideal placement would ensure that the hive receives morning sunlight while being partially shaded during the peak heat of the afternoon, particularly in regions with intense summer heat. This can be achieved by positioning the hive in an area where it is exposed to the east or southeast sun, with natural or artificial shading provided for the hotter part of the day. Trees, shrubs, or structures that cast a shadow during the afternoon can protect the hive from overheating, yet they should not obstruct the morning sunlight, creating a balanced environment for the bees.

Wind protection is another critical environmental factor to consider. Strong winds can cool the hive, increase the bees' energy consumption to maintain temperature, and in extreme cases, can topple hives. To mitigate this, hives should be placed near natural windbreaks such as hedges, bushes, or a row of trees. These barriers should be positioned to shield the hives from the prevailing wind direction, which varies depending on the geographic location. Alternatively, artificial windbreaks can be constructed using fencing or other materials that allow for some air flow to prevent stagnation of air around the hive while still reducing wind speed. The key is to ensure that the windbreak does not create a complete barrier that could lead to dampness or reduced air quality around the hive.

Rain coverage is equally important to protect the hive from water ingress, which can lead to mold growth and chill the bees. While the hive's construction typically includes an overhang to shed water away from the entrance, additional measures may be necessary in areas prone to heavy rainfall or flooding. Elevating the hive off the ground on a stand or pallet can prevent water from entering the hive and aid in air circulation. The chosen location should have good drainage to avoid water pooling around the hive base, which could attract pests or cause the wooden components of the hive to rot. In areas with significant rainfall, consider positioning the hive under a partial canopy that allows rain protection without restricting airflow or light. However, care must be taken to ensure that this canopy does not retain moisture or create a humid environment that could harm the hive.

The interaction of these environmental factors—sunlight, wind, and rain—must be carefully balanced to create an optimal setting for the hive. Each factor influences the others, and their collective management can significantly impact the health and productivity of the bee colony. By selecting a location that maximizes morning sunlight exposure, provides wind protection, and ensures adequate rain coverage without compromising ventilation, beekeepers can establish a strong foundation for their hives. This thoughtful approach to hive placement, grounded in an understanding of environmental influences, exemplifies the beekeeper's role in nurturing a conducive habitat for their bees, reflecting a commitment to the well-being of the colony and the broader ecosystem.

Nectar and Water Sources for Hives

Ensuring that your bees have access to abundant nectar and water sources is a cornerstone of successful beekeeping. This requires a strategic approach to selecting flora and water features that will support your hive throughout the year. When planning for nectar sources, prioritize a variety of plants that bloom in succession from early spring to late fall. This continuous bloom ensures that bees have a consistent source of nectar and pollen, which is vital for the colony's health and productivity. For example, planting crocuses, snowdrops, and willows can provide early spring forage, while asters and goldenrods offer valuable resources in the fall. It's advisable to research plants native to your region since these are often well-suited to local climate conditions and beneficial to native bee populations as well.

In addition to plant selection, consider the spatial arrangement of these resources. Clustering plants in groups rather than spacing them individually can make foraging more efficient for bees, reducing the energy they expend in collecting nectar and pollen. Also, incorporating trees and shrubs alongside smaller plants can create a layered foraging environment that mimics natural ecosystems, further enhancing forage availability.

Water sourcing is another critical aspect of hive location planning. Bees need water for cooling the hive, diluting stored honey, and feeding larvae. A natural water source, such as a pond, stream, or even a bird bath, should be within a quarter mile (about 400 meters) of the hive. If natural sources are not available, creating a bee waterer can be as simple as filling a shallow dish or tray with water and lining it with pebbles or twigs to provide landing spots for the bees. This setup prevents drowning, a common hazard for bees attempting to drink from open water sources. It's important to maintain and refill these artificial water sources regularly, especially during hot or dry periods, to ensure bees always have access to water.

Monitoring the health of the plants and water quality in your bee's environment is also crucial. Pesticides and other chemicals used in gardens and agriculture can be detrimental to bees, contaminating their water sources and nectar. Opting for organic gardening practices and encouraging neighbors to do the same can help create a safer foraging ground for your bees. Additionally, ensuring that water sources are clean and free from pollutants is essential for maintaining colony health.

Avoiding Risky Zones for Beekeeping

Navigating the complexities of avoiding risky zones for beekeeping requires a strategic approach to location selection, emphasizing the importance of minimizing exposure to pesticides and reducing disturbances from human activity. The first step involves conducting thorough research on potential sites, focusing on historical and current pesticide usage in the area. This can be achieved by consulting local agricultural extensions, environmental protection agencies, or community gardening associations, which often maintain records of pesticide applications and can offer insights into areas with minimal chemical usage. When evaluating these areas, prioritize regions that have implemented organic farming practices or integrated pest management (IPM) strategies, as these are less likely to introduce harmful chemicals into the environment that could adversely affect your bees.

In addition to chemical risks, it's crucial to assess the level of human and mechanical disturbances that could impact the hive. Areas close to busy roads, frequent foot traffic, or heavy machinery operations can stress the bees, leading to decreased productivity or even abandonment of the hive. To mitigate these risks, seek out locations that offer natural or artificial barriers, such as dense vegetation, earthen berms, or privacy fencing, which can serve as buffers between the hive and potential sources of disturbance. These barriers not only reduce noise and physical interference but also can provide a microclimate that further supports hive health by moderating temperature extremes and shielding the bees from strong winds.

Another factor to consider is the proximity to wild habitats or conservation lands, which are generally less affected by human activities and pesticide use. Establishing hives near these areas can provide bees with a richer, more diverse foraging ground, enhancing their nutritional intake and strengthening colony resilience. However, it's important to ensure that these areas are not subject to future development plans that could introduce the very risks you're aiming to avoid. Engaging with local environmental groups or land trusts can provide valuable information on protected areas and any potential threats that could arise.

When selecting a site, also take into account the potential for future changes in land use or agricultural practices that could introduce risks to your bees. Establishing a dialogue with neighboring landowners and engaging in community planning meetings can offer foresight into planned developments, pesticide application schedules, or changes in crop cultivation that may affect your chosen location. By being proactive and informed, you can anticipate and mitigate potential risks before they impact your hives.

Finally, consider the legal aspects of hive placement, including zoning laws, ordinances, and regulations that may restrict beekeeping activities in certain areas or require specific setbacks from property lines or dwellings. Compliance with these regulations not only ensures legal protection for your beekeeping venture but also can influence your site selection by identifying areas that are zoned for agricultural use or have bee-friendly policies in place.

By meticulously evaluating each of these factors—pesticide exposure, disturbances, proximity to natural habitats, potential for future risks, and legal considerations—beekeepers can make informed decisions that prioritize the safety and well-being of their bees. This comprehensive approach to site selection underscores the beekeeper's role in safeguarding their colonies against external threats, thereby contributing to the sustainability and success of their beekeeping endeavors.

Integrating Hives into Urban and Rural Settings

Integrating hives into urban and rural settings requires a nuanced understanding of the specific challenges and opportunities each environment presents. In urban areas, rooftops, balconies, and small backyards can become productive sites for beekeeping with the right approach. For rooftop hives, ensure there is easy access for regular maintenance and emergency situations. Use **non-penetrative roof mounts** to secure hives without damaging roofing materials. These mounts can be weighted with cinder blocks or similar to resist wind without drilling into the structure. For balconies, select compact hive models designed for small spaces, ensuring they are anchored securely to prevent tipping. In both cases, provide a **water source** such as a small container filled with pebbles and water to prevent bees from seeking water in neighbor's areas, potentially causing nuisance.

In rural settings, the focus shifts to leveraging the natural landscape for hive placement. Identify areas that offer a mix of sun and shade throughout the day. A **south-facing slope** is ideal, as it provides morning sunlight with natural drainage, reducing dampness around the hive. Surrounding trees can offer wind protection, but ensure they do not cast too much shade, especially in the morning. Establishing hives near **wildflower meadows** or **untreated crops** can provide abundant foraging resources. However, it's crucial to maintain a buffer zone from agricultural fields that may use pesticides. A **50 to 100-foot buffer** can minimize exposure to drift from chemical applications.

For both urban and rural beekeepers, engaging with the community is vital. In urban areas, this might mean informing neighbors about your beekeeping activities and addressing any concerns they have about bee safety. Educational outreach can help demystify bee behavior and highlight the benefits of urban beekeeping for local biodiversity and pollination. In rural areas, collaboration with local farmers and landowners can open up additional foraging grounds for bees and foster practices beneficial to pollinators, such as planting bee-friendly crops or hedgerows.

Landscaping around hives can also enhance integration into both settings. In urban areas, use potted plants and green roofs to create foraging opportunities right where the bees live. Native flowering plants that thrive in pots, like lavender and thyme, can attract bees and also beautify the space. In rural areas, planting a diversity of native flowers, shrubs, and trees around the hive site can provide a steady bloom sequence, ensuring bees have access to nectar and pollen throughout the growing season. This landscaping approach not only supports the hive but also contributes to local ecosystem health and biodiversity.

Pest management in both environments should emphasize non-chemical methods to maintain a healthy hive and environment. In urban areas, where space is limited, regularly inspect hives for signs of pests and diseases, and use mechanical barriers or natural predators to control pest populations. Rural beekeepers have the advantage of more space and can use strategies like crop rotation and diversified planting to naturally reduce pest pressures.

Finally, regardless of the setting, always ensure hives are **properly ventilated** to prevent overheating in summer and moisture accumulation in winter. Urban hives may require additional shade or ventilation modifications to deal with heat from surrounding buildings or paved surfaces. Rural hives benefit from natural breezes but may need windbreaks or insulation to protect against cold winter winds.

CHAPTER 18: ACQUIRING YOUR FIRST BEE COLONIES

Acquiring your first bee colonies is a pivotal step in your beekeeping journey, requiring careful consideration and preparation to ensure a successful start. The process involves deciding between purchasing packaged bees, nucleus colonies, or full hives, each with its own set of advantages and considerations.

Packaged Bees are a common choice for beginners due to their availability and cost-effectiveness. A typical package includes about 3 pounds of bees, which is approximately 10,000 bees, and a mated queen. When introducing packaged bees to a new hive, it's crucial to follow a step-by-step process to ensure they accept their new home and queen. This involves placing the queen cage between the frames in your hive, securely closing the hive, and then gently shaking the package to disperse the bees into the hive. Afterward, provide sugar syrup to help them start building comb and acclimatize to their new environment. It's important to check after a few days to ensure the queen has been successfully released and is laying eggs.

Nucleus Colonies (Nucs) offer a more established start, consisting of 4 to 5 frames of brood, honey, pollen, bees, and a laying queen. The advantage of a nuc is that it's a mini-hive with a queen that has already begun laying eggs, and the bees are already accustomed to her. To transfer a nuc into your hive, carefully move each frame to your hive box, maintaining their original order to preserve the brood nest's integrity. This method reduces the stress on the bees and allows for a smoother transition. Ensure your hive is ready with additional frames to fill out the box, providing space for the colony to grow.

Full Hives are an option for those looking to start with a larger colony. Purchasing a full hive involves acquiring a complete setup from another beekeeper, including the bees, frames, hive boxes, and possibly even honey stores. This option requires the least amount of initial work since the colony is already established. However, it's essential to inspect the hive thoroughly for diseases or pests and to understand the history of the colony's management before making the purchase. Transporting a full hive requires careful planning to secure the frames and close the hive entrance to prevent bees from escaping during transport.

Regardless of the option you choose, sourcing your bees from a reputable supplier is critical. Look for suppliers with positive reviews from other beekeepers, and if possible, visit their operation to see the conditions of their bees and ask about their breeding practices. Healthy bees are the foundation of successful beekeeping, so ensure the colonies show signs of vigor, the queen has a good laying pattern, and there are no visible signs of disease or pests.

After acquiring your bees, monitoring the colony's health and growth becomes your next focus. Regular inspections will help you identify any issues early on, such as signs of queen failure, disease, or pest infestations. Providing adequate food sources, either through natural forage or supplementary feeding, especially in the early stages, will support your colony's development.

In summary, acquiring your first bee colonies is a significant step that sets the stage for your beekeeping experience. Whether you choose packaged bees, nucleus colonies, or full hives, understanding the needs and best practices for each option will help you establish healthy, productive colonies. With careful preparation, observation, and management, your beekeeping journey will be rewarding and contribute positively to the environment by supporting these essential pollinators.

Reliable Sources for Buying Bee Colonies

Finding reliable sources for purchasing healthy and thriving bee colonies is essential for a successful beekeeping endeavor. When seeking out suppliers, prioritize those who are reputable within the beekeeping community and have a track record of providing vigorous, disease-free bees. Begin by consulting local beekeeping associations or clubs, as these groups often have insights into the best local suppliers. Members can share their experiences with different suppliers, offering a wealth of information on who provides the healthiest bees and the best customer service.

Another valuable resource is beekeeping forums and social media groups, where beekeepers from various regions share their experiences and recommendations. Look for suppliers who are consistently praised for the quality of their bees and their commitment to customer support. Suppliers who actively participate in these communities and are open to answering questions and providing advice are often a good choice, as they demonstrate an ongoing commitment to the beekeeping community.

When evaluating a potential supplier, inquire about their breeding practices and the origin of their bees. Suppliers who breed their own queens and manage their own apiaries are often more knowledgeable about the health and history of their colonies. Ask about the measures they take to ensure their bees are free from diseases and pests, such as routine health checks and treatment protocols. Suppliers should be transparent about their practices, including any treatments used to manage mite infestations or other health issues.

It's also important to consider the genetic diversity and adaptability of the bees. Suppliers who offer bees adapted to your local climate and environmental conditions can provide colonies that are more likely to thrive in your specific setting. Bees that are well-adapted to local conditions are generally more resilient and productive, reducing the need for intensive management and interventions.

Before finalizing a purchase, request to inspect the bees, if possible. This may not be feasible if you're ordering from a distant supplier, but local or regional suppliers may allow visits. During an inspection, look for signs of a healthy colony, such as active foraging behavior, a visible laying pattern from the queen, and the absence of pests or diseases within the hive. A reputable supplier should be willing to discuss any concerns you have and offer guarantees or support to address any issues with the bees after purchase.

By thoroughly researching suppliers, asking the right questions, and seeking advice from experienced beekeepers, you can identify reliable sources for purchasing healthy and thriving bee colonies. This careful selection process is a critical step in establishing a successful and sustainable beekeeping practice.

Introducing Bees to a New Hive Safely

When introducing bees to a new hive, the process must be approached with care to ensure the safety of the bees and the successful establishment of the colony. The following steps outline the specific techniques to be employed during this critical phase:

1. **Prepare the Hive**: Before the arrival of your bees, ensure that the hive is fully assembled and situated in its final location. The hive should have frames with foundation installed to give the bees a head start on comb building. It's advisable to place a feeder filled with sugar syrup inside the hive to provide an immediate food source for the bees upon their introduction.

2. **Queen Introduction**: The queen usually arrives in a separate queen cage. Before placing the queen cage between the frames, remove the cork or cap from the candy end of the cage, ensuring there's a candy plug that the worker bees will eat through to gradually release the queen. This slow release allows the worker bees to acclimate to the queen's pheromones, reducing the risk of rejection.

3. **Releasing the Bees**: If you're introducing packaged bees, carefully remove the can of syrup that comes with the package and set it aside. Then, gently tap the package to move the bees to the bottom, open the package, and remove the queen cage. After placing the queen cage, hold the package over the hive and gently shake it to guide the bees into the hive, onto the frames. Some beekeepers prefer to place the entire open package near the hive entrance, allowing the bees to enter the hive at their leisure.

4. **Feeding**: Install a feeder filled with sugar syrup to provide an essential food source for the bees. Early feeding helps the colony establish itself by encouraging comb building and giving the bees the energy they need to start foraging for nectar and pollen. A boardman feeder or an in-hive frame feeder works well for this purpose.

5. **Closing the Hive**: After the bees are introduced, close the hive gently but firmly. Ensure that the lid is secure and that there are no gaps for bees to escape, except for the hive entrance. It's crucial to reduce the entrance size using an entrance reducer to protect the new colony from potential robbers and pests while they are still weak.

6. **Observation**: Over the next few days, observe the hive from a distance to minimize disturbance. Look for signs of normal activity, such as bees entering and leaving the hive. After 3-7 days, you can perform a quick inspection to confirm that the queen has been released from her cage and is laying eggs. However, keep this inspection brief to avoid stressing the colony.

7. **Environmental Considerations**: Ensure that the hive is placed in an area with adequate sunlight, minimal wind, and protection from predators. The hive should have a clear flight path and be tilted slightly forward to prevent water accumulation inside the hive.

8. **Pest and Disease Management**: Implement an integrated pest management strategy from the beginning to monitor and control mite levels and other potential diseases. Early detection and management are crucial for the health and longevity of the colony.

By following these detailed steps, beekeepers can successfully introduce bees to a new hive, fostering a healthy and productive colony. Remember, patience and careful observation are key during the initial introduction period to ensure the colony's successful establishment and future growth.

Differences: Natural Swarms vs. Bred Queens

Understanding the key differences between acquiring bees through natural swarms and opting for professionally bred queens is crucial for beekeepers aiming to make informed decisions that align with their beekeeping goals and management practices. Natural swarms represent a traditional method of expanding a beekeeping operation or starting new colonies. This process involves capturing a swarm, which is a natural occurrence where a queen bee leaves the hive with a portion of the worker bees to form a new colony. In contrast, purchasing professionally bred queens involves selecting a queen from a breeder, who has raised queens specifically for their genetic traits, such as gentleness, productivity, and disease resistance.

One of the primary distinctions between these two methods lies in the predictability and control over the genetic makeup of the colony. Professionally bred queens come with known genetic backgrounds and specific traits that have been selected for, offering beekeepers the ability to tailor their colonies to specific needs or preferences. For instance, a beekeeper looking to enhance honey production might opt for a queen known for high productivity. On the other hand, natural swarms offer less predictability in terms of genetic traits, as the exact background and characteristics of the swarm's queen and her workers are generally unknown.

Another significant difference is the health and disease resistance of the colony. Professional breeders often ensure that their queens come from stock that is resistant to certain diseases and pests, such as Varroa mites, which can be a considerable advantage in maintaining colony health. These queens are usually accompanied by certification or assurance of their health status, reducing the risk of introducing diseases or pests into the beekeeping environment. Conversely, natural swarms might carry an unknown health status, potentially introducing diseases or pests to the beekeeper's existing colonies. Therefore, beekeepers capturing natural swarms must be diligent in monitoring and possibly treating new swarms for health issues.

The cost and availability of bees also differ significantly between these two methods. Capturing natural swarms typically involves minimal direct costs, making it an attractive option for beekeepers on a budget or those looking to expand their apiaries economically. However, the opportunity to capture swarms depends heavily on the season and local bee activity, which can be unpredictable. In contrast, professionally bred queens can be ordered and shipped at specific times, allowing beekeepers to plan their colony expansions or replacements with greater certainty. The cost of purchasing a queen and potentially shipping her can be higher than capturing a swarm, but the investment may be justified by the benefits of selected genetic traits and health assurances.

The impact on beekeeping practices and hive management is another area where differences emerge. Introducing a professionally bred queen to a new or existing colony requires careful management to ensure acceptance by the worker bees, including methods such as the slow release from a queen cage. This process allows the colony to acclimate to the pheromones of the new queen, reducing the risk of rejection. With natural swarms, the swarm has already accepted the queen, and the primary focus is on providing a suitable environment for the swarm to establish a new colony. However, beekeepers must still monitor these new colonies closely for signs of successful queen acceptance and laying patterns, as well as overall colony health.

CHAPTER 20: ECONOMIC ASPECTS OF BEEKEEPING

Exploring potential income streams in beekeeping is crucial for understanding the economic viability of this endeavor. One of the primary sources of income for beekeepers is the sale of **honey**. To maximize profits from honey, it's essential to harvest at the right time, ensuring the honey is of high quality and can command a better price in the market. This involves monitoring the nectar flow and ensuring that the honey is correctly extracted, filtered, and packaged. The packaging plays a significant role in marketing honey, with options ranging from glass jars to plastic containers, each with its labeling requirements to meet local health regulations and appeal to consumer preferences.

Another significant income stream comes from the sale of **beeswax**, a by-product of honey harvesting. Beeswax can be purified and molded into blocks or sheets for sale to candle makers, cosmetic companies, and hobbyists. The purification process involves melting the raw beeswax, filtering out debris, and possibly bleaching it to achieve a uniform color. Offering beeswax in various forms, such as organic or cosmetic grade, can cater to different market segments and increase its value.

Propolis and **royal jelly** are niche products with higher market values due to their perceived health benefits. Harvesting these requires specific techniques; for example, propolis can be scraped from the hive frames, and royal jelly is extracted from queen cells using a small, soft spatula. These products must be stored and packaged under controlled conditions to preserve their quality, often requiring refrigeration for royal jelly and airtight containers for propolis.

Pollen collection is another potential income source, using pollen traps at the hive entrance to collect pollen as bees return from foraging. This pollen must be dried and stored properly to prevent mold growth, making it a more labor-intensive product but one that can fetch a high price, especially when sold as a nutritional supplement.

For those looking to diversify further, offering **pollination services** to local farmers and orchards can be lucrative. This involves transporting hives to fields or orchards during bloom to enhance crop pollination, a service for which farmers are willing to pay. The logistics of hive transportation and ensuring the health of bees during the pollination service are critical aspects of this income stream.

Queen rearing and selling is a specialized field within beekeeping that requires knowledge of bee genetics and breeding techniques. A high-quality queen can command a premium price, especially if she comes from a line known for productivity, gentleness, or disease resistance. This venture requires careful planning and marketing, often involving shipping queens to customers, which adds a layer of complexity in terms of packaging and delivery.

Beekeeping classes and workshops offer an additional revenue stream, sharing knowledge with aspiring beekeepers. This can range from basic beekeeping courses to more advanced topics like queen rearing or disease management. Offering these classes requires a deep understanding of beekeeping, teaching skills, and the ability to engage with students of varying levels of knowledge and interest.

Lastly, **government incentives and support programs** can provide financial assistance for starting or expanding beekeeping operations. These may include grants for purchasing equipment, subsidies for conservation practices, or funding for research into sustainable beekeeping methods.

Navigating these programs requires understanding the application processes and eligibility criteria, which can vary widely by region.

Each of these income streams requires careful consideration of the costs involved, including initial investments, ongoing expenses, and the time required to manage each aspect of the business. Balancing these factors against the potential income and market demand is crucial for developing a sustainable and profitable beekeeping operation.

Beekeeping Costs: Setup and Ongoing Expenses

Understanding the **initial setup costs** of beekeeping is essential for beginners to budget effectively. The first significant expense is the **purchase of hives**. A standard Langstroth hive, which includes a bottom board, hive bodies or brood chambers, supers for honey storage, an inner cover, and a telescoping outer cover, can range from $100 to $250 depending on the materials (e.g., cedar is more expensive than pine) and whether the hive is pre-assembled or requires assembly.

Bee colonies can be acquired through packages, nucs (nucleus colonies), or full hives. A package, which includes approximately 10,000 bees and a queen, can cost between $100 to $150. Nucs, which are small, established colonies with a queen, frames, bees, brood, and honey, are slightly more expensive, ranging from $150 to $200. The cost of a full hive, although less common for initial purchases, can exceed $300 but provides the advantage of a fully functional colony from the start.

Protective gear is a must for any beekeeper to avoid stings and ensure comfort while working with bees. A full bee suit with a veil can cost between $100 to $200, gloves range from $15 to $30, and a pair of beekeeping boots can cost up to $50. While some opt for less expensive alternatives, such as simple veils or jackets, full protection is recommended for beginners.

Tools for hive management include a smoker, hive tool, bee brush, and frame gripper. A basic smoker costs between $30 to $50, a hive tool around $10, a bee brush less than $10, and a frame gripper about $15. These tools are essential for inspecting hives, managing frames, and calming bees during hive inspections.

For **feeding the bees**, especially in the early stages or during nectar dearths, an initial investment in feeders and food (sugar syrup or fondant) is necessary. Feeders cost between $5 to $20 depending on the type (e.g., entrance feeder, top feeder, frame feeder), and the cost of sugar for syrup can vary depending on local prices and the quantity needed.

Ongoing expenses include the replacement of queens, which may be necessary every 2 to 3 years or due to unexpected loss, costing between $20 to $40 each. **Hive maintenance** supplies, such as paint for wooden hives to protect them from the elements, can add to annual costs, though this is generally minimal. **Medications and treatments** for pests and diseases, such as Varroa mites and American foulbrood, represent another recurring cost. For example, Varroa mite treatments can range from $3 to $5 per hive for organic acids like oxalic acid to $20 to $30 for synthetic acaricides per treatment.

Harvesting equipment for extracting honey involves a one-time purchase for items like an extractor, which can be manual or electric, ranging from $150 for a small, manual model to over $500 for a larger, electric one. Buckets, filters, and jars for packaging honey also contribute to the costs, though these can vary widely based on scale and preferences.

Lastly, **educational resources** such as books, courses, or club memberships can be considered part of the ongoing investment in beekeeping knowledge. Costs here can range from minimal to several hundred dollars annually, depending on the level of engagement and learning desired.

By breaking down these costs, prospective beekeepers can better plan and budget for their new venture, ensuring they are prepared for both the initial investment and the ongoing expenses associated with maintaining healthy and productive hives.

Income Streams from Honey, Wax, and Propolis

Diving deeper into the economic aspects of beekeeping, it's essential to consider the various channels through which bee products can be marketed and sold. Beyond the primary income streams of honey, wax, and propolis sales, beekeepers can explore additional avenues to enhance their profitability.

When it comes to marketing honey, understanding the preferences of your target market is key. For instance, raw, unfiltered honey often appeals to health-conscious consumers and can fetch a higher price. Packaging can significantly influence customer perception and sales; glass jars with attractive labeling that includes information on the honey's origin, such as "Wildflower Honey from [Location]," can create a premium product. Selling honey in local farmers' markets, online platforms, or through community-supported agriculture (CSA) shares can open up direct channels to consumers who are willing to pay more for locally sourced, sustainable products.

Beeswax, with its versatile applications, presents another lucrative opportunity. Pure, filtered beeswax can be sold to candle makers, cosmetic companies, and artisans who value its natural qualities. Beekeepers can also venture into creating their own beeswax products, such as candles, lip balms, and skincare items. Crafting these products requires an initial investment in molds, ingredients like essential oils, and packaging materials. However, the added value of these finished goods can significantly increase profit margins. Marketing these products alongside honey at local markets or through online stores can attract a diverse customer base interested in natural, eco-friendly products.

Propolis, known for its medicinal properties, caters to a niche market that is growing in awareness and demand. Extracting and preparing propolis for sale involves careful collection and possibly creating tinctures or other health supplements. This process requires knowledge of safe handling and preparation techniques to preserve its beneficial properties. The market for propolis products is expanding, with opportunities to sell through health food stores, online wellness platforms, and at beekeeping or natural health conventions.

Pollen collection, while more labor-intensive, opens up another specialty product line. Fresh or dried bee pollen can be marketed as a nutritional supplement, appealing to health enthusiasts. The key to successful pollen sales lies in its quality and purity, necessitating meticulous collection, drying, and packaging processes to maintain its nutritional value. Engaging with local gyms, health food stores, and nutritionists to offer samples or educational materials about the benefits of bee pollen can help in establishing a market for this product.

In addition to product sales, beekeepers can generate income through educational endeavors. Hosting workshops on beekeeping, honey extraction, or crafting with hive products can attract individuals interested in learning these skills. These workshops can be marketed through local community centers, gardening clubs, and social media platforms dedicated to sustainability and natural living.

Collaborating with local businesses such as bakeries, breweries, and restaurants to supply honey or other bee products for their offerings can establish steady B2B income. This requires building relationships with business owners and understanding their needs, offering a consistent supply of quality products that can enhance their goods.

Lastly, renting out hives for pollination services to local farms and orchards provides a dual benefit of income generation and supporting local agriculture. This requires coordination with farmers to understand their crop cycles and ensuring that your bees are healthy and capable of providing effective pollination.

Each of these income streams involves specific considerations in terms of production, marketing, and sales strategies. Beekeepers should assess their capacity, resources, and interests to select the avenues that best fit their operation. By diversifying income sources, beekeepers can create a more resilient and sustainable business model that not only benefits their livelihood but also contributes to the health of the bee population and the broader ecosystem.

Government Incentives for Beginner Beekeepers

Accessing government incentives and support programs for beginner beekeepers can significantly reduce the financial burden of starting and maintaining a beekeeping operation. These programs often provide financial assistance in the form of grants, subsidies, or low-interest loans designed to promote sustainable agriculture practices, including beekeeping. To navigate these opportunities, it's essential to start by researching local, state, and federal agricultural departments that oversee such incentives. Websites like the United States Department of Agriculture (USDA) offer a wealth of information on available programs, including eligibility criteria and application processes.

One practical step is to subscribe to newsletters or follow social media accounts of relevant agricultural agencies to stay updated on new funding opportunities or changes in existing programs. Additionally, attending local beekeeping association meetings can provide insights from experienced beekeepers who may have successfully obtained government support in the past. These gatherings are invaluable for networking and gaining tips on preparing a compelling application.

When preparing to apply for an incentive or support program, meticulously gather all required documentation, which may include proof of land ownership or lease, a detailed business plan outlining your beekeeping project, and any environmental assessments if required. It's crucial to highlight in your application how your beekeeping operation will contribute to biodiversity, pollination services, or organic farming practices, as many programs favor projects that have a positive environmental impact.

For programs offering grants, understand that these funds do not need to be repaid, making them highly competitive. Your application should clearly articulate the uniqueness and sustainability of your beekeeping venture. On the other hand, if applying for low-interest loans, prepare to demonstrate financial stability and a solid plan for repayment.

Managing Financial Risks in Beekeeping

Managing financial risks in beekeeping, particularly those associated with hive losses or market fluctuations, requires a strategic approach that combines proactive management with informed decision-making. Beekeepers, whether novices or experienced, must understand that the financial health of their operations is as crucial as the health of their hives. To mitigate risks effectively, it's essential to adopt several key practices, each aimed at minimizing potential losses and ensuring the sustainability of the beekeeping venture.

Firstly, diversification is a fundamental strategy. Just as financial investors spread their investments across various assets to reduce risk, beekeepers can diversify their income sources. Beyond honey production, consider venturing into the sale of beeswax, propolis, royal jelly, and pollen. Additionally, offering pollination services or beekeeping workshops can provide alternative revenue streams that help stabilize income, especially when one source underperforms due to market conditions or production issues.

Secondly, maintaining healthy bee colonies is paramount to preventing hive losses. This involves regular monitoring for pests and diseases, such as Varroa mites and American foulbrood, and implementing integrated pest management strategies that balance effectiveness with environmental sustainability. Early detection and treatment can prevent the spread of issues that might lead to colony collapse. Investing in quality queens, ensuring genetic diversity within colonies, and practicing good nutrition and hive management techniques also contribute to stronger, more resilient hives.

Insurance is another critical tool for managing financial risk. Beekeeping operations can benefit from insurance policies that cover hive theft, vandalism, and natural disasters, which are unpredictable but can have devastating effects on production. Some policies may also offer compensation for colony collapse disorder, providing a financial safety net that can help beekeepers recover more quickly from significant losses.

Adopting a conservative approach to financial planning can also safeguard against market fluctuations. This involves setting aside a portion of profits during good years to create a buffer for leaner times. Monitoring market trends and adjusting production and sales strategies accordingly can help anticipate changes and make informed decisions about inventory management, pricing, and marketing efforts.

Lastly, building strong relationships with local beekeeping communities, agricultural extensions, and industry associations can offer invaluable support. These networks can provide early warnings about emerging threats, advice on best practices, and opportunities for collaborative marketing or bulk purchasing of supplies to reduce costs. Engaging with these communities fosters a culture of knowledge sharing and mutual support, which can be critical during challenging times.

By implementing these strategies, beekeepers can navigate the complexities of financial risk management in their operations. The goal is not only to survive the inevitable ups and downs of the market and environmental conditions but to thrive and grow, contributing to the sustainability and resilience of the beekeeping industry as a whole.

Part 6: Building and Managing Hives

CHAPTER 21: TYPES OF HIVES

Top-bar hives offer a more sustainable and natural approach to beekeeping, appealing to those looking to maintain a closer relationship with the natural behaviors and health of their bee colonies. Unlike the Langstroth hive, which is designed for maximum honey production, the top-bar hive prioritizes the comfort and natural living conditions of the bees. The hive consists of a horizontal trough with wooden bars laid across the top. Each bar serves as a foundation upon which the bees build their comb. This design allows for easy inspection and harvesting without the need for heavy lifting or disturbing the entire colony. The top-bar hive is well-suited for small-scale beekeepers who prefer a more hands-on approach to beekeeping.

When constructing a top-bar hive, it's essential to use durable, untreated wood to avoid chemicals that could harm the bees. Cedar and pine are popular choices due to their longevity and resistance to rot. The hive should be elevated off the ground on a stand to prevent moisture accumulation and to deter pests. The length of the hive is critical; too long, and the hive becomes difficult to manage, too short, and it won't provide enough space for a healthy colony. A length of 3 to 4 feet is generally recommended.

The top bars themselves should be precisely cut to fit snugly across the width of the hive body, preventing bees from building comb in areas that cannot be inspected or harvested. A gentle slope on the sides of the hive encourages the bees to build their comb hanging from the bars, facilitating easy removal. Each bar can be prepped with a small strip of beeswax to guide the bees in building straight comb.

Ventilation is another critical aspect of top-bar hive design. Adequate air flow is necessary to keep the hive dry and prevent mold. However, too much ventilation can chill the brood and stress the colony. Strategically placed vents or adjustable openings can help maintain the right balance.

The hive entrance should be positioned to face away from prevailing winds and should be just large enough for two or three bees to pass through at once. This helps the bees defend their hive against intruders. An overhang above the entrance provides shelter from the rain.

In terms of maintenance, the top-bar hive requires regular monitoring to ensure the colony is healthy and the comb is being built correctly. Since the bees build their comb naturally, there's a risk of cross-combing or uneven comb structures, which can complicate hive inspections and harvesting. Beekeepers must carefully manage the space within the hive, adding or removing bars as needed to guide the bees' construction efforts.

Harvesting honey from a top-bar hive is a simple process that involves removing the bars with fully capped honeycomb. Since the comb is cut away from the bar, this method is less intrusive and leaves the bees with enough honey stores. However, it's important to leave a sufficient amount of honey for the bees to survive the winter months.

Overall, the top-bar hive represents a bee-centric approach to beekeeping, emphasizing the health and well-being of the bee colony over maximum honey production. Its design and management practices align with sustainable and natural beekeeping philosophies, making it an attractive option for environmentally conscious beekeepers.

Langstroth Hives: Pros and Cons

Langstroth hives, named after Lorenzo Lorraine Langstroth who revolutionized beekeeping in the 19th century, are distinguished by their use of removable frames which the bees build their comb into. This innovation not only made it easier to inspect and manage bee colonies but also to harvest honey without destroying the hive. The design of Langstroth hives includes a series of boxes stacked vertically, each containing frames for the bees to build their comb. The standardization of hive dimensions allows for the interchangeability of parts and ease of expansion by adding more boxes, known as supers, on top.

One of the primary advantages of Langstroth hives is their scalability. As a beekeeper, you can start with a few boxes and expand vertically by adding more supers as the colony grows or as honey production increases. This flexibility makes it an attractive option for both hobbyists and commercial beekeepers. The removable frames also facilitate easy inspection for diseases and pests, crucial for maintaining colony health. By pulling out frames, beekeepers can check for signs of problems such as Varroa mites or brood diseases without much disturbance to the bees. Additionally, the extraction of honey is simplified with Langstroth hives. Frames can be removed and placed in an extractor, spinning out the honey without damaging the comb, allowing bees to reuse it, which is more efficient and less labor-intensive compared to traditional methods.

However, Langstroth hives are not without their disadvantages. The very structure that offers scalability and ease of inspection can also be cumbersome. The boxes, when full of honey, can be heavy, often weighing over 50 pounds each, making them difficult to lift and manage, especially for individuals with back problems or limited physical strength. This aspect of Langstroth beekeeping can be a significant drawback for some, potentially requiring additional equipment or assistance to manage the hives effectively.

Another potential downside is the initial cost. While the standardized dimensions of Langstroth hives mean that parts are readily available, the cost of purchasing multiple boxes, frames, and foundation can add up, especially for those just starting. Additionally, because the frames and foundations are often coated with wax or made from plastic to encourage the bees to build comb, there is an ongoing debate about the impact of these materials on bee health and honey quality.

Moreover, the design of Langstroth hives, while efficient for honey production, can be seen as less natural from the bees' perspective. The imposition of a predetermined frame structure for comb building can interfere with the natural comb building instincts of bees, potentially stressing the colony. Some natural beekeeping advocates argue that this can lead to a disconnection from the natural behaviors observed in wild bee populations, although many commercial and hobbyist beekeepers find the trade-offs acceptable given the benefits in terms of honey production and hive management.

In managing Langstroth hives, it's essential to monitor the space within the hive carefully. Overcrowding can lead to swarming, where a portion of the colony leaves to find a new home, reducing your bee population and potentially your honey yield. Conversely, too much space can lead to inefficient heating and cooling of the hive, stressing the bees. Thus, understanding how to manage the expansion and contraction of the hive space through the addition or removal of supers is a critical skill for successful Langstroth beekeeping.

Despite these considerations, Langstroth hives remain the most popular choice among beekeepers globally. Their design caters well to the needs of beekeeping, offering a practical balance between ease of management, honey production, and scalability. For those considering beekeeping, weighing these advantages and disadvantages in light of personal circumstances, physical capabilities, and beekeeping goals is crucial. Whether you're a beginner looking to start a hobby or a professional aiming to scale up production, understanding the nuances of Langstroth hive management will be key to your success in beekeeping.

Top-Bar Hives: Sustainable Beekeeping Choice

The simplicity of managing a top-bar hive extends beyond its construction and into the day-to-day interaction with the bees. For instance, inspecting a top-bar hive for health and productivity involves gently lifting each bar to view the comb, a process that minimally disturbs the bees and reduces stress on the colony. This method contrasts with more traditional approaches that often require more significant manipulation of the hive structure and can be more disruptive to the bees. Beekeepers find that this approach not only fosters a more harmonious relationship with their bees but also offers a deeper insight into the natural processes of the hive, encouraging a more intuitive form of beekeeping.

In managing pests and diseases, the top-bar hive's design facilitates a more natural approach. The absence of foundation in the comb construction allows bees to build cells to their preferred size, which some beekeepers believe can help naturally reduce issues with Varroa mites, as the mites prefer larger cells typically found in standard frame-based hives. Additionally, the ease of removing individual bars and combs allows for targeted management practices, such as removing comb sections affected by pests or diseases without having to disturb the entire hive structure.

The harvesting process in a top-bar hive is notably distinct and aligns with sustainable beekeeping principles. Instead of using extractors or other mechanical means to remove honey, beekeepers cut the comb directly off the bar, crush it, and strain it to separate the honey. This method is gentle on the bees and the environment, as it requires no electricity and minimal equipment. Moreover, it allows for the production of comb honey, a premium product that many consumers prefer for its natural presentation and purity. Beekeepers must be mindful, however, to leave enough honey in the hive to sustain the colony through the winter, a practice that underscores the sustainable ethos of taking only what is needed and ensuring the health and well-being of the bee community.

Another aspect of sustainability in top-bar beekeeping is the emphasis on local adaptation and resilience. By allowing bees to build comb naturally, they can adapt their living space to suit local environmental conditions and their specific needs, potentially leading to stronger, more resilient colonies. This adaptability is crucial in the face of changing climate conditions and challenges to bee health. Furthermore, top-bar hives can be made from locally sourced materials, reducing the carbon footprint associated with beekeeping and supporting local economies.

The design of the top-bar hive also contributes to a more ergonomic beekeeping experience. The horizontal layout eliminates the need to lift heavy boxes, a common requirement in traditional beekeeping that can lead to strain and injury. This accessibility makes beekeeping more inclusive, allowing individuals of all ages and abilities to participate in and contribute to the practice. It opens up the world of beekeeping to a broader audience, fostering a greater appreciation for bees and their vital role in our ecosystems.

Moreover, the top-bar hive's design simplicity extends to its maintenance. The natural materials used in construction, such as untreated wood, and the hive's design, which promotes good air circulation, help to prevent issues like mold and wood rot, ensuring the hive's longevity with minimal intervention. This durability further underscores the sustainable nature of top-bar beekeeping, as it reduces the need for frequent replacements and repairs, thereby minimizing waste and resource consumption.

In conclusion, the top-bar hive embodies the principles of sustainable and natural beekeeping through its design, management practices, and the ethos it promotes. By fostering a close, respectful relationship with bees, emphasizing the health and well-being of the colony, and minimizing environmental impact, top-bar hives offer a compelling choice for beekeepers committed to sustainability and the future of beekeeping.

Warré Hives: Tradition Meets Innovation

Warré hives, named after their inventor Abbé Émile Warré, embody a beekeeping philosophy that emphasizes minimal intervention and allowing bees to construct their hive in a manner that closely mimics their natural behavior in the wild. This approach is often referred to as "beekeeping for the bees," and it integrates traditional methods with modern innovations to support the health and productivity of the bee colony. The Warré hive is characterized by its vertical stacking of boxes, similar to the Langstroth system, but with a few key differences that cater to natural beekeeping practices.

The design of a Warré hive starts with the hive body, which consists of several boxes stacked on top of each other without frames or foundation, allowing bees to build their comb naturally from the top down. Each box is slightly smaller than the one below it, creating a conical shape that encourages the bees to move upward as they build comb and fill the hive with honey. This design facilitates the bees' natural thermoregulation and ventilation needs, promoting a healthy living environment within the hive.

To construct a Warré hive, begin with untreated wood to avoid any chemical contamination that could harm the bees. Cedar is a popular choice due to its durability and resistance to rot. Each box should measure approximately 20 inches by 16 inches, with a depth of 8 inches, although dimensions can vary based on local bee sizes and preferences. The top of each box features a set of bars, similar to the top-bar hive, onto which the bees will attach their comb. These bars are placed closely together to prevent the bees from building comb in areas that are not easily inspectable or harvestable.

One of the modern innovations in Warré beekeeping is the quilt box, a unique feature that sits atop the uppermost hive box under the roof. The quilt box is filled with natural insulating materials, such as wood shavings or straw, to regulate temperature and absorb moisture, creating an optimal microclimate within the hive. This feature is crucial for overwintering success, as it keeps the bees warm and dry during cold months.

The roof of the Warré hive is designed to overhang the sides of the top box, providing protection from the elements. It is often insulated or constructed with materials that reflect sunlight to prevent overheating in summer. The hive's entrance is located at the bottom of the lowermost box and is usually just wide enough to allow two or three bees to pass through at a time, optimizing hive defense and temperature control.

For beekeepers adopting Warré hives, monitoring and maintenance practices differ significantly from those used with Langstroth or top-bar hives. The philosophy behind Warré beekeeping discourages frequent inspections, promoting a hands-off approach to allow bees to thrive with minimal human interference. However, beekeepers should still monitor the hive's external indicators of health, such as bee activity at the entrance, and be prepared to intervene if problems are suspected.

Harvesting honey from a Warré hive typically involves removing the top box once it is full of honey and the bees have moved down to lower boxes. This method, known as "nadiring," is less disruptive to the bees and maintains the integrity of the brood nest. The removed box's comb can then be crushed and strained to extract the honey, or sections of comb can be cut out and packaged as comb honey, a delicacy highly valued by many consumers.

Incorporating modern innovations into Warré beekeeping, such as varroa mite monitoring techniques and organic treatments for pests and diseases, allows beekeepers to maintain the naturalistic ethos of Warré hives while addressing contemporary challenges. For example, integrating mesh floors can help manage varroa mite populations by allowing mites to fall out of the hive, reducing their impact on bee health.

By blending traditional beekeeping methods with modern innovations, Warré hives offer beekeepers a sustainable and bee-centric approach to hive management. This system not only supports the well-being of the bee colony but also fosters a deeper connection between the beekeeper and the natural world, aligning with principles of conservation and ecological stewardship.

CHAPTER 22: DIY HIVE CONSTRUCTION TECHNIQUES

When embarking on the journey of **DIY hive construction**, it's crucial to start with a comprehensive plan and a clear understanding of the materials and tools required. The first step is selecting the **type of hive** you wish to build, be it a Langstroth, top-bar, or Warré hive. Each has its unique dimensions, advantages, and construction challenges. For a **Langstroth hive**, you will need:

- **Untreated wood**: Cedar or pine are excellent choices due to their durability and resistance to rot. For a standard Langstroth hive, boards should be at least ¾ inch thick to provide adequate insulation and strength. You will need boards for the hive body, supers, bottom board, and cover. The exact quantity and dimensions will vary based on the hive design, but a typical Langstroth hive consists of two deep hive bodies and two to three medium supers.

- **Frame wire and foundation**: To support the bees in comb building. The wire should be stainless steel to resist rust, and the foundation can be made of wax or plastic, depending on your preference.

- **Nails or screws**: Galvanized or stainless steel to assemble the hive components. Screws offer more durability but require pre-drilling to avoid splitting the wood.

- **Exterior-grade glue**: To enhance the structural integrity of the hive.

- **Paint or sealant**: Only for the hive's exterior to protect the wood from the elements. Use a non-toxic, water-based product to ensure bee safety.

Tools needed for hive construction include a saw (hand saw or power saw for precision cutting), hammer or screwdriver, drill with various bits for pre-drilling screw holes and creating entrance holes, measuring tape, square, and a paintbrush or roller for applying the sealant or paint.

Construction Steps:

1. **Cut the Wood**: According to the dimensions of your chosen hive type. For a Langstroth hive, the standard deep body is 19 7/8" long by 16 1/4" wide by 9 5/8" high. Precision is key to ensure all parts fit together snugly.

2. **Assemble the Hive Bodies and Supers**: Apply glue to the joints before screwing or nailing them together for extra strength. Ensure the boxes are square by measuring diagonally from corner to corner; the measurements should be the same.

3. **Prepare the Frames**: If not using pre-assembled frames, construct the frames by nailing the four pieces together and adding the wire and foundation. The standard Langstroth frame size is 19" by 9 1/8" for deep frames and 19" by 6 1/4" for medium frames.

4. **Build the Bottom Board and Lid**: The bottom board serves as the hive's foundation, so it must be sturdy. A simple design includes a solid piece of wood with a rim to keep the hive bodies in place. The lid can be a flat cover with a rim around the edge to fit snugly over the top super or a telescoping cover with sides that extend down over the top super for added protection.

5. **Paint or Seal the Hive**: Apply at least two coats of paint or sealant to the hive's exterior surfaces, avoiding the inside where the bees will live. Allow ample drying time between coats.

Final Touches:

- **Install the Entrance Reducer**: This can be a simple piece of wood placed at the hive entrance to control access.
- **Add the Frames to the Hive Bodies**: Once the paint or sealant is dry, you can place the frames inside the hive bodies.
- **Set Up the Hive Stand**: Elevating the hive off the ground on a stand helps protect it from moisture and pests. The stand can be as simple as cinder blocks or a custom-built wooden stand.

By following these detailed steps and ensuring precision in measurements and assembly, you can build a durable and efficient beehive that will serve your beekeeping needs for years to come. Remember, the key to successful DIY hive construction lies in careful planning, selection of quality materials, and meticulous assembly.

Durable and Eco-Friendly Hive Materials

Selecting materials for your DIY hive construction not only involves considering durability and the ability to withstand the elements but also ensuring the materials are eco-friendly, supporting the overall health of the bees and the environment. When choosing wood, which is the primary material for hive construction, opt for locally sourced, untreated varieties to minimize the carbon footprint associated with transportation and to avoid chemical treatments that could harm the bees. Cedar and pine are excellent choices due to their natural resistance to decay and pests. Cedar, in particular, offers additional benefits with its aromatic properties that can deter insects and pests naturally without the need for chemical treatments.

For the hive's frames, consider using bamboo as an alternative to wood. Bamboo is a highly sustainable material due to its rapid growth rate and its natural strength and durability, making it an ideal choice for supporting the comb structure within the hive. Ensure that any metal components, such as screws or hive tools, are made of stainless steel to prevent rust and degradation over time, which can compromise the hive's structural integrity and potentially introduce harmful substances into the hive environment.

The use of natural finishes on the hive exterior is crucial for protecting the wood from weathering while ensuring the safety of the bees. Linseed oil, derived from the flax plant, is a renewable resource that provides a durable, water-resistant finish without the use of volatile organic compounds (VOCs) found in many conventional wood treatments. Apply multiple thin coats of linseed oil, allowing adequate drying time between applications, to create a protective barrier that enhances the wood's natural beauty and longevity.

In constructing the hive's foundation, consider incorporating materials that improve ventilation and moisture management within the hive. A mesh bottom board made of galvanized steel allows for adequate air circulation and helps to keep the hive dry, reducing the risk of mold and mildew development. It also facilitates the removal of varroa mites, a common pest, by allowing them to fall through the mesh and out of the hive. Ensure the mesh is of appropriate gauge to prevent the bees from passing through while allowing mites and debris to exit.

For insulating the hive during colder months, use natural fibers such as wool or hemp. These materials are not only effective insulators but also breathable, helping to regulate temperature and humidity levels within the hive. Additionally, they are renewable resources that biodegrade at the end of their useful life, reducing waste and environmental impact.

When applying paint or sealants to the hive exterior, choose products that are labeled as bee-friendly or non-toxic to ensure they do not off-gas harmful chemicals that could affect the health of the bee colony. Water-based acrylic paints are a good option, offering a wide range of colors and a durable finish without the harmful solvents found in oil-based paints.

By carefully selecting materials and finishes that prioritize durability, sustainability, and the well-being of the bees, beekeepers can construct hives that support the health of their colonies and the environment. These choices reflect a commitment to responsible beekeeping practices that contribute to the sustainability of bee populations and the ecosystems they support.

Step-by-step guide to assembling a Langstroth hive.

Materials and tools

- Pre-cut Langstroth hive kit (includes bottom board, hive bodies or supers, frames, inner cover, and telescoping outer cover)

- Wood glue

- Hammer

- Nails or screws (usually provided with the kit)

- Paint or wood preserver (optional for exterior)

- Level

- Square

- Protective gear (gloves, safety glasses)

Step-by-step instructions

1. **Prepare your workspace** by laying out all the materials and tools. Ensure you have a flat and stable surface to work on.

2. **Assemble the bottom board**: This is the base of your hive. If your kit includes a reversible bottom board, decide which side will face up based on your climate (the shallow side for warmer climates and the deeper side for cooler climates).

3. **Build the hive bodies and supers**: Apply wood glue to the joints of each corner for added stability. Fit the pieces together, making sure they form right angles by using a square. Secure the corners with nails or screws.

4. **Assemble the frames**: Each frame will consist of a top bar, bottom bar, and two side bars. Frames usually come with pre-cut grooves for easy assembly. Glue and nail/screw these together. Ensure they are square.

5. **Install the foundation**: If your frames didn't come with foundation (wax or plastic that encourages the bees to build straight comb), now is the time to add it. Slide the foundation into the groove of the top bar of each frame.

6. **Paint or treat the exterior of the hive bodies and bottom board**: Use a paint or wood preserver suitable for outdoor use. Do not paint the inside of the hive or the frames, as this could harm the bees. Allow to dry completely.

7. **Stack the hive bodies on the bottom board**: Place your first hive body or super on the bottom board. Ensure it sits flush and square.

8. **Add the frames to the hive body**: Carefully insert the assembled and foundation-equipped frames into the hive body. They should hang securely from the top bar slots.

9. **Place the inner cover on top of the last super**: The inner cover provides insulation and helps regulate the hive's internal environment.

10. **Finish with the telescoping outer cover**: This cover goes on top of the inner cover and provides additional protection from the elements. It should overlap the edges of the top super.

11. **Ensure the hive is level**: Using a level, check that the hive is perfectly horizontal. This encourages the bees to build their comb straight and helps with water drainage.

Troubleshooting

- If the hive components do not fit snugly together, double-check that all pieces are correctly oriented and that no excess glue is obstructing the joints. Sanding down any rough edges can also help.

- Should the frames feel loose within the hive body, ensure that you have used the correct number of frames and that they are evenly spaced.

- If the paint or wood preserver is taking too long to dry, check that you have applied it thinly and evenly. Avoid painting in humid conditions.

Care and maintenance

- Regularly inspect the hive for signs of wear or damage, especially after severe weather conditions.

- Repaint or reapply wood preserver to the hive's exterior as needed to maintain its resistance to the elements.

- Keep the area around the hive clear of tall grass and debris to discourage pests.

By following these detailed steps, you will have assembled a Langstroth hive ready for your bees. This hive type is popular for its modularity and ease of inspection and honey harvesting, making it an excellent choice for both beginner and experienced beekeepers.

Designing Flexible Top-Bar Hives

Designing and personalizing top-bar hives involves a thoughtful approach to ensure they meet the needs of both the beekeeper and the bees, focusing on flexibility and ease of use. The top-bar hive, characterized by its horizontal structure and individual bars from which bees build their comb, offers a more natural environment for bees and a simpler harvesting process for beekeepers. To maximize these benefits, every aspect of the hive's design and personalization must be considered with precision.

Firstly, select the wood for your top-bar hive with care. Western red cedar is highly recommended for its natural resistance to rot and its insulating properties, which help maintain a stable internal temperature. Each top bar should be approximately 19 inches long to accommodate the natural comb size of bees, with a width that varies depending on local bee species, typically between 1.25 to 1.5 inches. The bars should be fitted with a starter strip of wax or a wooden guide to encourage the bees to build straight comb along the bar.

The hive body's length can vary, but a standard size is around 4 feet, accommodating around 28 to 30 top bars. This size balances the space needed for a healthy colony's growth and the manageability of the hive for the beekeeper. The sides of the hive body should slope inward at a 30-degree angle, mimicking the natural shape of comb and discouraging the bees from attaching their comb to the hive walls.

Ventilation is critical in a top-bar hive to prevent moisture accumulation, which can lead to mold and disease within the colony. Drill small ventilation holes along the upper sides of the hive body, ensuring they are equipped with mesh screens to keep pests out. Additionally, the hive's entrance should be located at one end of the hive body, sized to allow two to three bees to pass simultaneously, optimizing defense and ventilation.

The roof design is equally important for protecting the hive from the elements. A sloped or gabled roof with an overhang provides rain protection and helps regulate the hive's internal temperature. Insulating the roof with natural materials such as wool felt can further stabilize the temperature, particularly in colder climates. For ease of access, design the roof to be easily removable or hinged for simple opening.

Personalization of the hive can extend to its aesthetic and functional aspects. Painting the hive with non-toxic, water-based paint not only protects the wood but also allows the beekeeper to integrate the hive into the garden's visual landscape. Colors should be chosen carefully; light colors reflect sunlight to keep the hive cool in summer, while darker colors can be beneficial in colder regions by absorbing heat.

Incorporating observation windows can greatly enhance the beekeeping experience, allowing for non-intrusive monitoring of the colony. These windows should be placed on one side of the hive and covered with a removable darkened cover to prevent light from disturbing the bees.

For those beekeepers interested in sustainability and craftsmanship, personalizing the top-bar hive offers an opportunity to create a hive that supports the health of the bee colony, simplifies management practices, and integrates beautifully into its environment. By focusing on the details of construction and design, from the selection of materials to the incorporation of features that enhance the well-being of the bees, beekeepers can craft a hive that is not only functional but also a testament to their commitment to beekeeping and the natural world.

Maintaining Handmade Hives

Maintaining handmade hives is crucial to ensure their longevity and functionality, providing a safe and productive environment for the bee colony. Regular maintenance checks and timely repairs can prevent larger issues, ensuring that the hive remains a conducive home for bees for many years. The following detailed tips and practices are essential for the upkeep of your handmade hives.

Firstly, conduct seasonal inspections of the hive's structural integrity. Before the onset of spring, thoroughly examine the hive for any signs of wear, such as cracks, warping, or loose joints. These issues can compromise the hive's ability to keep moisture and pests out, directly impacting the health of the bee colony. Use a high-quality wood filler to patch minor cracks and apply additional layers of non-toxic, water-based sealant to areas showing signs of water damage. For loose joints, apply exterior-grade glue and clamp the joint until the glue sets completely, usually 24 hours. If screws or nails were used in the original construction, check if they need tightening or replacement.

The roof of the hive requires special attention, as it protects the colony from the elements. Ensure the roof is securely attached and that its waterproof coating is intact. Reapply a layer of eco-friendly, water-resistant paint or sealant annually, or as needed, to prevent water ingress. If the roof is damaged beyond simple repairs, consider constructing a new one using the same durable and bee-friendly materials recommended for the original build.

Ventilation and moisture control within the hive are critical for bee health. Inspect the ventilation holes or mesh screens to ensure they are not blocked by debris or propolis, which bees use as a sealant. Use a soft brush to gently clean these areas, maintaining proper air flow. Additionally, check the hive's bottom board, especially if it's a mesh design for varroa mite control, to ensure it's intact and securely in place. Replace the mesh if it's torn or damaged to keep mites and other pests out.

Pests and diseases pose significant threats to bee colonies. Regularly inspect the hive for signs of infestation or illness, such as the presence of varroa mites, wax moths, or foulbrood disease. Early detection is key to managing these issues effectively. Implement integrated pest management strategies, such as mechanical mite removal or the use of natural treatments, to address infestations without resorting to harsh chemicals.

The hive's exterior also requires maintenance to ensure it withstands weather conditions and blends into its surroundings without stressing the bees. If you've painted the hive, check annually for chips or fading and apply a fresh coat of paint as needed to protect the wood and maintain the aesthetic appeal. Choose light colors to reflect sunlight and keep the hive cooler in hot climates, or darker colors for warmth in cooler regions.

Finally, the interior of the hive, including frames and comb, should be monitored for cleanliness and structural integrity. Replace old or damaged frames to prevent the spread of diseases and ensure the bees have ample space for honey storage and brood rearing. During inspections, gently clean any excess propolis or wax from the hive components to keep the interior tidy and functional.

By following these detailed maintenance tips, beekeepers can significantly extend the life of their handmade hives, ensuring they remain a vibrant and healthy home for their bee colonies. Regular upkeep not only supports the wellbeing of the bees but also enhances the beekeeper's experience, fostering a deeper connection with their environment and the fascinating world of beekeeping.

CHAPTER 23: PROPER HIVE PLACEMENT

Ensuring your hive is placed in an optimal location is crucial for the health and productivity of your bee colony. The right placement can significantly reduce stress on the bees, enhance their ability to regulate temperature, and improve their access to food sources. When selecting a spot for your hive, consider the following detailed aspects to ensure the best possible environment for your bees.

Sunlight Exposure: Bees thrive in areas that receive morning sunlight, especially during cooler months. Position your hive so that it faces east or southeast. This orientation takes advantage of the morning sun to warm the bees, encouraging them to start their foraging activities early. However, it's also essential to provide shade during the hottest part of the day in warmer climates. A location that offers morning sunlight but becomes shaded in the afternoon is ideal. You can achieve this by placing the hive near deciduous trees that lose their leaves in winter, allowing more sunlight during colder months.

Wind Protection: Wind can cool a hive significantly, making it harder for bees to maintain the necessary warmth inside. Choose a spot that is naturally protected from prevailing winds by natural landforms, buildings, or vegetation. If such protection isn't available, consider erecting a windbreak. A hedge or a fence made of natural materials can serve this purpose without obstructing airflow entirely. Ensure the windbreak is positioned so that it deflects winds above the hive, rather than creating turbulent air currents around it.

Water Source: Bees need a reliable water source close to the hive, especially during hot weather. A natural water source within a quarter mile (about 400 meters) of the hive is ideal. If natural sources are not available, create a bee-friendly water station. This can be as simple as a shallow dish filled with pebbles and water, allowing bees to land on the pebbles to drink without the risk of drowning. Regularly check and refill the water station, especially during dry spells.

Avoiding Pesticides: The location should be in an area where pesticide use is minimal or nonexistent. Exposure to pesticides is a significant risk to bee health and can lead to colony collapse. Before setting up your hive, observe the surrounding area for agricultural activities and speak with neighboring property owners about their pesticide use. Opt for locations that are isolated from commercial farming or areas known for organic or low-impact gardening practices.

Accessibility: Your hive should be easily accessible for management and observation, but not so close to human activity that bees become a nuisance or hazard. A distance of at least 10 feet from pathways or property lines is a good rule of thumb, allowing bees to fly out of their hive at an elevation above human height. Ensure the path to the hive is clear of obstructions and safe to walk on, even when carrying equipment or during adverse weather conditions.

Ground Stability and Drainage: The hive should be placed on stable ground that does not collect water. Poor drainage can lead to damp conditions in the hive, promoting the growth of mold and attracting pests. If the ideal location has poor drainage, consider elevating the hive on a stand or platform. This elevation also helps to protect the hive from ground-dwelling predators and facilitates easier access for beekeeping activities.

Integration into the Environment: Finally, consider how the hive integrates with the surrounding environment. The goal is to create a harmonious balance where bees can contribute to pollinating local flora while also gathering the nectar and pollen they need to thrive. Planting bee-friendly flowers, shrubs, and trees near the hive can support this balance, providing bees with a diverse and abundant source of food throughout the growing season.

By carefully considering these factors when placing your hive, you can create a supportive and productive environment for your bee colony. Each aspect, from sunlight exposure and wind protection to water access and environmental integration, plays a crucial role in the health and well-being of your bees. Proper hive placement is a foundational step in successful beekeeping, setting the stage for a thriving colony that can contribute to local ecosystems and yield the rewards of beekeeping, from honey to pollination services.

Maximizing Hive Success on Your Property

When identifying the ideal spot on your property for hive placement, it's imperative to consider **visibility and interaction** with the bees. The hive should be positioned in a location where it can be easily seen from a distance, allowing for unobtrusive observation of the bees' activity patterns. This can help in early detection of potential issues such as overcrowding or signs of disease. A clear line of sight to the hive from a frequently used area of your property, such as a window from your home or a garden seating area, ensures that you can enjoy watching the bees and also stay informed about their well-being.

Noise levels in the surrounding environment should also be taken into account. Bees are sensitive to vibrations and loud sounds, which can cause them stress. Avoid placing hives near busy roads, loud machinery, or frequent foot traffic. A tranquil spot that is removed from noise pollution provides a peaceful environment for the bees and reduces the risk of them becoming agitated or aggressive.

The **orientation** of the hive entrance is another critical factor. In addition to facing east or southeast to catch the morning sun, ensure the entrance is not directly exposed to prevailing winds, which could chill the bees as they enter and exit the hive. A slight tilt forward of the hive can also aid in water drainage, preventing water from pooling at the entrance and making it difficult for bees to access their hive.

Consideration of the **local flora** is essential for providing bees with a rich source of nectar and pollen. Survey your property and the surrounding area for the availability of flowering plants throughout the seasons. Planting a variety of bee-friendly flowers, shrubs, and trees can greatly enhance your bees' diet and health, encouraging a more productive hive. Ensure there's a succession of blooms from early spring through late fall to provide continuous forage for your bees.

Legal requirements and **neighborhood considerations** should not be overlooked. Check local zoning laws regarding beekeeping, as some areas may have restrictions on hive placement relative to property lines or public spaces. It's also courteous to inform neighbors of your beekeeping endeavors, addressing any concerns they might have about bee activity near their property. In some cases, positioning hives in a less visible area or incorporating a bee-friendly barrier, like a tall hedge, can mitigate potential issues and maintain good relations with neighbors.

Security measures against wildlife and pets should be implemented to protect your hives. A secure fence or enclosure can deter bears, raccoons, and domestic animals, preventing them from disturbing or damaging the hives. Electric fencing is particularly effective in areas prone to bear activity. Additionally, ensure the hive stands are sturdy and stable, capable of supporting the hive's weight even in adverse weather conditions, and elevated enough to keep the hives out of reach from skunks and rodents.

In summary, the selection of a hive location on your property involves a comprehensive assessment of environmental factors, bee health considerations, legal compliance, and community relations. By meticulously evaluating each of these aspects, you can ensure a supportive and productive environment for your bee colony, contributing to its success and the overall health of the local ecosystem.

Deterring Predators from Hives

To effectively deter predators such as raccoons or bears from accessing hives, beekeepers must employ strategic measures that safeguard their colonies without harming the natural environment or the animals. Predators, primarily drawn to hives by the scent of honey and bee larvae, can cause significant damage if not properly managed. Here are detailed strategies for protecting hives from these common threats:

Physical Barriers: The most effective way to prevent bears from accessing bee hives is by installing **electric fencing**. A setup with multiple strands of wire, positioned at varying heights (starting from about 8 inches off the ground to 3 feet high), ensures that bears, regardless of their size, receive a deterring shock upon attempting to reach the hives. It's crucial to maintain a minimum voltage of 5,000 volts on the fence to effectively deter bears. For areas with heavy snowfall, consider installing taller posts to keep the wires at an effective height year-round. Regular checks to ensure the electric fence is free of vegetation and debris that could ground the charge are essential for maintaining its effectiveness.

Raccoon Deterrence: Raccoons, adept climbers and smaller in size, require a different approach. Using **hive stands** with smooth, metal poles can prevent raccoons from climbing up to the hives. Coating the poles with a slippery substance, such as grease, further enhances their effectiveness. Additionally, securing hive components with **ratchet straps** or **hive locks** can prevent raccoons from prying open the hives. These straps or locks should be made of durable materials, such as heavy-duty plastic or metal, to withstand the elements and the strength of raccoons.

Chemical Repellents: While physical barriers are more effective and environmentally friendly, certain situations may call for the use of **chemical repellents**. These should be used as a last resort and selected carefully to ensure they are not harmful to bees, other wildlife, or the environment. Natural repellents, such as capsaicin-based sprays, can be applied around the perimeter of the beekeeping area, rather than directly on the hives, to deter bears and raccoons without harming the bees.

Sanitation and Hive Maintenance: Minimizing attractants is key to deterring wildlife. Regularly removing any dead bees or comb from around the hive area reduces the scents that attract predators. Ensuring that honey is not leaking from the hives and repairing any damages promptly can also decrease the likelihood of bear and raccoon visits.

Community Awareness and Reporting: In areas with frequent bear sightings, participating in community wildlife awareness programs can be beneficial. Reporting sightings and engaging in collective prevention efforts can help manage the presence of bears and racoons in residential areas. Collaboration with local wildlife authorities can provide additional resources and support for implementing effective deterrents.

Protecting Hives from Harsh Weather

Protecting hives from harsh weather conditions, such as heavy rain or wind, requires careful planning and execution to ensure the safety and well-being of the bee colony. The first line of defense against rain involves the construction and positioning of the hive roof. A roof made from durable, water-resistant materials such as galvanized metal or heavy-duty plastic sheeting is essential. The roof should extend several inches beyond the hive body on all sides to prevent water from seeping into the hive. Ensuring the roof is slightly sloped facilitates water runoff, reducing the risk of water pooling and eventually infiltrating the hive.

For additional protection against rain, especially in areas prone to heavy downpours, consider installing a hive cover or awning that provides extended coverage. This can be made from a waterproof tarp secured over a frame constructed around the hive. The frame should be sturdy enough to withstand wind while allowing ample air circulation to prevent condensation inside the hive, which could lead to mold growth and weaken the bees' health.

Wind protection involves not only strategic placement of the hive within the landscape but also the potential addition of windbreaks. In areas with strong prevailing winds, positioning the hive near a natural windbreak such as a dense row of trees, a hedge, or a solid fence can significantly reduce wind speed and the chilling effect on the hive. For man-made windbreaks, ensure there is enough distance from the hive to avoid creating turbulent airflows that could stress the bees. Approximately 10 to 15 feet away from the hive is a good starting point, but this may vary based on the height and density of the windbreak.

To secure the hive against strong winds, use heavy-duty straps or anchors to fasten the hive components together and to the ground. Ground anchors can be driven into the soil to provide a secure point to which straps can be attached, preventing the hive from toppling over. For added stability, consider placing bricks or large stones on the hive's roof, but ensure these do not impede the roof's ability to shed water.

In colder climates, where winter storms bring both wind and snow, insulating the hive is crucial. Using insulation boards cut to fit the exterior of the hive, or wrapping the hive in an insulating blanket designed for beekeeping, can help maintain a more stable internal temperature. The entrance should be reduced to a small opening to minimize heat loss while still allowing bees to exit for cleansing flights on warmer days. However, it's vital to ensure that the hive's ventilation is not compromised by the insulation, as proper air flow is necessary to control moisture levels inside the hive.

Lastly, regular maintenance checks are vital to ensure that the protective measures remain effective. After severe weather events, inspect the hive for any signs of damage or water ingress and make repairs as needed. Checking the stability of the hive's anchoring system, the condition of the roof and awning, and the effectiveness of windbreaks can prevent future weather-related issues.

By adopting these strategies, beekeepers can significantly mitigate the risks posed by harsh weather conditions, providing a safer environment for their bee colonies to thrive. The key is to anticipate potential weather challenges and implement preventative measures proactively, ensuring the longevity and productivity of the hive through the changing seasons.

Ventilation for Hive Health and Moisture Control

Ensuring proper ventilation within the hive is crucial for maintaining bee health and preventing moisture buildup, which can lead to mold growth and increase the risk of diseases within the colony. Ventilation helps regulate the temperature inside the hive, removes excess humidity, and allows for the evaporation of water from nectar as bees produce honey. Here are detailed strategies and techniques to achieve optimal ventilation in bee hives.

First, consider the design and placement of the hive's entrance. The entrance should be located at the bottom of the hive front, allowing cooler air to enter. This positioning takes advantage of the bees' natural tendency to circulate air by fanning their wings, pushing warm, moist air out through the top of the hive. The size of the entrance can be adjusted seasonally, with a larger opening in summer for increased air flow and a smaller one in winter to retain heat while still allowing for adequate ventilation.

In addition to the entrance, the use of screened bottom boards is an effective method for enhancing ventilation. Unlike solid bottom boards, screened versions allow air to flow freely through the bottom of the hive, helping to remove excess moisture and keep the hive dry. This is particularly beneficial in humid climates or during rainy seasons. Screened bottom boards also provide the added benefit of helping to control varroa mite populations, as mites fall through the screen and are unable to return to the hive.

Another technique involves the strategic placement of small ventilation holes or notches at the top of the hive bodies or supers. These openings, often located under the hive cover or in the uppermost box, allow warm, moist air to escape. When creating these ventilation holes, ensure they are small enough to prevent robbers or pests from entering the hive, typically no larger than ⅜ inch in diameter. It's also important to position these holes in a way that does not create direct drafts on the cluster of bees during colder months, as this could chill the bees.

The use of an inner cover with a built-in ventilation slot can further improve air circulation within the hive. The slot allows warm air to escape while preventing direct exposure to the elements. During winter, the inner cover can be flipped to reduce the ventilation space, helping to conserve heat while still allowing moisture to escape.

In regions with extreme temperature variations, the addition of insulation to the hive can help regulate internal temperatures and reduce condensation. However, it's vital to balance insulation with ventilation. Insulating materials should be breathable, such as natural wool or specially designed beekeeping insulation boards, to avoid trapping moisture inside the hive.

Regular monitoring and maintenance of the hive's ventilation system are essential. Check the entrance, screened bottom board, ventilation holes, and inner cover periodically to ensure they are clear of debris, dead bees, or propolis that could block air flow. Adjustments may be necessary based on seasonal changes, weather conditions, and the health and size of the bee colony.

CHAPTER 24: MONITORING HIVE HEALTH

Monitoring hive health is a critical aspect of beekeeping that requires vigilance and a systematic approach to ensure the well-being of the colony. One of the first steps in this process involves **regularly inspecting the hive's interior** for signs of disease, pests, and the overall condition of the bees. This includes looking for **irregular patterns in the brood**, such as scattered or missed cells which could indicate issues like queen failure or disease. **Brood cells should be compact and uniform**, with larvae visible at various stages of development.

Another key indicator of hive health is the **presence of pests such as Varroa mites**. These pests can be identified by closely examining the bees and the brood. Varroa mites are small, reddish-brown parasites that feed on the bodily fluids of honey bees. **A sugar roll test or alcohol wash** can be performed to assess the mite load within the colony. For the sugar roll test, gently shake a sample of bees in a jar with powdered sugar; the mites will detach and can be counted to estimate infestation levels. The alcohol wash involves submerging bees in alcohol, which kills the bees but also the mites, allowing for an accurate count. While the alcohol wash is more accurate, the sugar roll is less invasive and does not kill the bees.

Observing the behavior of bees at the hive entrance is another non-invasive method to gauge hive health. Healthy colonies will show a steady stream of worker bees departing and returning, indicating foraging activity. The entrance should also be monitored for **signs of robbing**, where bees from other hives attempt to steal honey, which can stress or weaken the colony. Robbing behavior includes fighting at the hive entrance and an unusually high number of bees hovering outside the hive.

Hive hygiene is a critical component of maintaining health. A healthy bee colony will efficiently remove dead bees and debris from the hive, an activity that should be observable during inspections. Accumulation of dead bees inside the hive or at the entrance can be a sign of disease or poor colony health.

The condition of the comb also provides insights into the health of the hive. Combs should be well-structured, with cells being filled with brood, pollen, or honey in an organized manner. **Moldy or discolored combs** can indicate moisture problems or disease within the hive and should be addressed immediately.

Honey stores are another important indicator of hive health, especially approaching winter. A healthy hive will have ample stores of honey to sustain the colony through periods when foraging is not possible. Inspect the hive in late summer or early fall to ensure there are sufficient honey reserves. If honey stores are low, supplemental feeding may be necessary to prevent starvation.

Water availability is crucial for hive health, especially in hot climates or during dry spells. Bees require water for cooling the hive and for diluting honey to feed to the brood. Ensure there is a clean, reliable water source near the hive to support the colony's needs.

In summary, monitoring hive health involves a combination of visual inspections, behavioral observations, and proactive management practices. Regular and thorough inspections are essential for early detection of issues, allowing for timely interventions to maintain colony health and productivity. By understanding and responding to the signs of a healthy or distressed hive, beekeepers can support the well-being of their bees and ensure the sustainability of their beekeeping endeavors.

Signs of a Healthy Hive

Recognizing signs of a healthy hive is paramount for beekeepers of all experience levels. A thriving colony is not only a sign of successful beekeeping but also an indicator of the overall health of the ecosystem surrounding the hive. One of the most evident signs of a healthy hive is the presence of active bees. Activity levels can vary depending on the time of day and weather conditions; however, a healthy hive typically exhibits a bustling entrance with bees coming and going, foraging for nectar and pollen. Observing the entrance during peak hours of activity, usually mid-morning to early afternoon on sunny days, provides insight into the colony's foraging efficiency and population health. Bees should appear purposeful in their movements, with foragers returning to the hive laden with pollen visibly stored in their corbiculae, or pollen baskets, indicative of successful foraging.

Another critical aspect to monitor is the cleanliness and structure of the comb. Healthy colonies maintain impeccably clean and well-organized comb structures. The comb should contain a pattern of brood in the center, surrounded by rings of pollen and then honey towards the outer edges. This arrangement reflects the colony's natural organization and prioritization for brood rearing, food storage, and ease of access for the queen to lay eggs. The brood pattern itself is a key health indicator; capped brood cells should be uniform in appearance, with a slightly convex cap that is tan to dark brown in color. Irregularities in the brood pattern, such as scattered or sunken caps, may indicate health issues within the hive that require further investigation.

Inspecting the comb also involves checking for signs of diseases or pests. Healthy hives manage to keep pests at bay through their grooming behaviors and the maintenance of propolis barriers. A lack of pest presence on the comb, such as Varroa mites or hive beetles, further confirms the effectiveness of the colony's internal defenses and overall health. Additionally, the comb should be free of mold or mildew, which can signify ventilation issues within the hive that need to be addressed to prevent respiratory stress on the bees.

The vibrancy and behavior of the bees themselves are also telling. Healthy bees exhibit a vigorous demeanor, with workers performing their respective duties diligently. Nurse bees should be seen tending to the brood, cleaners removing debris from the hive, and foragers processing nectar and pollen. The queen's presence is crucial; spotting her regularly, along with evidence of fresh egg-laying, signifies a productive and well-functioning colony. A good laying pattern, where eggs are deposited singly in the center of each cell, reflects the queen's health and the hive's ability to support brood rearing.

Monitoring these aspects of hive health not only ensures the well-being of the bee colony but also empowers beekeepers to make informed management decisions. Addressing any deviations from these signs of health promptly can mitigate potential issues, ensuring the colony's longevity and productivity.

The presence of **pollen diversity** in the hive is another indicator of a healthy bee colony. Pollen, brought back by foragers from various plant sources, should display a range of colors when stored within the comb cells. This diversity not only signifies a rich foraging environment but also ensures a balanced diet for the bees, crucial for their nutrition and immune system strength. Observing the types and colors of pollen can also provide insights into the foraging patterns and preferences of the colony, as well as the health of the local ecosystem.

Honey stores are a critical measure of hive health, particularly as the seasons change. A healthy hive will have significant reserves of honey, which appear as capped cells of a light to dark golden color, depending on the floral sources. These reserves are essential for the colony's survival through winter or periods of scarcity. The pattern and extent of honey storage, with adequate space left for the queen to lay eggs, indicate a well-balanced hive management by the bees.

Ventilation and moisture control within the hive play a significant role in maintaining colony health. A healthy hive will have mechanisms in place, such as propolis-coated screens or strategically placed vents, to regulate airflow and humidity. Proper ventilation ensures that the internal environment of the hive remains conducive to bee health, preventing the growth of harmful fungi and bacteria that can thrive in damp conditions.

The **sound of the hive** is an often-overlooked aspect of bee health monitoring. A healthy colony emits a steady, calm buzz that indicates contentment and productivity. In contrast, a high-pitched whining or erratic buzzing can signal distress or disorder within the hive, such as queenlessness or an imminent swarm. Listening to the hive can provide beekeepers with immediate feedback on the colony's state of well-being.

Seasonal behaviors also offer clues to the hive's health. For instance, in preparation for winter, a healthy colony will reduce its population, concentrate honey stores, and seal cracks with propolis to insulate the hive. Observing these behaviors can reassure beekeepers of the colony's readiness for the colder months.

Finally, **record-keeping** is an invaluable tool in monitoring hive health. Detailed records of hive inspections, including observations of bee behavior, comb condition, pest presence, and honey stores, can help beekeepers track the health of their colony over time. This historical data can reveal patterns, identify potential problems early, and inform management decisions to support the hive's health and productivity.

By paying close attention to these signs and maintaining vigilant observation, beekeepers can ensure their hives remain vibrant and healthy. Each aspect of hive health, from the activity of the bees to the condition of the comb and the sounds of the colony, contributes to a comprehensive understanding of the hive's well-being. This holistic approach to monitoring enables beekeepers to foster strong, productive colonies that can withstand challenges and contribute positively to the ecosystem.

Detecting Foulbrood and Chalkbrood Diseases

Detecting foulbrood and chalkbrood diseases early in bee colonies is crucial for maintaining hive health and preventing the spread of these destructive conditions. Foulbrood, caused by the bacteria Paenibacillus larvae, presents in two forms: American foulbrood (AFB) and European foulbrood (EFB). AFB is highly contagious and lethal, leading to the death of larval bees and resulting in a characteristic foul smell. EFB affects the larvae before the capping stage, causing them to appear twisted and discolored, but is less deadly and can sometimes be overcome by a strong colony. Chalkbrood, a fungal infection caused by Ascosphaera apis, leads to mummified larvae that resemble chalky, white pieces inside the comb.

To detect these diseases, beekeepers must perform regular, detailed inspections of their hives, focusing on the brood patterns and the appearance of the larvae. For American foulbrood, one of the most definitive tests is the "rope test": using a small stick or similar tool, gently touch a larva suspected of being infected. If the larva is infected, it will stretch out in a ropey, sticky thread when lifted. This symptom is a clear indication of AFB. Additionally, the presence of sunken, greasy-looking cappings or perforated cappings with a dark color can indicate AFB. For European foulbrood, look for irregular brood patterns, discolored (yellow or brown) larvae that appear melted down in the cells, and a lack of the characteristic foul odor associated with AFB.

Chalkbrood detection involves identifying mummified larvae, which are typically hard, chalk-like, and can be white or black in color, depending on the stage of the infection. These mummies can sometimes be found outside the hive entrance, as worker bees attempt to remove the infected larvae to prevent the spread of the disease within the colony. The presence of these mummies in or around the hive is a clear sign of a chalkbrood outbreak.

Preventative measures for these diseases include maintaining strong and healthy colonies, as stressed or weak hives are more susceptible to infections. Ensuring good ventilation within the hive can help reduce moisture and prevent the growth of the chalkbrood fungus. Regularly replacing old combs with new ones can help prevent the accumulation of pathogens that cause foulbrood. Additionally, beekeepers should practice good hygiene, sterilizing tools and equipment between hives to prevent the spread of disease.

In the case of American foulbrood, because of its highly contagious nature and the lack of a cure, infected colonies often must be destroyed to prevent the spread of the disease to other hives. For European foulbrood and chalkbrood, treatments may include antibiotic therapy (only for EFB and under veterinary guidance to prevent antibiotic resistance), increasing hive ventilation, and providing supplemental feeding to strengthen the colony. However, the use of antibiotics is strictly regulated and should only be considered after consultation with a bee health specialist.

The early detection and management of these diseases are critical steps in ensuring the health and productivity of bee colonies. Beekeepers should remain vigilant, conducting regular inspections and taking immediate action upon the first signs of disease. By doing so, they can protect their hives and contribute to the overall health of the beekeeping community.

Integrated Pest Management for Varroa Mites

Controlling varroa mites, a significant threat to bee colonies worldwide, requires a comprehensive and strategic approach known as Integrated Pest Management (IPM). IPM is a multifaceted strategy that incorporates various techniques to manage pest populations at levels that do not cause significant harm to bees, beekeepers, and the environment. This method emphasizes the importance of monitoring mite levels within the hive, understanding the life cycle of varroa mites, and applying control measures that are both effective and sustainable.

The first step in an effective IPM strategy for varroa mites is regular and accurate monitoring. Beekeepers can employ several techniques to assess the mite load within their colonies. One common method is the powdered sugar roll, which involves collecting a sample of bees in a jar, coating them with powdered sugar, and then shaking the jar to dislodge the mites from the bees. The number of mites that fall off provides an estimate of the infestation level within the hive. Another method is the alcohol wash, which is more accurate but lethal to the sampled bees. It involves submerging the bees in alcohol, which kills the bees but also separates the mites, allowing for an accurate count. These monitoring techniques should be performed regularly, especially during the late summer and fall when mite populations tend to peak.

Understanding the life cycle of the varroa mite is crucial for effective control. Varroa mites reproduce on the developing bee brood, with a preference for drone brood because of its longer development time. Breaking the mite's life cycle involves strategic interventions such as drone brood removal, where frames specifically designed to attract drone brood are inserted into the hive and then removed once capped, effectively trapping and removing the mites that have infested these cells.

Cultural controls form another pillar of IPM, focusing on beekeeping practices that naturally reduce mite populations. These include maintaining strong, healthy colonies through proper nutrition and hive management, as healthier bees are more capable of resisting and surviving mite infestations. Additionally, beekeepers can select bee strains known for their mite resistance or tolerance. Breeds such as the Russian honey bee and hygienic strains of bees have shown an innate ability to detect and remove infected brood, a behavior that naturally suppresses mite populations.

Mechanical controls also play a role in IPM. These can include the use of screened bottom boards, which allow mites to fall out of the hive, reducing their numbers. The use of mite traps or sticky boards placed beneath the hive can help monitor mite fall and indicate the effectiveness of other control strategies being employed.

The integrated approach to managing varroa mites emphasizes the use of chemical controls as a last resort, due to the potential for mite resistance and the risk of contaminating hive products. When chemical controls are necessary, options include organic acids such as formic acid and oxalic acid, which have shown effectiveness in reducing mite loads without leaving harmful residues in the hive. These treatments, however, must be carefully timed and applied according to label instructions to minimize harm to the bees.

By combining regular monitoring with a deep understanding of the varroa mite's life cycle and employing a mix of cultural, mechanical, and, when necessary, chemical controls, beekeepers can implement an effective IPM strategy. This approach not only controls mite populations but also supports the overall health and sustainability of bee colonies.

Biotechnological advancements have introduced another dimension to the IPM strategy against varroa mites, focusing on non-invasive methods to bolster bee immunity and disrupt mite reproduction. One such innovation involves the use of RNA interference (RNAi) technology, which targets specific mite genes critical for their survival and reproduction, without affecting the bees. This method, still in research and development stages, promises a future where beekeepers can control mite populations using highly specific, environmentally friendly treatments that do not contribute to chemical resistance or residue issues.

In addition to these strategies, the importance of community-level management practices cannot be overstated. Varroa mites can travel between colonies and apiaries through robbing behavior and drifting bees. Therefore, coordinated treatment efforts among beekeepers in the same region can significantly enhance the effectiveness of individual IPM plans. Sharing knowledge and resources, such as bulk purchasing of treatment supplies or organizing community-wide treatment days, ensures a synchronized approach to mite control, reducing the overall mite load in the area and minimizing the risk of re-infestation.

Furthermore, the role of beekeeping associations and extension services in educating and supporting beekeepers in IPM practices is vital. Workshops, webinars, and field days provide platforms for disseminating the latest research findings, demonstrating new techniques, and fostering a community of practice that encourages the adoption of sustainable beekeeping methods.

To complement these efforts, record-keeping is an essential practice within the IPM framework. Detailed records of mite monitoring results, treatment types, dates, and outcomes, as well as notes on colony health and productivity, enable beekeepers to refine their IPM strategies over time. This historical data becomes invaluable in understanding the dynamics of varroa populations in relation to specific management practices and environmental conditions, allowing for more precise and effective interventions.

Lastly, the adoption of IPM strategies for varroa mite control requires a commitment to ongoing learning and adaptation. As new research emerges and the context within which beekeeping occurs evolves, so too must the strategies employed to ensure the health and viability of bee colonies. Engaging with the beekeeping community, participating in citizen science projects, and staying informed about advancements in bee health research are all ways in which beekeepers can contribute to the collective effort to combat varroa mites and ensure the sustainability of beekeeping practices.

By embracing an integrated approach that combines regular monitoring, understanding of the varroa mite life cycle, application of cultural, mechanical, and selective chemical controls, along with community coordination and education, beekeepers can effectively manage varroa mite populations. This holistic strategy not only mitigates the impact of varroa mites on bee colonies but also promotes the overall health of the beekeeping ecosystem, ensuring the resilience and productivity of bee populations for future generations.

Identifying Hive Stressors and Triggers

Identifying stressors and environmental triggers that lead to hive abandonment requires a meticulous and observant approach. Hive abandonment, or absconding, is a phenomenon where the entire bee colony leaves the hive due to various stress factors. This can be devastating for beekeepers, leading to the loss of a colony and its productive capabilities. To address this issue, beekeepers must first understand the common stressors that can trigger this behavior.

One significant stressor is the presence of pests and parasites within the hive. Varroa mites, for example, are a common parasite that weakens bees by feeding on their bodily fluids and transmitting diseases. Beekeepers should regularly check for varroa mites using methods such as the powdered sugar roll or alcohol wash mentioned previously. If mite levels are high, appropriate integrated pest management strategies must be employed. This could include mechanical methods like drone brood removal or chemical treatments with organic acids, ensuring that these are applied in a way that minimizes harm to the bees.

Another critical factor is the hive's internal environment. Bees are highly sensitive to their living conditions, requiring adequate ventilation, temperature control, and humidity levels to thrive. A hive that becomes too hot, cold, or damp can prompt bees to abandon it. To prevent this, beekeepers should ensure proper hive placement, ideally in a location that receives morning sun and is protected from prevailing winds. Insulation may be added to the hive for temperature regulation, particularly in colder climates, while ensuring that ventilation is not compromised. Screened bottom boards can aid in managing both ventilation and mite control by allowing mites to fall out of the hive.

Chemical exposure is another environmental trigger for hive abandonment. Bees exposed to pesticides or herbicides can exhibit disorientation and may not return to the hive, leading to colony weakening and potential abandonment. Beekeepers should advocate for responsible pesticide use in their communities and consider planting bee-friendly forage to provide safe nutrition sources away from treated crops.

Food scarcity can also lead to absconding. A hive that is not able to collect enough nectar and pollen will be stressed and may decide to leave in search of a better location. Ensuring that hives have access to diverse floral resources throughout the foraging season is essential. This may involve planting a variety of bee-friendly plants that bloom at different times, providing a continuous food supply. Additionally, supplementary feeding with sugar syrup or pollen patties can be used during periods of scarcity, though this should be done carefully to avoid dependency or health issues related to poor nutrition.

Water availability is another often-overlooked factor. Bees need a consistent water source not just for consumption but also for regulating hive temperature through evaporative cooling. Providing a nearby water source, such as a shallow water dish with stones or floating debris for bees to land on, can prevent the colony from relocating to be closer to water.

Finally, beekeepers should be mindful of their own interactions with the hive. Rough handling, frequent disturbances, and improper use of smoke can stress the colony, contributing to the risk of absconding. Adopting gentle beekeeping practices, minimizing hive inspections to necessary checks, and using smoke judiciously can help maintain colony calmness and reduce stress.

By addressing these stressors and environmental triggers with careful, informed management practices, beekeepers can significantly reduce the risk of hive abandonment. Regular monitoring for pests and diseases, maintaining optimal hive conditions, ensuring access to food and water, and practicing gentle beekeeping are all critical steps in supporting the health and stability of bee colonies. Through such diligent care, beekeepers not only prevent absconding but also contribute to the resilience and productivity of their bees, fostering a thriving beekeeping operation and supporting the broader ecosystem.

CHAPTER 25: EXPANDING THE HIVE

Expanding the hive is a critical step for beekeepers aiming to increase their colony's size and productivity. This process involves adding frames or supers, splitting hives to create new colonies, and introducing new queens to ensure colony stability. Each of these steps requires careful planning and execution to minimize stress on the bees and ensure the success of the expansion.

Adding Frames or Supers: When a colony outgrows its current space, it becomes necessary to add more room to prevent overcrowding and swarming. To determine the right time for expansion, inspect the hive regularly for signs of crowding, such as comb being built between frames or at the hive entrance. Before adding new frames or supers, ensure they are compatible with your existing hive setup. For Langstroth hives, this means checking that the dimensions and frame styles match. When adding a super, place it directly above the current top super of the hive, using a queen excluder if you wish to keep the queen from laying eggs in the honey supers. It's best to add these during the early morning or late evening when most bees are in the hive, reducing the disruption to their foraging activities.

Splitting Hives: Splitting a hive involves dividing an existing colony into two, with one part keeping the old queen and the other receiving a new queen. This is typically done in the spring when colonies are growing rapidly. To split a hive, first, select a strong colony with plenty of brood and resources. Then, carefully move several frames of brood, honey, and pollen, along with a substantial number of bees, into a new hive box. It's crucial to ensure that one of the splits contains the old queen, while the other will require a new queen. Introduce a new queen to the queenless split by placing her in a queen cage within the hive, allowing the bees to slowly acclimate to her over several days. Monitor both splits closely for signs of acceptance and healthy brood patterns, feeding them sugar syrup if necessary to support their growth.

Introducing New Queens: Requeening a hive can revitalize an aging colony or replace a lost queen. When introducing a new queen, first, remove the existing queen from the hive. Then, place the new queen, still in her cage, between the frames in the center of the brood nest. The cage should have a candy plug that the worker bees will gradually eat through, releasing the queen into the hive over several days. This slow release gives the colony time to get used to her pheromones and accept her as their new queen. Check the hive after a week to confirm that the queen has been released and is laying eggs. If the colony does not accept the queen, they may ball around her and kill her, in which case, a new introduction may be necessary.

Throughout the expansion process, maintain vigilant monitoring of the hive's health, population dynamics, and resource levels. Regular inspections will help identify any issues early, allowing for timely interventions to keep the colony thriving. Additionally, keeping detailed records of each hive's expansion history, queen introductions, and any challenges encountered provides valuable insights for future management decisions.

Adding Frames and Supers for Hive Growth

As the bee colony grows and flourishes, the beekeeper must remain vigilant in observing the signs that indicate it's time to expand the hive by adding frames or supers. This expansion is not just about providing more space; it's about ensuring the health, productivity, and well-being of the colony. The decision to add frames or supers is critical and should be based on a combination of visual cues, bee behavior, and seasonal timing.

One of the first indicators that it might be time to add more space is when you observe that the bees have started to draw comb on 80% to 90% of the existing frames. This suggests that the colony is running out of room for the queen to lay eggs, for workers to store pollen and nectar, and for the processing and storage of honey. It's essential to provide additional frames before the bees feel too constrained, as overcrowding can trigger swarming, where the queen and a significant portion of the workers leave to find a new home, significantly reducing your hive's population and productivity.

Another sign is the observation of bees clustering outside the hive entrance, a behavior known as "bearding." While bearding can also indicate overheating, if it occurs in conjunction with the hive frames being mostly filled, it's likely a sign that the hive is too crowded. Before adding supers or frames, ensure that the hive's temperature and ventilation are adequately managed to rule out overheating as the primary cause of bearding.

Seasonal timing also plays a crucial role in deciding when to add supers, especially for honey production. In anticipation of major nectar flows, such as those from clover, acacia, or citrus blooms, adding supers in advance can give bees the space they need to store the incoming abundance of nectar. This proactive approach helps maximize honey harvests and prevents the colony from becoming honey-bound, where the bees fill every available cell with honey, leaving no room for the queen to lay eggs.

When adding frames to a brood box or supers for honey storage, it's vital to do so in a manner that minimizes disruption to the bees. For Langstroth hives, this often means employing the checkerboarding technique, where new frames are alternated with drawn frames, encouraging the bees to draw out the new frames more quickly. When adding a super, placing it directly on top of the brood boxes or the existing super, and ensuring it's the correct size and type for your hive configuration, is crucial. For example, using a medium super on a hive that has been using deep boxes for brood may disrupt the colony's internal architecture.

The materials of the frames and supers should match the existing hive construction to maintain consistency for the bees. Frames should be equipped with foundation, either wax or plastic, embossed with a honeycomb pattern to encourage the bees to build comb. When introducing new frames or supers, checking for proper alignment and spacing is essential to facilitate the bees' movement and work within the hive.

In conclusion, understanding the right time and method for expanding your hive by adding frames or supers is a nuanced aspect of beekeeping that requires attention to detail, observation, and timing. By providing your bees with the space they need to grow and store resources effectively, you can support a healthy, productive colony that contributes to the ecosystem through pollination and provides valuable hive products such as honey, beeswax, and propolis.

Splitting Hives for New Colonies and Overcrowding Prevention

Splitting hives, a crucial technique for managing bee populations and preventing overcrowding, involves dividing an existing colony into two or more groups, each of which will grow into a full colony on its own. This method not only helps in increasing the number of hives but also in controlling swarming, a natural process where bees leave an overcrowded hive to start a new one. To execute a hive split effectively, detailed planning and understanding of bee biology are essential. The process typically begins in the spring, aligning with the bees' natural reproductive cycle and ensuring that both the original and new colonies have ample time to establish themselves before the colder months.

First, assess the parent hive for splitting readiness. A strong, healthy colony with at least two brood boxes full of bees, brood at various stages, and sufficient stores of honey and pollen is a good candidate. The presence of fresh eggs indicates that the queen is still productive, a critical factor for the success of the split. Selecting frames for the new hive involves choosing ones with eggs, larvae, capped brood, honey, and pollen. This variety ensures that the new colony has all the necessary resources to raise a new queen and sustain itself. Typically, three to five frames from the parent hive, depending on its strength and the time of year, are moved to a new hive box.

When transferring frames, it's important to maintain the brood nest's integrity. Arrange the frames in the new hive in a similar order to how they were in the original hive, with brood in the center surrounded by frames of honey and pollen. This arrangement mimics the natural structure of a hive, providing optimal conditions for brood rearing and making it easier for the bees to regulate temperature and humidity within the new hive.

Introducing a new queen to the split is the next critical step. There are two common methods: purchasing a mated queen or allowing the bees to raise their own queen from the eggs present on the transferred frames. If introducing a purchased mated queen, it's vital to do so carefully to ensure she is accepted by the new colony. The queen usually comes in a queen cage with a candy plug, which the worker bees will eat through, gradually releasing her into the hive over a few days. This slow introduction gives the bees time to get accustomed to her pheromones, increasing the chances of acceptance.

If the decision is to let the colony raise its own queen, ensure that there are eggs or very young larvae present on the frames you're transferring. Bees can only raise a new queen from eggs or larvae less than three days old. The workers will select several larvae to rear as potential queens by feeding them royal jelly, resulting in the development of new queens. Once the first new queen emerges, she will typically eliminate any other developing queens before going on her mating flight.

After the split, monitor both the original and new hives closely. Check for signs of queen acceptance in the new hive and ensure the original hive is recovering well from the split. It may be necessary to feed both colonies, especially the new split, to help them build up their strength. Providing a 1:1 sugar syrup can stimulate comb building and give the bees the energy they need to establish the new colony.

Introducing New Queens for Colony Stability

Introducing a new queen to a hive is a delicate process that requires precision and understanding of bee behavior to ensure colony stability. The success of this introduction can significantly impact the health and productivity of the hive. Here's a detailed breakdown of how to introduce a new queen to your bee colony.

Acquiring a New Queen: First, source your new queen from a reliable breeder. The queen should come in a queen cage, often accompanied by a few attendant bees and a candy plug sealing the entrance. Ensure the queen is healthy and active before proceeding with the introduction.

Preparing the Hive: Before introducing the new queen, remove the existing queen from the hive. This may require carefully inspecting each frame for her presence. Once the old queen is removed, wait 24 hours before introducing the new queen to give the hive time to realize they are queenless, which increases the chances of the new queen's acceptance.

Introducing the Queen Cage: Place the queen cage between two central frames in the brood chamber, ensuring the screen faces outward so worker bees can access it. The candy plug should be at the top of the cage to prevent dead attendant bees from blocking the exit. If the weather is warm and the hive is calm, consider removing the cork from the candy end immediately. In cooler temperatures or if the hive seems agitated, wait a day before exposing the candy plug to slow the release process.

Monitoring and Assistance: After placing the queen cage, close the hive and monitor without disturbing it for at least three to five days. This period allows the worker bees to chew through the candy plug and gradually release the queen at a pace that lets her pheromones become familiar to the colony, promoting acceptance.

Confirming Queen Release: Carefully inspect the hive after five to seven days to ensure the queen has been released from her cage. Look for signs of acceptance, such as the queen moving freely among the bees and the presence of new eggs. If the queen has not been released, carefully remove the remaining candy plug or mesh covering, taking care not to harm the queen.

Post-Release Observation: Once the queen is released, continue to monitor the hive from a distance, minimizing disturbances that could stress the colony and affect the queen's acceptance. Look for normal bee behavior and signs of a healthy, laying queen, such as consistent egg-laying patterns and calm worker bees.

Troubleshooting: If the colony does not accept the queen, evidenced by aggressive behavior towards her or a lack of new eggs, a second introduction attempt may be necessary. This might involve acquiring a new queen and repeating the introduction process, ensuring any potential stressors within the hive are addressed.

Record Keeping: Document the queen introduction process, including the source of the queen, the date of introduction, and any observations regarding the hive's acceptance or rejection of the new queen. This information can be invaluable for future management decisions and understanding the dynamics of your bee colonies.

By adhering to these detailed steps and maintaining a close observation of the hive's behavior, beekeepers can successfully introduce new queens, ensuring the stability and productivity of their colonies. This process, while intricate, is essential for the health of the bee community and the success of beekeeping endeavors.

PART 7: SEASONS AND BEEKEEPING

CHAPTER 26: SPRING

Spring is a pivotal time for beekeepers, marking a period of renewal and activity within the hive. As the weather warms, flowers begin to bloom, providing bees with a rich source of nectar and pollen. This abundance triggers a surge in hive activity, with bees working tirelessly to collect these vital resources. Beekeepers must be vigilant during this season, ensuring their hives are well-prepared to capitalize on the spring bloom. The following steps outline essential spring beekeeping tasks, focusing on inspection, feeding, swarm prevention, and hive expansion.

Inspect the Hive Thoroughly: Early spring is the ideal time for a comprehensive hive inspection. Look for signs of a healthy queen, such as fresh eggs and a consistent brood pattern. Assess the population size; a strong colony should have multiple frames covered in bees. Check food stores, ensuring there's enough honey and pollen to support the colony until nectar flow is consistent. Inspect for diseases or pests, such as Varroa mites, and take appropriate action if necessary.

Feed if Necessary: If food stores are low, feeding the colony may be necessary to prevent starvation. A 1:1 sugar syrup can stimulate brood rearing and give bees the energy they need until natural nectar sources become abundant. Pollen patties can also be added to supplement the bees' protein intake, crucial for brood development.

Monitor for Swarm Signs: Spring's increase in resources and population growth can lead to swarming, where half of the colony leaves with the old queen to form a new hive. To prevent swarming, regularly check for queen cells along the bottom of frames, a sign that the colony is preparing to swarm. If queen cells are found, consider creating a controlled split to manage the colony size.

Expand the Hive: As the colony grows, adding space is crucial to accommodate the increasing population and prevent overcrowding, which can also lead to swarming. Add additional boxes or supers as needed, ensuring there is enough room for the queen to lay and for workers to store nectar and pollen. This is also the time to replace any old or damaged frames to maintain a healthy environment for the bees.

Ensure Adequate Water Sources: Bees need water for cooling the hive and diluting honey to feed brood. Ensure there is a clean, reliable water source near the hive to prevent bees from venturing too far or visiting neighboring swimming pools or water features.

Plant Bee-Friendly Flowers: Enhance your garden or the area around your hives with plants that bloom early in the spring. This not only provides your bees with additional forage options but also supports other local pollinators. Consider planting a mix of native flowers, shrubs, and trees to offer a variety of nectar and pollen sources.

Prepare Equipment for the Honey Flow: Clean and prepare supers, frames, and extraction equipment in anticipation of the spring honey flow. Having everything ready in advance ensures you can add storage space to the hive promptly and harvest honey at the optimal time without stress.

By following these detailed steps, beekeepers can ensure their hives thrive in spring, setting a solid foundation for a productive season. Regular inspections and proactive management during these critical months support the health and growth of the colony, ultimately contributing to successful honey production and hive expansion.

Preparing Hives for Active Season

Ensuring that your hives are ready for the active season involves a meticulous cleaning and inspection process that goes beyond a simple visual check. This step is crucial for the health and productivity of your bee colony. Start by choosing a warm, calm day when bees are more likely to be out foraging, reducing the stress on the hive during inspection and cleaning.

Cleaning the Hive: Begin with the removal of any debris that has accumulated over the winter months. This includes dead bees, wax cappings, and propolis. Use a hive tool to gently scrape away these materials from the frames and the bottom board. Be careful not to damage the comb, especially if you're planning to reuse the frames. If you find mold, which can occur due to poor ventilation during colder months, lightly scrape the affected area and expose the frame to sunlight, as UV rays are a natural disinfectant. For severe cases, consider replacing the frame to avoid health issues in the colony.

Inspect wooden components for signs of rot or damage. The hive's exterior, especially, can suffer from weather exposure. If you find any structural weaknesses, repair or replace these parts to ensure the hive remains secure against predators and the elements. For hives painted for protection, check if a new coat is necessary and apply a bee-safe paint or sealant.

Inspecting the Hive: Inspection goes hand-in-hand with cleaning. Your goal is to assess the health of the colony and the condition of the hive's interior. Look for a healthy brood pattern, which indicates a productive queen. A spotty brood pattern or the presence of multiple eggs in a single cell could signal problems, such as a failing queen or laying workers.

Check for pests and diseases during your inspection. Varroa mites are a common threat and can be spotted on bees or within the brood cells. Utilize a magnifying glass for a closer look if necessary. For Varroa management, consider mechanical methods like drone comb removal or chemical treatments that are safe and approved for use in beekeeping. Also, be on the lookout for signs of American Foulbrood, a serious bacterial disease characterized by a distinct odor and ropey larval remains. If detected, follow state agricultural guidelines for dealing with infected hives.

Frame Management: Evaluate the condition of the wax combs. Over time, combs can become dark and filled with cocoon remnants, reducing their effectiveness for brood rearing and honey storage. Plan to rotate out old combs, aiming to replace them every 3-5 years. This not only improves hive hygiene but also helps manage diseases and pests.

Feeding and Nutrition: Post-cleaning, assess the need for supplemental feeding. Early spring can still have cold snaps that reduce foraging opportunities. If natural nectar and pollen sources are scarce, provide a 1:1 sugar syrup feeder and consider pollen substitutes to ensure the colony has the necessary nutrients for brood rearing.

Ventilation: Proper ventilation is essential to prevent moisture buildup, which can lead to mold and cold stress. Ensure the hive's ventilation system, whether it's a notched inner cover or screened bottom board, is clear of debris and functioning. Adjustments may be needed depending on local climate conditions.

Record Keeping: Throughout this process, keep detailed records of your observations, actions taken, and any supplies used. This information is invaluable for tracking the health of your colony and planning future management activities.

By meticulously cleaning and inspecting your hives in preparation for the active season, you're setting the stage for a healthy, productive colony. This proactive approach helps mitigate potential issues before they become serious, ensuring your bees are in the best possible condition to take advantage of the spring nectar flow.

Post-Winter Hive Health Inspections

After the meticulous cleaning and inspection tasks have been completed, the next critical step in ensuring the vitality of your bee colony during spring involves conducting a thorough post-winter health assessment. This detailed examination is pivotal for identifying any underlying issues that may have developed over the colder months, which could impact the colony's ability to thrive in the upcoming active season.

Assessing Brood Health and Patterns: Begin by examining the brood patterns across the frames. Healthy brood patterns are compact and cover most of the frame, indicating a strong and productive queen. Look for uniformity in the age of the larvae within each cell, as erratic brood patterns could suggest issues such as disease or an underperforming queen. Utilize a magnifying glass to inspect for any signs of brood diseases like European Foulbrood or Sacbrood virus, characterized by discolored or sunken cappings and larvae that appear melted or twisted within their cells.

Varroa Mite Load Evaluation: Varroa mites are one of the most significant threats to bee health, and their management should be a priority during your spring inspection. Employ a standardized method such as the powdered sugar roll or alcohol wash to estimate the mite load within your colony. Collect a sample of approximately 300 bees from the brood area, apply the testing method, and count the number of mites present. This will give you an infestation rate per 100 bees, guiding your decision on whether treatment is necessary. Remember, early detection and management of Varroa mites are crucial to prevent their detrimental impact on colony health.

Nutritional Status Check: Evaluate the colony's food stores by inspecting frames for remaining honey and pollen. A healthy colony should have ample stores to sustain themselves until the first nectar flow of spring. If stores are low, continue feeding a 1:1 sugar syrup and consider adding pollen patties to ensure the bees have enough resources for brood rearing. This not only supports the current population but also encourages the queen to increase her laying rate, essential for building up colony strength.

Queen Viability and Performance: The queen's health and performance are central to the colony's success. Observe if the queen is present, and note her behavior and the pattern of eggs she lays. A well-performing queen lays a consistent pattern of eggs, with one egg per cell, predominantly in the center of the cells. If the queen is not visible, or if there are signs of poor laying patterns, consider whether requeening is necessary to ensure the colony's productivity and health.

Pest and Disease Surveillance: Beyond Varroa mites, be on the lookout for other pests such as small hive beetles or wax moths, which can take advantage of weakened colonies. Check for any signs of diseases like Nosema or chalkbrood, which can be more prevalent in the spring. Utilize specific diagnostic tools or kits available for beekeepers to test for these ailments if you suspect their presence.

Structural Integrity and Space Requirements: Finally, ensure that the hive's physical structure is in optimal condition to support the colony's growth. Check that all frames are intact and that there is enough space for the queen to lay and for workers to store incoming nectar and pollen.

As the colony expands, be prepared to add additional boxes or frames to prevent overcrowding and promote healthy colony development.

By conducting a thorough post-winter inspection and addressing any identified issues, beekeepers can set their colonies on a path toward a successful and productive season. This proactive approach to hive management in the spring is a cornerstone of successful beekeeping, ensuring that colonies are healthy, queens are productive, and the beekeeper is prepared for the challenges and rewards of the coming year.

Early Nutrition for Honey Production and Growth

Early nutrition is paramount for stimulating honey production and growth in the spring. As beekeepers, ensuring your colonies have access to adequate nutrition during this critical period can significantly impact their productivity and health. Here's a detailed breakdown of steps and considerations for providing early nutrition to your bees.

Supplemental Feeding: Early in the spring, natural nectar sources may not yet be abundant. Providing a **1:1 sugar syrup** solution mimics nectar and can help stimulate the queen to lay eggs and support the growth of the colony. Use a feeder that minimizes drowning risk, such as an entrance feeder or a top feeder, to supply this syrup. It's essential to monitor the consumption rate and replenish the feeders as needed.

Pollen Substitutes: Pollen is crucial for protein, which is necessary for brood rearing. In early spring, if natural pollen is scarce, offering a pollen substitute can provide the necessary protein to support the developing brood. Pollen patties, available from beekeeping suppliers, can be placed directly inside the hive near the brood chamber. Ensure that the patties are accessible to the bees but protected from moisture to prevent mold.

Micro-Nutrient Supplements: Some beekeepers opt to add micro-nutrient supplements to the sugar syrup or pollen substitutes to further support bee health. These supplements can include vitamins and minerals that promote hive vitality. However, it's important to use products specifically designed for bees to avoid harm.

Water Availability: Bees need water not just for cooling the hive but also for diluting stored honey to feed to the larvae. Ensure there is a clean, shallow water source near the hive. Adding small stones or floating debris to the water can help prevent bees from drowning.

Early Blooming Plants: Planting or encouraging the growth of early blooming flowers can provide a natural and sustainable source of nectar and pollen for your bees. Crocus, snowdrop, and willow are among the first plants to bloom in the spring and can significantly benefit bees emerging from winter clusters. Research and plant native flora that thrives in your region for the best results.

Monitoring Hive Activity: Regularly inspect the hive to assess the health of the colony and the effectiveness of your feeding strategy. Look for signs of active brood rearing and an increase in foraging behavior. Adjust your feeding practices based on the hive's progress and the availability of natural food sources.

Clean and Functional Feeders: Ensure that feeders are clean and free from mold or contamination. A dirty feeder can spread disease throughout the colony. Regularly inspect and clean your feeders, replacing syrup with fresh batches and removing any debris or dead bees.

Gradual Transition to Natural Foraging: As natural food sources become more abundant, gradually reduce supplemental feeding to encourage bees to forage. This transition is crucial for the development of a healthy foraging behavior among bees, ensuring they utilize natural sources of nutrition, which are ultimately beneficial for their development and the environment.

By meticulously providing early nutrition through these methods, beekeepers can support their colonies' growth and productivity, setting a strong foundation for a successful year of beekeeping. This proactive approach to early spring management underscores the importance of understanding and catering to the nutritional needs of bees, ensuring a robust start to the beekeeping season.

CHAPTER 27: SUMMER BEEKEEPING ESSENTIALS

During the summer months, beekeepers face a unique set of challenges and opportunities as the hive's activity reaches its peak. The primary focus shifts to **monitoring hive activity**, **managing heat stress**, and **preparing for honey harvest**. The following detailed guidelines will assist beekeepers in navigating the summer season effectively.

Monitoring Hive Activity: Regular inspections are crucial during the summer to ensure the health and productivity of the hive. Look for signs of overcrowding, which can lead to swarming. If more than 80% of the frames are covered in bees, consider adding more space or creating a new split to prevent the colony from swarming. Check for fresh eggs, larvae, and capped brood, as these indicate a healthy, laying queen. Also, monitor the hive for pests such as Varroa mites and small hive beetles, which can proliferate during the warm months. Employ sticky boards or alcohol washes to assess Varroa mite levels and take appropriate action if thresholds are exceeded.

Managing Heat Stress: Bees work hard to maintain the hive's temperature at around 95°F for brood rearing. During hot weather, ensure the hive has adequate ventilation. This can be achieved by slightly propping open the hive's top cover with small sticks or spacers to allow heat to escape or by ensuring the bottom board is open and unobstructed. Additionally, provide a water source near the hive so bees can cool the hive through evaporation. A shallow dish with stones or marbles for the bees to land on will prevent drowning.

Preparing for Honey Harvest: The timing of honey harvest is critical and should be based on the nectar flow in your area. Once the majority of the frames in the super are capped, it's time to harvest. This usually occurs in late summer but can vary depending on local floral sources. Before harvesting, ensure you have the necessary equipment ready, such as a bee brush, extractor, and containers for the honey. Remove the frames from the hive and gently brush off any bees. Use a honey extractor to remove the honey from the frames. Strain the honey through a fine mesh to remove any wax particles before bottling.

Supplementary Feeding: If nectar sources become scarce during the late summer, monitor the hive's food stores. Be prepared to provide supplementary feeding, especially if you've harvested a significant amount of honey. A 1:1 sugar syrup can stimulate foraging behavior and support the colony if natural food sources are insufficient.

Water Management: Bees require more water during the summer for cooling and brood rearing. Ensure there's a consistent and clean water source near the hive. Adding a water station with floating platforms like cork or small twigs can help prevent bees from drowning while allowing them to collect water safely.

Pest and Disease Management: Summer is an active season for pests and diseases. Regularly check for signs of brood diseases such as American Foulbrood or European Foulbrood. Increase vigilance for pests like the small hive beetle, which thrives in warm weather and can quickly ruin honey stores and comb. Use traps or chemical controls as necessary, following manufacturer instructions and considering the impact on the hive and honey.

Record Keeping: Maintain detailed records of each inspection, noting the health of the queen, brood patterns, pest and disease presence, honey stores, and any interventions you perform. This information is invaluable for tracking the colony's health over time and making informed decisions.

By following these summer beekeeping essentials, beekeepers can ensure their colonies are healthy, productive, and well-prepared for the coming seasons. Regular inspections, proactive management of heat stress and pests, and timely honey harvesting are key to a successful beekeeping season.

Monitoring Hive Activity and Managing Heat Stress

To effectively monitor hive activity during peak nectar flow, beekeepers should employ a structured approach to inspecting the hive. This involves systematically checking for signs of robust nectar collection and storage, ensuring that the bees are efficiently processing and capping honey. Look for frames that are predominantly filled and capped with honey, indicating strong nectar flow and active foraging. The presence of uncapped nectar suggests ongoing foraging and nectar processing. Use a **bee escape board** to gently clear bees from the supers for a more thorough inspection without causing stress to the colony.

During these inspections, it's crucial to assess the hive's ventilation needs to manage heat stress effectively. High temperatures can lead to overheating, which not only affects the bees' health but can also cause honey to ferment. To enhance ventilation, consider adding a **screened bottom board** or employing **propolis traps** as a dual-purpose tool for mite management and increased air flow. Another method is to slightly offset the hive's supers or to introduce spacers between frames, allowing for better air circulation.

Providing shade can also be beneficial during the hottest parts of the day. Positioning the hive so that it receives morning sunlight but is shaded during the peak afternoon heat can help maintain an optimal temperature. However, ensure that the hive still has adequate sun exposure to prevent dampness and to discourage pests.

Water management becomes increasingly important during summer. Establish a **water station** near the hive with shallow containers filled with pebbles or twigs to prevent bees from drowning. This not only helps the bees cool the hive but also meets their hydration needs without forcing them to travel long distances, which can affect their energy levels and productivity.

To manage heat stress, observe the bees' behavior around the hive entrance. Bees that are fanning at the entrance or along the inner walls are working to regulate the hive temperature. If this behavior is observed along with bees clustering outside the hive (bearding), it may indicate that internal temperatures are too high, and additional ventilation or cooling methods should be considered.

Regular monitoring of hive activity and environmental conditions allows for timely interventions to support the colony's health and productivity during the summer. Adjustments to hive management practices based on these observations can mitigate the risks associated with heat stress and ensure a successful nectar flow season. Keep detailed records of hive inspections, noting any changes in bee behavior, brood patterns, honey stores, and the effectiveness of heat stress management strategies. This documentation will serve as a valuable reference for optimizing hive conditions and bee health year over year.

Harvesting Honey for Quality and Quantity

Harvesting honey involves meticulous timing and attention to detail to ensure the maximum quality and quantity of the harvest. The process starts with the observation of the hive's supers, where honey is stored. Beekeepers must wait until at least 80% of the honeycomb is capped with wax, a sign that the honey is ripe and moisture content is low enough to prevent fermentation. This typically occurs in late summer but can vary based on local climate and floral sources. Using a bee brush gently to remove any bees from the frames, or employing a bee escape board the day before extraction to minimize the bee population in the super, can facilitate a smoother harvesting process.

Once the frames are removed from the hive, the wax caps can be sliced off using a heated uncapping knife or an uncapping fork, exposing the honey. This step requires a steady hand and a keen eye to remove just enough wax without damaging the delicate comb beneath, as preserving the comb structure allows bees to refill it more efficiently, saving energy and time for the colony.

The uncapped frames are then placed in a honey extractor, a centrifugal force device that spins the frames, forcing the honey out of the comb while leaving the structure intact. For small-scale beekeepers, a manual extractor is sufficient, but larger operations may benefit from an electric model to handle higher volumes. The honey is spun out onto the sides of the extractor and then drains through a spigot at the bottom. It's crucial to balance the frames inside the extractor evenly to prevent it from becoming unbalanced during operation.

After extraction, the honey should be strained through a fine mesh filter to remove any remaining wax particles, bee parts, or other debris. This step is essential for ensuring a clean, pure final product. The honey can then be funneled into sterilized containers. Glass jars are preferred for long-term storage as they do not impart any flavors to the honey and allow its natural color to shine through. It's important to leave the honey to settle for a few days, as air bubbles introduced during the extraction and jarring process will rise to the top, allowing for a clearer product.

Labeling the jars with the harvest date and floral source, if known, provides valuable information for consumers and is a requirement for selling honey. Small-scale beekeepers selling directly to consumers can benefit from highlighting the unique characteristics of their honey, such as being raw, local, or derived from specific plant sources, which can differentiate their product in the marketplace.

During the entire harvesting process, beekeepers must wear appropriate protective gear, including gloves and a veil, to prevent stings. The area around the extractor should be kept clean and free from honey spills to avoid attracting bees or other insects.

Finally, after the honey has been harvested, beekeepers should assess the hive's food stores and consider providing supplementary feeding to ensure the colony has enough resources to sustain itself, especially if a significant amount of honey was removed. A 2:1 sugar syrup can be provided as a substitute for the harvested honey, helping the bees to prepare for the upcoming winter months.

By adhering to these detailed steps and considerations, beekeepers can efficiently harvest honey while maintaining the health and productivity of their hives, ensuring a sustainable operation that can continue to produce high-quality honey year after year.

Managing Summer Pest Issues in Beekeeping

Addressing pest issues, particularly wax moths, during the active summer months is crucial for maintaining hive health and productivity. Wax moths, including the Greater Wax Moth (*Galleria mellonella*) and the Lesser Wax Moth (*Achroia grisella*), pose significant threats to bee colonies. They thrive in warmer temperatures, making summer an ideal time for their populations to grow. These pests lay their eggs in the cracks and crevices of bee hives, and their larvae feed on the wax, honey, and pollen stored within the combs, causing extensive damage and potentially leading to colony collapse.

To effectively manage wax moth infestations, beekeepers should adopt a multi-faceted approach:

1. **Regular Hive Inspections**: Conduct thorough inspections of your hives every two weeks during the summer. Look for signs of wax moth activity, such as webbing, larvae, and damaged comb. Early detection is key to preventing widespread infestation.

2. **Hygiene and Maintenance**: Keep your apiary clean. Remove old, unused combs and equipment from the vicinity of active hives. Wax moths are attracted to weak or abandoned hives and leftover wax. By eliminating these attractants, you reduce the risk of infestation.

3. **Physical Removal**: If wax moth larvae are detected during inspections, physically remove them and any affected comb. This can be done using a hive tool to scrape off the larvae and webbing. Dispose of the infested materials properly to prevent re-infestation.

4. **Freezing Infested Frames**: For frames that are infested but salvageable, freezing is an effective method to kill wax moth eggs and larvae. Place the affected frames in a freezer at 0°F (-18°C) for at least 24 hours. After freezing, allow the frames to return to room temperature before reintroducing them to the hive.

5. **Use of Moth Traps**: Moth traps can be placed around the apiary to capture adult wax moths before they have a chance to lay eggs. These traps typically use a pheromone or a light source to attract and capture the moths. Regularly check and empty these traps throughout the summer months.

6. **Maintaining Strong Colonies**: Strong, healthy bee colonies are less susceptible to wax moth infestations. Ensure your colonies have a good queen, are well-fed, and are free of other stressors. A strong bee population can effectively patrol and clean the hive, removing wax moth eggs and larvae before they become a problem.

7. **Chemical Controls**: In severe cases, chemical controls may be necessary. Use products specifically designed for wax moth control in bee hives, and follow the manufacturer's instructions carefully. Always prioritize non-chemical methods to minimize risk to the bees.

8. **Storage of Comb and Equipment**: When storing comb and beekeeping equipment, ensure it is in a well-ventilated, dry area. Consider using moth repellents or securely sealing equipment in airtight containers to prevent moth infestation during storage.

CHAPTER 28: AUTUMN BEEKEEPING

As the **autumn** season approaches, the focus of beekeeping shifts towards preparing the bees for the colder months ahead. This preparation is crucial for the survival of the hive, and several steps must be taken to ensure the bees are adequately prepared.

Assessing the hive's honey reserves is the first step in autumn beekeeping. It's essential to check that there are enough honey stores to sustain the colony through the winter. A strong colony requires approximately 60 to 90 pounds of honey. To evaluate the honey stores, inspect the frames in the upper boxes of the hive, looking for fully capped honey. If the reserves are low, you may need to feed the bees sugar syrup in the fall to boost their stores.

Providing supplementary feeding becomes necessary if natural nectar sources are scarce. A 2:1 ratio of sugar to water creates a thick syrup that mimics the consistency of honey, providing the bees with a vital energy source. Place the feeder inside the hive or at the hive entrance to allow easy access for the bees. Monitor the consumption regularly and refill as needed to ensure the bees have enough food.

Checking the queen's health is another critical task in autumn. A healthy, productive queen is essential for a strong colony in the spring. Observe the brood pattern for uniformity and the presence of eggs, which indicates the queen is still laying. If the queen is underperforming or missing, consider introducing a new queen before the winter sets in, ensuring the colony has the leadership it needs.

Preparing the hive for winter involves several maintenance tasks to protect the bees from cold and moisture. First, reduce the hive entrance to prevent mice and other pests from entering. This also helps the bees defend their hive against intruders. Next, ensure the hive has proper ventilation to prevent moisture buildup inside the hive, which can be more deadly to the bees than the cold. A small ventilation hole at the top of the hive can help moisture escape.

Insulating the hive can help maintain a stable internal temperature. Wrapping the hive in a breathable material like tar paper helps keep the bees warm without trapping moisture. Be sure to leave the entrance and ventilation holes uncovered to ensure airflow.

Pest and disease management must continue into the autumn. Diseases like Nosema can be particularly problematic in cooler temperatures. Administer treatments for Varroa mites if the infestation level is high, following the recommended guidelines to avoid harming the bees. Removing any dead bees from the hive and ensuring the hive is clean and free from debris can also help prevent disease and pest issues.

Finally, **monitoring the hive** throughout the autumn is vital. Regular checks, though less frequent than in the summer, can help identify any issues early on. Pay attention to the weight of the hive, which can indicate how much honey the bees have stored. A sudden decrease in weight might mean the bees are consuming their reserves too quickly, signaling a need for supplementary feeding.

Assessing Honey Reserves for Winter Survival

To accurately assess the hive's honey reserves, beekeepers should conduct a thorough inspection of the hive's frames, particularly focusing on the upper boxes where honey is predominantly stored.

This evaluation is critical as a strong colony requires approximately **60 to 90 pounds** of honey to sustain itself through the winter months. Here's a step-by-step guide to ensure a precise assessment:

1. **Frame Inspection**: Begin by gently removing the frames from the upper boxes of the hive. Look for frames that are fully capped with wax, indicating that the honey within is ripe and moisture content is low, ideal for long-term storage. Use a soft brush or a puff of smoke to calmly remove any bees from the frames for a clearer view.

2. **Weighing the Frames**: To estimate the weight of honey in the hive, select a representative sample of capped honey frames. A fully capped Langstroth frame can weigh approximately **5 to 7 pounds**. Weigh the frame using a scale and multiply this by the number of similar frames in the hive to estimate the total honey reserves. Remember, the weight includes both the honey and the frame, so you may need to subtract the weight of an empty frame for accuracy.

3. **Visual Inspection for Empty Spaces**: Not all frames may be fully capped or filled with honey. Inspect for any empty spaces or partially filled frames. This can indicate that the bees have not stored enough honey and may need supplementary feeding.

4. **Calculating Required Reserves**: Based on the estimated total weight of honey, compare this against the **60 to 90 pounds** requirement. If the hive has less than this, consider feeding the bees with a 2:1 sugar syrup to boost their reserves before winter.

5. **Monitoring Consumption Rates**: Keep in mind that bees will start consuming their honey reserves as soon as colder weather sets in. Regularly weighing the hive, using a hive scale, throughout the autumn can help you monitor the rate of consumption. A sudden drop in weight could indicate that the bees are consuming their reserves faster than expected, necessitating additional feeding.

6. **Supplementary Feeding**: If supplementary feeding is required, place the feeder inside the hive or at the hive entrance. Ensure the feeder is secure and easily accessible to the bees. Monitor the feeder regularly, refilling as necessary to ensure the bees have enough food leading into the winter.

7. **Record Keeping**: Maintain detailed records of your observations, including the weight of honey reserves, the number of frames inspected, and any supplementary feeding undertaken. This information will be invaluable for managing the hive through the winter and preparing for the following year.

By following these detailed steps, beekeepers can accurately assess their hive's honey reserves, ensuring that the colony has sufficient resources to survive the winter. This careful preparation is a key component of successful beekeeping, supporting the health and productivity of the bee colony in the long term.

Supplementary Feeding with Sugar Syrup

Supplementary feeding with sugar syrup is a critical intervention for beekeepers to ensure their colonies have sufficient energy reserves to survive the colder months. The preparation of the syrup and its delivery to the bees require careful attention to detail to mimic the consistency and sweetness of natural honey as closely as possible. A 2:1 ratio of sugar to water by weight is recommended to achieve a thick syrup that closely resembles the viscosity of honey. This ratio provides the bees with an energy-dense feed, crucial for building up their reserves.

To prepare the syrup, start by measuring out the sugar and water using a kitchen scale to ensure accuracy. For example, to make a gallon of syrup, you would need approximately 11.5 pounds of sugar and 5.75 pounds of water. Heat the water in a large pot until it is warm but not boiling, then gradually add the sugar, stirring continuously until it is completely dissolved. Avoid overheating the mixture as this can produce compounds that are harmful to bees.

Once the syrup is prepared, allow it to cool to room temperature before feeding it to the bees. This can be done using various types of feeders, such as an entrance feeder, which attaches to the front of the hive, or an internal feeder, which sits inside the hive. The choice of feeder depends on the beekeeper's preference, the size of the colony, and the weather conditions. Entrance feeders are easy to monitor and refill but can attract robbing from other colonies and pests. In contrast, internal feeders are protected from the elements and robbing but require opening the hive for refills, which can stress the bees.

When placing the feeder, ensure it is secure and will not tip over, spilling the syrup. If using an entrance feeder, place it at the hive entrance and adjust the entrance reducer to prevent robbing. For internal feeders, place the feeder above the brood chamber where the bees can easily access it without significant disruption to the colony structure.

Monitoring the consumption of the syrup is crucial. Refill the feeders as needed, typically every few days, depending on the size of the colony and the feeder. As the weather cools and the bees consume more of their honey stores, the frequency of feeding may increase. Keep detailed records of how much syrup is consumed and how often the feeders are refilled. This information will help in planning for future seasons and understanding the specific needs of your colonies.

In addition to providing energy, supplementary feeding is an opportunity to administer medications or supplements that can help the bees maintain their health through the winter. However, any additives should be used judically and according to the manufacturer's instructions to avoid harming the colony.

As the autumn progresses and temperatures continue to drop, gradually reduce the frequency of syrup feeding to encourage the bees to cluster for warmth and slow their consumption of stores. This gradual reduction helps the colony transition smoothly into winter, reducing the risk of overconsumption of their winter reserves too early in the season.

Queen Bee Health and Replacement

Ensuring the health of the queen bee is paramount as autumn transitions into winter, as her well-being directly impacts the colony's productivity and survival. The queen's ability to lay eggs and the overall health of the brood are clear indicators of her condition. A healthy queen should be actively laying eggs, and the pattern of the brood should be dense and uniform, covering most of the frame. This demonstrates not only her fertility but also the hive's ability to maintain its temperature and protect its young, crucial abilities in the colder months.

If upon inspection, the brood pattern appears spotty or the number of eggs is significantly reduced, this may indicate that the queen's health is declining or that she is aging and becoming less productive. Additionally, if no new eggs are observed, it could mean the queen is missing. In such cases, immediate action is required. Introducing a new queen to the hive before winter ensures that the colony will not be left without a leader during the critical overwintering period.

When selecting a new queen, consider the timing and method of introduction carefully. It's advisable to introduce a new queen when the bees are still actively foraging and there's enough time for her to be accepted and to start laying eggs before the colony clusters for winter. The new queen should ideally be introduced in a queen cage with a candy plug, allowing the bees to gradually eat through the candy and release the queen at their own pace, usually within a few days. This method helps improve acceptance rates by giving the bees time to adjust to her pheromones.

During the introduction period, monitor the hive closely for signs of acceptance or rejection. Acceptance is indicated by the bees' calm behavior around the queen cage and the eventual release of the queen. Rejection may be observed if the bees are aggressively clustering around the queen cage or attempting to sting through the cage. If rejection occurs, removing the queen and consulting with a more experienced beekeeper or a beekeeping association for further advice may be necessary.

Once the new queen is released, continue to monitor the hive, paying close attention to the brood pattern and the queen's egg-laying activity. It may take a few weeks for the new queen to reach her full laying potential. Successful introduction of a new queen will be evident through a revitalized brood pattern and increased activity within the hive, ensuring the colony's strength going into winter.

In addition to queen health, ensure that the hive's environment supports the new queen's acceptance and productivity. This includes maintaining adequate stores of honey and pollen, providing insulation and ventilation to regulate the hive's temperature, and protecting the hive from pests and diseases. These steps, combined with careful monitoring and management, will help secure the hive's health and vitality through the winter and into the next season, allowing the beekeeper to enjoy the rewards of a thriving bee colony.

CHAPTER 29: WINTER BEEKEEPING STRATEGIES

Preventing moisture buildup within the hive is crucial during the winter months as excess moisture can be more detrimental to the bees than the cold itself. Moisture in the hive can lead to mold growth and chill the bees, potentially leading to their demise. To effectively manage moisture, beekeepers should ensure that their hives are well-ventilated. One method to achieve this is by slightly propping up the hive's outer cover with small sticks or spacers, allowing moist air to escape while keeping the interior insulated. Additionally, incorporating an upper entrance can facilitate better air circulation, drawing damp air out of the hive.

Another strategy involves the use of moisture boards or quilts placed above the top bars of the uppermost box. These boards are designed to absorb condensation that rises with the warm air generated by the cluster of bees. Made from absorbent materials such as wood shavings or natural fibers, moisture boards can hold a significant amount of moisture, keeping it away from the bee cluster. It is essential to check these boards periodically throughout the winter and replace them if they become saturated.

Insulation plays a pivotal role in maintaining stable temperatures within the hive and preventing the walls from becoming cold enough to cause condensation. Wrapping the hive with a breathable insulation material, such as tar paper or a specially designed hive wrap, can help keep the bees warm while allowing moisture to escape. When applying insulation, it's important to leave the hive's entrance and upper ventilation points unobstructed to maintain airflow.

Monitoring hive activity periodically without disturbing the colony is another vital aspect of winter beekeeping. On warmer winter days, when temperatures rise above 50 degrees Fahrenheit, bees may take cleansing flights. This natural behavior is an opportunity for beekeepers to briefly check the hive's exterior for signs of activity and ensure the entrance is clear of dead bees or debris, which could block the entrance and trap moisture inside.

To safeguard the hive from harsh winter winds, positioning the hive with its entrance facing away from prevailing winds can reduce wind chill, helping to keep the interior warmer. Strategic placement of windbreaks, such as bales of straw or a fence, can also provide additional protection from cold winds, further insulating the hive.

Lastly, ensuring proper hive health before winter sets in is critical. A strong, healthy colony with a productive queen, adequate honey stores, and minimal pest and disease issues stands the best chance of surviving the winter. Beekeepers should have treated for Varroa mites and other pests in the late summer or early fall to reduce the colony's stress and prevent winter losses. By combining these strategies, beekeepers can help their hives emerge from winter ready for the active season ahead, maintaining the health and productivity of their bee colonies.

Insulating Hives for Winter Stability

Insulating hives for winter stability requires a methodical approach to ensure that bees remain active and healthy during the colder months. The primary goal is to maintain a stable internal temperature within the hive, preventing the colony from freezing while also avoiding excessive moisture buildup which can lead to mold and disease. To achieve this, beekeepers must select the right materials and apply them in a way that enhances the hive's natural defenses against the cold.

The first step in insulating a hive is to choose insulation materials that are breathable, such as natural wool, hemp, or specially designed synthetic wraps that allow moisture to escape while keeping the cold out. These materials have the dual purpose of retaining warmth generated by the bee cluster and allowing any excess moisture to wick away, thus preventing condensation on the interior walls of the hive.

When applying insulation, start by wrapping the hive body, ensuring that the material covers the sides, back, and front of the hive, but leaves the entrance unobstructed to allow for ventilation and bee access. It is crucial to secure the insulation so it stays in place even in windy conditions, using straps, ropes, or even a hive cover that fits snugly over the insulation.

For the top of the hive, a moisture-absorbing quilt or board can be placed directly above the top bars, underneath the roof. This quilt should be made of a material that can hold moisture away from the colony, such as wood chips, straw, or a commercial moisture board designed specifically for beekeeping. This layer acts as a buffer, capturing any condensation that forms as warm air from the cluster meets the cold surface of the hive lid.

In addition to insulation, providing adequate ventilation is key to managing moisture. Even in winter, bees need fresh air and a way to expel damp, CO_2-rich air from the hive. This can be achieved by ensuring the hive's upper ventilation holes are clear and by slightly raising the hive's lid to create a small ventilation gap. However, care must be taken not to create drafts that could chill the bees. Some beekeepers use a small piece of wood or foam as a spacer to lift the lid just enough to allow air flow without significantly lowering the internal temperature of the hive.

Another aspect to consider is the hive's location. Placing hives in a spot that receives winter sunlight can help raise the internal temperature naturally during the day. Additionally, shielding the hive from prevailing winds with a windbreak can significantly reduce the cooling effect of the wind on the hive's exterior.

Finally, monitoring the hive throughout the winter is essential. On warmer days, when it is safe to open the hive briefly, checking the moisture levels and the condition of the insulation can help beekeepers make necessary adjustments. If the insulation is wet or the moisture board is saturated, they should be replaced to ensure the hive remains dry and warm.

By carefully selecting insulation materials, applying them in a way that complements the hive's design, and balancing insulation with adequate ventilation, beekeepers can effectively maintain stable temperatures within their hives during freezing months. This careful preparation helps ensure that the bee colony remains strong and healthy, ready to emerge in the spring fully capable of resuming their vital role in pollination and honey production.

Monitoring Hive Activity Without Disturbance

Monitoring hive activity during the winter months requires a delicate balance to ensure the well-being of the colony without causing undue stress or exposure to cold temperatures. The key to successful winter monitoring lies in the subtle observation of external indicators and the judicious use of technology to assess the hive's internal conditions.

One effective method for monitoring without opening the hive is to observe the entrance for signs of bee activity. On warmer days, bees may venture outside for short cleansing flights. The presence of dead bees near the entrance can also provide clues to the hive's health, indicating normal cleansing behavior or, in larger numbers, potential issues within the hive. Careful removal of debris and dead bees from the entrance using a soft brush or a small hand tool can help maintain clear access for the bees without the need to open the hive.

Another non-intrusive technique involves placing a hand gently on the side of the hive and feeling for warmth, or carefully listening for the hum of bee activity within. These signs can indicate that the colony is alive and maintaining the necessary temperature for survival. However, it's important to perform these checks quickly to minimize heat loss through the hive walls.

Advancements in beekeeping technology have also provided tools such as infrared thermometers or thermal imaging cameras, which can be used to check the temperature inside the hive without opening it. These devices can detect heat patterns and clusters of bee activity, offering insights into the colony's size and health. When using such technology, aim the device at the middle sections of the hive, where the cluster is likely to be located during colder months, and compare readings over time to monitor for significant changes.

For beekeepers who employ hive scales, the weight of the hive can serve as an indicator of honey stores consumption. A gradual decrease in weight suggests that bees are consuming their stores as expected, while a rapid decline might indicate a problem, such as robbery by other animals or insufficient honey reserves requiring supplementary feeding.

Additionally, installing an entrance reducer during the winter can aid in monitoring efforts. Not only does it help conserve heat and protect the hive from intruders, but it also restricts the entrance area, making it easier to observe bee activity and health indicators at the hive's entrance.

It's crucial to remember that every interaction with the hive, no matter how minimal, can influence the colony's energy expenditure. Therefore, all monitoring activities should be conducted with the utmost care, ideally during the warmer part of the day to minimize cold air intrusion. By combining these non-invasive monitoring techniques, beekeepers can gather valuable information about their hives' health through the winter months, ensuring that the colonies remain strong and ready for the active season ahead.

Preventing Moisture Buildup in Hives

Preventing moisture buildup within the hive is critical, especially during the winter months when the combination of cold air and moisture can create a lethal environment for bees. A key strategy in managing moisture is the strategic use of **moisture boards or quilts**, which was previously mentioned. However, the material and placement of these boards are crucial for optimal effectiveness. For instance, using a moisture board made of a porous material such as untreated wood pulp can absorb moisture effectively. Positioning this board directly above the bees but under the hive cover ensures that moisture rising from the cluster of bees is captured before it condenses on the hive's interior surfaces.

Another effective method is the installation of a **propolis trap** or a **ventilation rim**. A propolis trap, typically used to encourage bees to fill gaps with propolis, can also serve as an additional barrier to moisture when placed beneath the hive's lid. The small gaps in the trap allow moisture to escape while keeping the warmth inside. On the other hand, a ventilation rim, a shallow frame placed between the top super and the hive cover, can create a space for moist air to exit the hive. Drilling small holes on the sides of this rim enhances airflow without creating a draft, especially when combined with an upper entrance.

In regions where winter precipitation is significant, **protecting the hive's exterior** from direct contact with rain or snow is vital. Employing a hive stand that elevates the hive off the ground prevents moisture from wicking up into the hive from the wet ground. Additionally, ensuring the hive's outer cover is sloped or peaked encourages water to run off rather than pool on the top, which could seep into the hive. Covering the hive with a waterproof but breathable material, such as a canvas tarp, allows for moisture to escape while keeping the hive dry. It's important that this cover extends over the edges of the hive but does not block the hive's entrance or ventilation points.

Winter wraps, specifically designed for beehives, offer dual benefits of insulation and moisture control. When selecting a winter wrap, opt for materials that are breathable to prevent moisture accumulation within the hive. Some wraps come with a reflective outer layer to maximize sunlight absorption, raising the hive's temperature during the day. Securing these wraps with straps ensures they stay in place even in windy conditions but are loose enough to not compress the hive's natural ventilation pathways.

Feeding practices also play a role in moisture management. Feeding bees with fondant or dry sugar instead of syrup in the late fall reduces internal hive moisture. If syrup feeding is necessary, it should be completed early enough in the fall so that bees can process and reduce the moisture content of the syrup before winter sets in.

Lastly, the role of **regular hive checks** cannot be overstated. On milder winter days, a quick visual inspection of the hive's exterior for signs of moisture, such as ice or mold formation around the entrance, can indicate an issue. Additionally, listening for the bees' activity through the hive wall can provide assurance that the colony is alive and generating heat, which is a natural byproduct of their movement and metabolic processes.

By implementing these detailed strategies, beekeepers can effectively manage moisture within the hive, ensuring their colonies remain healthy and strong throughout the winter months. Each step, from the material choice for moisture boards to the strategic feeding of bees, plays a crucial role in creating a stable and conducive environment for overwintering bee populations.

Part 8: Harvesting Hive Products

CHAPTER 31: HONEY

Identifying different types of honey, such as clover, acacia, and wildflower varieties, is crucial for both the beekeeper and the consumer. Each variety of honey has its unique flavor, color, and nutritional properties, largely influenced by the source of the nectar the bees have foraged. Clover honey, for instance, is typically light in color and mild in flavor, making it a versatile sweetener for a wide range of foods and beverages. Acacia honey, on the other hand, is known for its clear, almost transparent color and delicate floral taste. It's often preferred by those who appreciate a less sweet, more refined flavor profile. Wildflower honey, derived from the nectar of various species of flowers, can vary greatly in taste and color depending on the specific mix of wildflowers in bloom at the time of production. This variety is often darker and can offer a more complex flavor profile, reflecting the diversity of its floral sources.

Safe and efficient techniques for honey extraction are vital for preserving the quality of the honey and ensuring the health of the bee colony. The most common method of extraction involves the use of a centrifugal extractor, which spins the honey out of the frames, leaving the comb intact for bees to refill. This method is preferred for its efficiency and the fact that it minimizes waste and damage to the hive. Before extraction, frames must be uncapped, either manually with a knife or with an automatic uncapping machine, to remove the thin layer of wax that bees use to seal the honey into the cells. It's important to maintain a clean and controlled environment during extraction to prevent contamination and ensure the purity of the honey. Temperature control is also a key factor, as too much heat can degrade the enzymes and antioxidants that give honey its health benefits.

Best practices for storing honey to retain flavor and prevent spoilage include keeping it in airtight containers at room temperature, away from direct sunlight. Honey's natural acidity and low moisture content make it resistant to bacteria and spoilage, but it can absorb moisture and odors from its environment, which can affect its quality and shelf life. Glass jars are often recommended for storage as they do not impart any flavors to the honey and allow its natural color to shine through. It's also important to ensure that the containers used for storing honey are dry and free from contaminants before filling. Honey can crystallize over time, especially at cooler temperatures, but this is a natural process and does not indicate spoilage. Crystallization can be reversed by gently warming the honey in a water bath, being careful not to overheat and damage the honey's natural properties.

Labeling and marketing honey effectively requires a clear understanding of the unique characteristics of each honey variety and the preferences of the target market. Honey labels should provide consumers with essential information, including the type of honey, its floral source, and any unique qualities it possesses. Transparency about the origin and processing methods can also appeal to consumers who value sustainability and ethical production practices. Marketing strategies might include offering tastings at local farmers' markets, creating attractive packaging that highlights the honey's natural beauty, and educating consumers about the benefits of different honey varieties. Online marketing and social media can also be powerful tools for reaching a wider audience, allowing beekeepers to share stories about their bees and the production process, which can enhance the perceived value of their honey products.

For beekeepers aiming to maximize the appeal and marketability of their honey, understanding consumer preferences is key. This involves not only showcasing the distinct flavors and benefits of each honey variety but also emphasizing the artisanal and sustainable aspects of beekeeping.

Presenting honey in attractive, eco-friendly packaging can significantly enhance its appeal, with options such as reusable glass bottles or jars being particularly popular among environmentally conscious consumers. Additionally, including recipes or suggestions for use on the label can inspire buyers to incorporate honey into their daily lives in new and exciting ways.

To ensure the longevity and quality of honey, beekeepers should adopt meticulous harvesting and processing practices. This begins with the careful timing of honey collection, ideally when the majority of the nectar flow has ceased, and the honey is ripe, indicated by the bees capping the cells. The extraction process should be done gently to preserve the integrity of the comb and the honey's natural enzymes. After extraction, honey should be strained through a fine mesh to remove any impurities without overheating or excessively filtering, which can diminish its nutritional value and flavor profile.

Quality control is another critical aspect of honey production. Conducting regular taste tests and laboratory analyses can help beekeepers monitor the flavor, purity, and safety of their honey, ensuring it meets high standards. This not only protects consumers but also builds trust and credibility for the beekeeper's brand. Beekeepers can also consider obtaining certifications, such as organic or non-GMO, to further validate their commitment to quality and sustainable practices.

Educational outreach is an invaluable tool for beekeepers looking to engage with their community and market their honey. Hosting workshops, farm visits, or beekeeping demonstrations can provide hands-on learning experiences for consumers, fostering a deeper appreciation for beekeeping and the unique qualities of artisanal honey. These interactions offer an excellent opportunity to share knowledge about the importance of bees to our ecosystem, the challenges they face, and how responsible beekeeping practices contribute to environmental conservation.

In conclusion, successful honey production and marketing require a combination of passion, knowledge, and strategic planning. By focusing on quality, sustainability, and consumer education, beekeepers can create a strong demand for their products while supporting the health and vitality of their bees. With the right approach, honey can be more than just a sweet treat; it can be a catalyst for positive change, promoting environmental awareness and the preservation of our precious pollinators.

Types of Honey: Clover, Acacia, Wildflower

When assessing the **flavor profiles** of different honey varieties, it's essential to understand the influence of the bees' nectar source. **Clover honey**, for example, is one of the most common and widely available types of honey in the United States. It comes from bees that predominantly collect nectar from clover plants. This honey is characterized by its light, creamy color and mild, sweet flavor, making it a favorite for use in teas, baking, and as a general sweetener. The subtle flavor of clover honey makes it an excellent base for flavored honeys, infused with fruits or spices.

Acacia honey stands out due to its exceptionally clear and pure appearance. Sourced from the nectar of the acacia tree's flowers, it boasts a delicate floral taste with hints of vanilla. Its low sucrose content and high fructose level give it a smooth, liquid consistency that rarely crystallizes. Acacia honey's light and clean flavor profile pairs well with delicate foods, such as yogurt or fine cheeses, without overpowering them.

Wildflower honey, also known as polyfloral honey, is made from the nectar of various species of flowers and plants that are blooming at the same time. This results in a complex, variable taste and color that can change from season to season, even from the same hive. Wildflower honey can range from light and fruity to dark and rich, depending on the mix of wildflowers in bloom. This variety is particularly favored for its potential health benefits, including a higher antioxidant content due to the diverse sources of nectar.

To **taste test** these honey varieties accurately, use a clean spoon for each type and cleanse your palate between tastings with water. Observe the honey's color and consistency, then smell it to identify any floral or fruity notes before tasting a small amount. Let the honey dissolve on your tongue to fully appreciate its flavor profile and aftertaste.

For those interested in **pairing honey with foods**, consider the following guidelines:

- **Clover honey**'s mildness complements a wide range of foods, from oatmeal and yogurt to more savory dishes like glazed ham.

- **Acacia honey** works beautifully with light desserts, fresh cheeses, and as a sweetener for herbal teas, enhancing without overwhelming.

- **Wildflower honey**'s robust flavors stand up well to stronger, more flavorful foods. It can be drizzled over blue cheese, used in baking rye bread, or as a glaze for roasted meats.

When **storing honey**, always keep it in airtight containers to prevent moisture absorption which can lead to fermentation. Glass jars are preferable as they do not impart any odors or flavors to the honey. Store honey in a cool, dry place away from direct sunlight. If crystallization occurs, gently warm the honey in a water bath to return it to its liquid state, being careful not to overheat and degrade its quality.

For beekeepers and enthusiasts looking to explore the nuanced world of honey, understanding these differences is just the beginning. Experimenting with honey from different floral sources can offer a rich palette of flavors and benefits, enhancing culinary experiences and providing a deeper connection to the natural world of bees and their vital role in our ecosystem.

Honey Extraction Techniques

Moving beyond the initial steps of honey extraction, it's crucial to delve into the nuances of both manual and automated extraction methods to ensure the process is both safe for the bees and efficient for the beekeeper. Manual extraction, while more labor-intensive, offers a hands-on approach that can be especially appealing to hobbyist beekeepers or those with smaller operations. This method typically involves the use of a manual extractor, which is a device that spins the honey out of the frames using hand power. To prepare for manual extraction, beekeepers must first uncap the honeycomb cells. This can be done using a specialized uncapping knife or fork, which gently removes the wax cap without damaging the comb. The frames are then placed inside the extractor, and the beekeeper manually turns a handle to spin the frames, using centrifugal force to pull the honey out of the comb and down the sides of the extractor where it can be collected.

The key to efficient manual extraction is to ensure even loading of the frames within the extractor. This balance allows for a smoother spinning process and prevents potential damage to the extractor or the frames. Additionally, maintaining a steady rhythm while turning the handle maximizes the amount of honey extracted and minimizes the physical effort required. After extraction, the honey should be allowed to settle in a bucket or tank equipped with a spigot at the bottom. This settling process helps to remove air bubbles and small wax particles, resulting in a clearer honey product. The honey can then be drawn off from the bottom of the tank, ensuring that the cleanest possible honey is collected for bottling.

Automated extraction systems, on the other hand, are ideal for larger operations or beekeepers looking to streamline their honey processing. These systems often incorporate automated uncapping machines, which precisely remove the wax caps from the frames with minimal waste. The frames are then automatically loaded into an extractor, which uses electric power to spin the honey out. One of the significant advantages of automated systems is their ability to handle a larger volume of frames simultaneously, significantly reducing the time and labor required for extraction.

When setting up an automated extraction line, it's important to consider the flow of the process from uncapping to bottling. The equipment should be arranged in a logical sequence that minimizes the need to move frames and honey by hand, thus reducing the risk of contamination or spillage. Many automated extractors also feature variable speed controls, allowing the beekeeper to adjust the extraction speed based on the type of honey and the condition of the comb. This level of control can help to prevent damage to the comb, which is particularly beneficial when planning to return the frames to the hive for refilling by the bees.

Regardless of the method chosen, cleanliness and temperature control are paramount throughout the extraction process. All equipment should be thoroughly cleaned and sanitized before use to prevent the introduction of bacteria or other contaminants into the honey. Additionally, maintaining an ambient temperature of around 95°F (35°C) during extraction can make the honey less viscous and easier to handle, while preserving its natural enzymes and flavor. This temperature can be achieved through the use of a heated room or a honey warming cabinet.

Finally, the choice between manual and automated extraction methods depends on the scale of the beekeeping operation, budget, and personal preference. While manual extraction offers a more tactile and involved experience, automated systems can save time and labor, allowing beekeepers to process larger quantities of honey with greater efficiency. Regardless of the method, the goal remains the same: to safely and effectively harvest the honey while preserving its quality and the health of the bee colony.

Storing Honey: Best Practices

To maintain the **flavor** and **prevent spoilage** of honey, it's essential to focus on the **storage environment**. Honey should be stored in a **cool, dry place** away from direct sunlight. Direct sunlight can increase the temperature of the honey, leading to potential degradation of its natural enzymes and flavor. A pantry or a cupboard away from heat sources like stoves or ovens is ideal.

The choice of **container** is crucial for preserving honey's quality. **Glass jars** with tight-fitting lids are preferred as they are impermeable to air and moisture, which can lead to fermentation if the water content of the honey increases. Glass does not interact with the honey, ensuring that its flavor remains unaltered. Ensure the jar is thoroughly cleaned and completely dry before use to avoid introducing moisture into the honey.

If honey crystallizes, a natural process that does not indicate spoilage, it can be returned to liquid form by gently warming the container in a **water bath**. The water should be heated to no more than 104°F (40°C) to avoid damaging the honey's delicate flavors and enzymes. Place the glass jar of honey in the warm water, and let it sit, stirring occasionally until the crystals dissolve. Avoid using a microwave for warming, as it can unevenly heat the honey and potentially overheat it.

For larger quantities of honey, **food-grade plastic buckets** with airtight lids can be used for storage. However, it's important to ensure that the plastic is compatible with food use and does not leach any chemicals into the honey over time. Buckets should be kept in a cool, dark place to maintain optimal conditions.

Labeling the honey with the **date of harvest** is a good practice, as it helps keep track of its age. While honey does not spoil if stored properly, its flavor and aroma might evolve over time. Knowing the harvest date can help in managing the rotation of stock and ensuring that older honey is used first.

For those who prefer to **store honey in larger containers**, transferring a smaller amount into a glass jar for daily use can minimize the exposure of the bulk honey to air and potential contaminants. This practice can help maintain the quality of the honey while still allowing convenient access.

Humidity control in the storage area is another factor to consider. High humidity environments can lead to moisture absorption, increasing the risk of fermentation. Using a dehumidifier in particularly damp areas can help protect the quality of the honey.

In summary, proper storage of honey involves protecting it from excessive heat, light, and moisture. By selecting the appropriate containers, maintaining a cool and dry environment, and managing humidity, beekeepers and honey enthusiasts can ensure their honey retains its natural flavor, aroma, and beneficial properties for an extended period.

Labeling and Marketing Honey for Consumer Appeal

To effectively **label and market honey** to meet diverse consumer preferences, it's essential to understand the specifics that appeal to different segments of the market. Begin by ensuring that your honey labels are **clear, informative, and compliant** with local and federal regulations. This includes listing the **net weight** in pounds or ounces, **origin** (if required by local laws), **producer contact information**, and any **relevant certifications** such as organic, non-GMO, or fair trade.

Highlighting the **type of honey** (e.g., clover, acacia, wildflower) prominently on the label can attract consumers looking for specific flavors or health benefits associated with certain varieties. For consumers interested in health and wellness, emphasize any **unique properties** of your honey, such as antioxidant levels, antibacterial qualities, or if it's raw and unfiltered, which preserves its natural enzymes and nutrients.

In terms of marketing, **digital platforms** offer a powerful tool to reach a broad audience. Create engaging content that tells the story of your beekeeping practices and the quality of your honey. Use **social media** to share behind-the-scenes looks at your beekeeping operations, the natural environment of your bees, and the harvesting and bottling process. This transparency builds trust and connects consumers emotionally to your product.

Email marketing can be used to keep your audience informed about new product releases, special promotions, or educational content about the benefits of honey. Tailoring your messages based on consumer preferences and purchase history can increase engagement and loyalty.

For local market penetration, consider partnering with **farmers' markets, local food co-ops, and independent grocery stores**. Offering tastings can significantly boost interest and sales, as it allows potential customers to experience the quality and flavor profile of your honey firsthand.

Lastly, consider **collaborating with local chefs, bakeries, and cafes** to feature your honey in their products. This not only broadens your market reach but also associates your honey with high-quality, artisanal food experiences, appealing to foodies and culinary enthusiasts.

By focusing on clear labeling, leveraging digital marketing, engaging with your local community, and creating strategic partnerships, you can effectively market your honey to cater to a wide range of consumer preferences.

CHAPTER 32: ROYAL JELLY

Royal jelly, a milky secretion produced by worker bees, plays a pivotal role in the nutrition of developing larvae and the queen bee. Harvesting royal jelly requires precision and care to ensure the health of the colony is not compromised. The process begins with the preparation of artificial queen cells. Beekeepers use these cells to encourage worker bees to produce royal jelly. Once the cells are prepared and larvae are introduced, worker bees begin the secretion of royal jelly to feed the larvae. After a few days, these cells are rich in royal jelly and ready for harvest.

The harvesting process involves carefully removing the queen cells from the hive and extracting the royal jelly using a small, sterile spatula or syringe. It is crucial to perform this task gently to avoid damaging the cells or injuring the larvae. The collected royal jelly should be immediately cooled to preserve its freshness and nutritional properties. Optimal storage conditions involve refrigeration at temperatures between 36°F to 40°F. For long-term storage, freezing royal jelly at temperatures below 0°F can significantly extend its shelf life without diminishing its quality.

The nutritional composition of royal jelly makes it a sought-after product, rich in proteins, lipids, vitamins, and minerals. Its potential health benefits, including anti-inflammatory and antioxidant properties, make it a valuable addition to dietary supplements and natural cosmetics. When incorporating royal jelly into products, it is essential to maintain a controlled environment to prevent contamination. The use of clean, sterilized equipment cannot be overstated, as it ensures the royal jelly remains pure and safe for consumption or application.

For beekeepers looking to market royal jelly, compliance with local health regulations and standards is mandatory. Labeling should accurately reflect the product's contents and storage instructions to inform consumers effectively. As with any bee product, transparency about the harvesting and processing methods enhances consumer trust and supports the product's authenticity.

In summary, the harvesting and processing of royal jelly require meticulous attention to detail, from the preparation of queen cells to the storage and labeling of the final product. By adhering to best practices and regulatory standards, beekeepers can ensure the quality and safety of royal jelly, making it a sustainable and valuable addition to their beekeeping endeavors.

Health Benefits of Royal Jelly

Royal jelly, often hailed as nature's superfood, is a substance of complex chemical structure produced by the young nurse bees. It serves as the primary food for the queen bee throughout her life, as well as for all the bee larvae in their first few days. This creamy, acidic secretion from the glands on the heads of young worker bees is rich in proteins, vitamins, minerals, and fatty acids, making it a potent supplement for humans. The composition of royal jelly includes water, proteins, sugar, fats, vitamins, salts, and free amino acids. Its unique blend of nutrients contributes to its various health benefits and nutritional value.

Proteins, the most significant component of royal jelly after water, constitute about 12-15% of its substance and include all the essential amino acids, making it a complete protein source. These proteins are crucial for cell growth, tissue repair, and immune system function. The lipid content in royal jelly, which ranges from 3-6%, includes highly beneficial fatty acids that are not synthesized by the human body. Among these, 10-Hydroxy-2-decenoic acid (10-HDA) is exclusive to royal jelly and is believed to have anti-inflammatory and antibacterial properties.

Vitamins in royal jelly are plentiful, with a particularly high content of B vitamins, including thiamine (B1), riboflavin (B2), pantothenic acid (B5), niacin (B3), folic acid (B9), and biotin (B7). These vitamins play vital roles in energy production, DNA repair, and the maintenance of blood cells, skin, and brain function. The presence of trace minerals such as calcium, copper, iron, phosphorus, potassium, silicon, and sulfur further enhances its nutritional profile, supporting bone health, nerve function, and metabolic processes.

Royal jelly's health benefits are as diverse as its composition. It has been traditionally used to boost immunity, improve skin health, and enhance vitality and overall well-being. Its antimicrobial properties make it beneficial in fighting bacterial and fungal infections. The antioxidants present in royal jelly, including flavonoids and phenolic acids, help combat oxidative stress and may reduce the risk of chronic diseases such as heart disease and cancer. Furthermore, the neurotrophic effects of royal jelly have shown promise in improving cognitive function and protecting against neurodegenerative diseases.

For those considering adding royal jelly to their diet, it's available in various forms, including fresh jelly, capsules, and powders. Fresh royal jelly has the highest potency but requires refrigeration and has a limited shelf life. Capsules and powders offer convenience and longer shelf life but may have reduced bioavailability. When choosing a royal jelly product, look for one that specifies the content of 10-HDA, as this is a marker of the product's quality and potency.

Incorporating royal jelly into one's diet should be done with consideration of dosage and potential allergies. While royal jelly is generally safe for most people, it can cause allergic reactions in individuals sensitive to bee products. Starting with a small dose and gradually increasing it can help minimize the risk of adverse reactions. Consulting with a healthcare provider before beginning any new supplement regimen is always recommended, especially for those with pre-existing health conditions or those taking medications.

Royal jelly's blend of nutrients offers a natural way to support various aspects of health, from enhancing immune function to improving skin condition and potentially aiding in the management of certain health conditions. Its use as a dietary supplement underscores the importance of bees not only to our ecosystem but also to human health and nutrition.

Harvesting Royal Jelly Safely

Harvesting royal jelly without harming the colony is a delicate process that requires precision, patience, and a deep understanding of bee behavior and hive dynamics. The first step involves the use of artificial queen cells, which beekeepers introduce into the hive to stimulate the production of royal jelly. These cells mimic the natural environment where the queen bee would lay her eggs. To ensure the health and safety of the bees, it's crucial that these artificial cells are made from materials that are non-toxic and biocompatible with the bees, such as food-grade plastic or wax.

Once the artificial queen cells are in place, a very young larva, usually less than 24 hours old, is carefully transferred into each cell. This task requires a steady hand and a specialized grafting tool, designed to move the larvae without causing them harm. The presence of these larvae in the artificial cells prompts the worker bees to start producing royal jelly and depositing it into the cells to feed the larvae.

After a few days, typically between 3 to 4, the cells are filled with royal jelly. At this stage, the beekeeper must carefully remove the cells from the hive. This is done with minimal disturbance to the bees, often by gently smoking the hive to calm the bees before the cells are removed. The use of smoke should be moderate, as excessive smoke can stress the colony.

The extraction of royal jelly from the cells is the next critical step. Using a small, sterile spatula or a syringe, the beekeeper delicately scoops or suctions the royal jelly out of each cell. This tool must be sterilized between each use to prevent contamination of the royal jelly. It's imperative that this process is done gently to avoid damaging the artificial cells or injuring any larvae that remain inside. The larvae are usually returned to the hive after the royal jelly is extracted, where they can be reared as workers or removed if the cell is to be reused.

Immediately after extraction, the royal jelly is placed into pre-sterilized, airtight containers to prevent contamination and spoilage. Given its perishable nature, the royal jelly is then stored in a refrigerator or freezer, as previously mentioned, to preserve its freshness and nutritional properties until it's ready for use or sale.

Throughout this entire process, the beekeeper's actions must be guided by a commitment to the well-being of the bee colony. This includes monitoring the hive's health before, during, and after the royal jelly harvest to ensure that the production of royal jelly has not negatively impacted the colony. Factors such as the proportion of cells removed, the frequency of harvest, and the overall health and population of the hive must be carefully balanced to sustain the colony's vitality.

Moreover, beekeepers should adopt an integrated approach to hive management that supports the natural behaviors and needs of the bees. This includes providing adequate space for the bees to grow and thrive, ensuring a diverse and plentiful food source, and protecting the hive from pests and diseases through natural and non-invasive methods. By prioritizing the health of the bee colony, beekeepers can harvest royal jelly in a way that is not only productive but also sustainable and ethical, contributing to the preservation of these vital pollinators for future generations.

Storage Techniques for Freshness and Quality

Once the royal jelly has been harvested and placed into pre-sterilized, airtight containers, the focus shifts to maintaining its freshness and quality through proper storage techniques. The delicate nature of royal jelly necessitates a controlled environment to prevent degradation of its nutritional and medicinal properties. The primary goal is to minimize exposure to factors that can accelerate spoilage, such as heat, light, and moisture.

The optimal storage temperature for royal jelly is between 36°F to 40°F, which can typically be achieved in a standard refrigerator. This temperature range slows down the activity of enzymes and microorganisms that can cause the royal jelly to spoil. When storing royal jelly in a refrigerator, it's crucial to place it in the coldest part, often at the back of the lower shelves, away from the door. This location is less affected by temperature fluctuations that occur when the refrigerator door is opened frequently.

For those intending to store royal jelly for extended periods, freezing is an advisable option. Freezing royal jelly at temperatures below 0°F effectively halts enzyme activity and microbial growth, significantly extending its shelf life without diminishing its quality. When freezing royal jelly, it's essential to use containers that are both airtight and freezer-safe. These containers should be made of materials that do not crack or degrade at low temperatures. Additionally, leaving a small space at the top of each container before sealing helps accommodate the slight expansion that occurs when liquids freeze, preventing the container from bursting.

When preparing royal jelly for storage, whether in the refrigerator or freezer, using airtight containers is non-negotiable. Exposure to air not only increases the risk of contamination but can also lead to oxidation, which degrades the quality of the royal jelly. Glass containers with tight-fitting lids are ideal for this purpose, as they do not impart any odors or flavors to the royal jelly and provide an excellent barrier against air. Plastic containers can be used as well, provided they are of food-grade quality and designed to prevent the leaching of chemicals.

Labeling each container with the date of harvest and the date of storage is a good practice that helps in managing the inventory of royal jelly. This practice ensures that older stocks are used first and helps in monitoring the storage duration. It is generally recommended to consume fresh royal jelly within 6 to 8 months when stored in a refrigerator and up to 12 to 24 months when frozen, although it can remain safe beyond these periods if stored properly.

To further protect royal jelly from degradation, it's advisable to shield it from light, especially direct sunlight. Light can induce photooxidation, leading to the loss of some nutritional components. Storing royal jelly in opaque containers or wrapping transparent containers in aluminum foil are effective ways to block light.

In summary, the careful handling and storage of royal jelly are critical steps that ensure the preservation of its nutritional and medicinal qualities. By maintaining the correct temperature, using appropriate containers, minimizing exposure to air and light, and adhering to best practices for freezing and refrigeration, beekeepers and consumers can enjoy the full benefits of royal jelly. These storage techniques, grounded in understanding the biological and chemical properties of royal jelly, underscore the importance of meticulous attention to detail in the post-harvest handling of this remarkable bee product.

CHAPTER 33: PROPOLIS

Propolis, a resinous mixture produced by honeybees from substances collected from tree buds, sap flows, or other botanical sources, serves multiple purposes within the hive, primarily as a building material and an antimicrobial agent. The bees use propolis to seal unwanted open spaces in the hive, smooth out internal walls, and protect the colony from diseases and parasites. For beekeepers and enthusiasts, understanding how to harvest propolis efficiently and safely without disrupting the hive is crucial. The process begins with the installation of a propolis trap, a plastic grid or mesh that fits into the top of a hive frame. When bees fill the gaps in the trap with propolis, it can then be removed and the propolis collected. The optimal time to install a propolis trap is during warmer months when bees are most active in collecting and using propolis, as the substance is more pliable and easier to harvest in warmer temperatures.

To ensure a successful harvest, beekeepers should choose a propolis trap made from a flexible material that can be easily bent or manipulated to release the propolis without breaking it. Silicone or soft plastic grids are ideal. The trap should be placed directly above the brood chamber where bees are most likely to deposit propolis. After installation, it's important to monitor the hive and remove the trap for harvesting before it becomes overfilled, which typically occurs within 4 to 6 weeks during peak activity.

Once removed from the hive, the propolis trap should be placed in a freezer for at least a few hours. Freezing makes the propolis brittle, allowing it to be easily snapped or scraped off the trap. Using a plastic scraper or a similar non-metal tool is recommended to avoid damaging the trap. The collected propolis should then be stored in airtight containers in a cool, dark place to preserve its quality. Contaminants such as bee parts or wood chips should be removed from the propolis. This can be achieved by gently breaking the frozen propolis into smaller pieces over a mesh screen, allowing the pure propolis to fall through while retaining the larger debris.

The processing of propolis for use in various products involves careful consideration of its intended application. For health supplements, propolis can be tinctured in alcohol to extract its beneficial compounds. This involves combining propolis with a food-grade alcohol at a ratio that depends on the desired strength of the tincture, typically ranging from 1:5 to 1:10 propolis to alcohol by weight. The mixture should be kept in a dark glass container and shaken daily for 1 to 2 weeks before straining out the solid particles. The resulting tincture can be stored in a cool, dark place for up to a year.

For topical applications, such as creams or ointments, propolis can be infused into a carrier oil, such as coconut or olive oil. This is done by gently heating the oil and propolis mixture in a double boiler for several hours at a temperature not exceeding 122°F to avoid degrading the propolis's beneficial properties. After cooling, the mixture is strained, and the infused oil is ready for use or further formulation into skincare products.

Beekeepers aiming to produce propolis for sale or personal use must ensure the purity and quality of their product. This includes not only the careful harvesting and processing of propolis but also maintaining healthy, disease-free colonies from which to collect it. Sustainable beekeeping practices contribute to the production of high-quality propolis while supporting the health and longevity of the bee population.

Bee Propolis: Production and Hive Uses

Beyond the harvesting and processing stages, propolis holds a revered spot in both the hive and human applications due to its remarkable properties. Bees meticulously gather resin from tree buds and other botanical sources, which they then mix with their saliva and beeswax to create propolis. This substance is not only a testament to the bees' industrious nature but also to their innate wisdom in maintaining a sterile environment. Within the hive, propolis acts as a natural sealant, filling cracks and smoothing surfaces, thereby fortifying the hive's structure against external threats. Its application by bees extends to the embalming of invaders that are too large for them to remove, preventing decomposition and potential disease outbreaks within the hive. This insight into the bees' use of propolis underscores its significance as an antimicrobial and preservative agent.

The beekeeper's approach to propolis extraction must be both methodical and respectful of its importance to the hive. The installation of a propolis trap is a minimally invasive technique that encourages bees to deposit propolis in a designated area, making collection straightforward and sustainable. This method ensures that the hive's integrity and the bees' health are not compromised. The timing of propolis collection is crucial; removing the trap too early or too late can affect the quality and quantity of propolis harvested. Monitoring the hive's activity and environmental conditions provides beekeepers with the necessary cues for optimal harvest times.

Once collected, the propolis undergoes a purification process to remove impurities. This step is vital for ensuring the highest quality of propolis for use in various products. The freezing and breaking method not only facilitates the removal of debris but also preserves the propolis's bioactive compounds. The meticulous nature of this process reflects the beekeeper's role in maintaining the purity and potency of propolis, mirroring the bees' dedication to their hive's well-being.

The transformation of raw propolis into consumable and topical forms involves a careful balance of traditional knowledge and scientific understanding. The creation of propolis tinctures and infusions requires precision in ratios and temperatures to extract its beneficial compounds without diminishing its efficacy. This alchemical process, rooted in both art and science, highlights the beekeeper's skill in harnessing the natural world's gifts. The choice of carrier oils for propolis infusions, for example, is not arbitrary; it is informed by the oil's ability to complement the propolis's properties, enhancing its application in skincare and health products.

In preparing propolis for sale or personal use, beekeepers must navigate the complexities of product formulation with an eye towards sustainability and ethical practices. This extends to the selection of packaging materials that protect the propolis's integrity while minimizing environmental impact. The labeling of propolis products is not merely a regulatory requirement but an educational opportunity. It allows beekeepers to share the story of propolis, from hive to home, fostering a deeper appreciation for bees and their role in our ecosystems.

The stewardship of beekeepers over propolis collection and processing is a testament to their commitment to sustainable beekeeping and the well-being of the hive. By adopting techniques that prioritize the health of the bees and the quality of the propolis, beekeepers play a crucial role in preserving this ancient practice for future generations. Through their efforts, the remarkable properties of propolis continue to benefit both the hive and humanity, bridging the gap between nature's ingenuity and human health and wellness.

Extracting Propolis Safely and Efficiently

After the propolis has been harvested using a propolis trap and frozen to make it brittle, the next steps involve cleaning and preparing the propolis for various uses. The cleaning process is essential to remove any impurities and ensure the propolis is of the highest quality for its intended application.

Cleaning Process: Begin by breaking the frozen propolis into smaller pieces over a clean, fine mesh screen. This screen should have openings small enough to allow propolis to pass through while catching larger debris, such as bee parts or wood chips. It's advisable to perform this task over a clean, white sheet or surface to easily identify and remove any unwanted materials. For this purpose, a mesh screen with openings of about 1/16 inch (approximately 1.6 mm) works well.

Once the propolis has been broken down and sifted, inspect the material on the white surface for any remaining contaminants. Using a pair of tweezers, carefully remove any visible debris. This meticulous inspection ensures that the final product is as pure as possible, enhancing its quality for therapeutic or cosmetic applications.

Drying and Storage: After cleaning, spread the propolis pieces on a clean, dry surface or a non-stick mat to air dry. This step is crucial to remove any moisture that may have been introduced during the cleaning process. Moisture in propolis can lead to mold growth or degradation of its beneficial properties. The drying area should be well-ventilated, away from direct sunlight, and at room temperature. Depending on the humidity levels, drying can take anywhere from 24 to 48 hours.

Once dry, store the propolis in airtight containers made of glass or food-grade plastic. Glass is preferred for long-term storage as it does not interact with the propolis and maintains its purity. Label each container with the date of processing to keep track of freshness. Properly stored, dry propolis can last for several years without losing its potency.

Grinding for Use: For many applications, propolis needs to be in a powdered form. Once it is dry, you can grind the propolis using a coffee grinder or a mortar and pestle. It's important to use a grinder dedicated to propolis or other non-food substances to avoid contamination. Grind the propolis until it reaches a fine powder, which can then be used for making tinctures, infusions, or incorporated into homemade cosmetics or health products.

Safety Precautions: Throughout the propolis processing, wearing gloves can protect your skin from the sticky residue and potential allergens. Additionally, working in a well-ventilated area is advisable to avoid inhaling fine propolis dust during grinding, which can irritate the respiratory system.

By following these detailed steps for extracting, cleaning, drying, storing, and grinding propolis, beekeepers and enthusiasts can safely and efficiently prepare propolis for a wide range of uses. This process not only maximizes the yield from each harvest but also ensures the highest quality of the final product, preserving the remarkable properties of propolis for health and wellness applications.

Propolis in Natural Medicine and Wellness

Propolis, with its complex composition and broad spectrum of bioactive compounds, has found a significant place in natural medicine and wellness products. Its application ranges from oral health care to skin treatments, each leveraging propolis's antimicrobial, anti-inflammatory, and antioxidant properties. When incorporating propolis into wellness products, the method of preparation and the concentration of propolis play crucial roles in the efficacy of the final product.

For oral health care products, such as toothpaste and mouthwashes, propolis acts as a natural antiseptic that helps in reducing plaque buildup and preventing gum diseases. To create a propolis-infused toothpaste, one can start by mixing finely ground propolis powder with a base of calcium carbonate and a natural thickener like xanthan gum. The propolis powder should be sieved through a fine mesh to ensure its smooth incorporation into the paste, avoiding any gritty texture. A concentration of 0.5% to 1% propolis by weight is effective for such applications. Essential oils, such as peppermint or spearmint, can be added for flavor and additional antibacterial properties. The mixture is then homogenized to ensure even distribution of propolis throughout the toothpaste.

In skincare, propolis is valued for its healing, soothing, and regenerative properties. For a propolis cream or lotion, the initial step involves creating an infusion of propolis in a carrier oil, as previously described. This infused oil is then blended with an emulsifying wax to form the cream base. The concentration of propolis-infused oil in the cream can range from 1% to 5%, depending on the desired potency. Adding vitamin E oil not only acts as a natural preservative but also synergizes with propolis to enhance skin healing. The cream should be gently heated to allow the emulsifying wax to dissolve properly, followed by cooling and continuous stirring to achieve a smooth consistency.

For dietary supplements, propolis is often prepared as a tincture. The propolis-to-alcohol ratio is critical and should be adjusted based on the intended use. A 1:10 ratio yields a tincture suitable for general health support, enhancing immune function, and providing antioxidant benefits. The mixture must be stored in a dark glass container, shaken daily, and then strained to remove solid particles. The final tincture can be administered directly under the tongue or diluted in water. Dosage recommendations vary, but starting with a few drops and gradually increasing, based on tolerance and effects, is a prudent approach.

In the realm of wound care, propolis has been formulated into salves and ointments for its antimicrobial and healing properties. To prepare a propolis salve, begin by infusing propolis into a carrier oil, followed by blending with beeswax to achieve the desired consistency. The ratio of infused oil to beeswax can be adjusted to make the salve more or less firm. A typical formulation might include 1 part beeswax to 4 parts propolis-infused oil by weight. The mixture is gently heated until the beeswax melts, then poured into containers to solidify. This salve can be applied to cuts, scrapes, and other minor skin injuries to promote healing and prevent infection.

When developing products with propolis, it's essential to consider the source and quality of the propolis used. Ethically sourced, high-quality propolis will yield a more effective product. Additionally, individuals should be advised to patch test topical products for allergies and to consult healthcare professionals before incorporating new supplements into their regimen, especially if they have existing health conditions or are taking other medications. The integration of propolis into natural medicine and wellness products offers a bridge between ancient apicultural wisdom and modern health practices, providing a natural alternative for those seeking to enhance their health and well-being through the gifts of the hive.

Propolis Processing Techniques

Once the propolis has been transformed into a fine powder or infused into a carrier oil, the next step is preparing these products for sale or personal use. This involves packaging, labeling, and understanding the best practices for storage to maintain the propolis's potency and effectiveness.

Packaging: For powdered propolis, use small, airtight containers that are easy to open and close. Glass jars with screw-top lids are ideal because they do not react with propolis and prevent air and moisture from entering, which could degrade the product. For propolis tinctures or infused oils, dark glass bottles with dropper caps are recommended. The dark glass minimizes light exposure, preserving the bioactive compounds in propolis. Ensure that all containers are sterilized before use to prevent contamination.

Labeling: Each product should have a clear, informative label. Include the product name (e.g., "Propolis Powder" or "Propolis Tincture"), the net weight or volume, usage directions, and any cautionary statements, such as advising a patch test for allergies before use. It's also helpful to add a "packed on" or "best before" date to inform customers of the product's freshness. If you claim any health benefits, ensure they comply with local regulations to avoid legal issues.

Storage Recommendations for End Users: On the label or accompanying pamphlet, provide storage instructions to help users maintain the product's quality. Advise storing powdered propolis in a cool, dry place away from direct sunlight and moisture. Tinctures and infused oils should be kept in their dark glass bottles and stored in a cabinet or pantry to avoid light degradation. Inform customers that, when stored properly, propolis products can last for several years but that checking the product's appearance and smell is a good practice to ensure it remains usable.

Creating an Informative Insert: Including an insert or pamphlet inside the product packaging can provide customers with additional valuable information. This could cover the benefits of propolis, suggested uses, detailed storage instructions, and any safety precautions. Providing educational content can enhance customer satisfaction and promote the responsible use of propolis products.

Online Resources: If you sell your products online or have a website, consider creating a digital resource center where customers can find detailed articles, research studies, and FAQs about propolis. This not only establishes your brand as a knowledgeable authority on propolis but also supports your customers in getting the most out of their purchases.

By meticulously preparing propolis for sale or personal use, you ensure that the end product retains its beneficial properties and meets the expectations of your customers or personal needs. Proper packaging, informative labeling, and providing detailed usage and storage instructions are key to achieving this goal.

CHAPTER 34: BEESWAX

Beeswax, a natural substance produced by honeybees, serves as a foundation for constructing honeycomb cells. The bees secrete wax scales from glands on their abdomens, which are then chewed and molded to form the hexagonal cells we recognize as honeycomb. This remarkable material has been utilized by humans for centuries, not only for its role in the hive but also for its versatile applications in crafting, cosmetics, and household products. The process of collecting, purifying, and utilizing beeswax is a testament to the sustainable practices inherent in beekeeping.

Collecting Beeswax: The first step in harvesting beeswax is to remove the honeycomb frames from the hive. Once the honey has been extracted, the remaining comb, which contains both honey and beeswax, is collected. It's important to use a gentle heat source, such as a solar wax melter or a double boiler, to melt the comb without burning the wax. The melted wax is then poured through a fine mesh strainer or cheesecloth to remove impurities such as propolis, bee parts, and other debris.

Purifying Beeswax: After the initial melting and straining, the beeswax may still contain some impurities and water. To further purify it, the wax can be melted again and allowed to sit undisturbed. As it cools, impurities will either float to the top or sink to the bottom, depending on their density. The clean, molten beeswax can then be carefully poured off, leaving the contaminants behind. For additional purity, this process can be repeated as necessary.

Molding Beeswax: Once purified, the beeswax is ready to be formed into usable blocks or shapes. Pour the molten wax into molds of your choice, such as silicone baking molds or custom beeswax molds. Allow the wax to cool and harden completely before removing it from the molds. These beeswax blocks can be stored indefinitely in a cool, dry place until ready for use.

Uses for Beeswax: Beeswax's applications are vast and varied. In crafting, it's used to make candles, which burn cleaner and longer than paraffin-based candles. For cosmetics, beeswax is a key ingredient in lip balms, lotions, and salves due to its moisturizing properties and natural barrier against environmental irritants. In the home, beeswax can be used to waterproof leather, polish wood, and even as a natural lubricant for squeaky doors or drawers.

Beeswax Candles: To create beeswax candles, melt purified beeswax in a double boiler and prepare your molds or containers with wicks secured. Once the wax is fully melted, you can add essential oils for fragrance if desired. Carefully pour the molten wax into the prepared molds or containers, ensuring the wick remains centered. Allow the candles to cool and harden for several hours or overnight before trimming the wick to the appropriate length.

Cosmetic Applications: For lip balms, melt beeswax with carrier oils such as coconut or almond oil, and add essential oils or flavorings as desired. Pour the mixture into lip balm tubes or tins and let them solidify. For lotions or salves, the process is similar, with beeswax melted alongside carrier oils and butters, such as shea or cocoa butter, before being whipped or poured into containers to cool.

Household Uses: Beeswax can be applied directly to wooden furniture or leather goods as a polish or conditioner. Melt the wax and apply it with a soft cloth, then buff to a shine. For a natural lubricant, a small piece of beeswax can be rubbed along the surface of drawers or window tracks to reduce friction.

In conclusion, beeswax is not only a byproduct of beekeeping but a valuable resource in its own right. Its versatility in applications from crafting to cosmetics and household uses demonstrates the sustainable and beneficial nature of beekeeping. By following these detailed steps for collecting, purifying, and utilizing beeswax, beekeepers and enthusiasts can fully appreciate and make use of this natural treasure.

Collecting and Purifying Beeswax

Once the beeswax has been collected and the initial purification process through melting and straining has been completed, further refinement is necessary to ensure the beeswax is of the highest quality for use. This next stage involves a meticulous approach to remove the remaining impurities and water content that could compromise the integrity and aesthetic of the final product. The beeswax, now in a liquid state after the initial purification, should be transferred to a clean, heat-resistant container, such as a stainless steel pot, which allows for controlled reheating. It's crucial to maintain a gentle heat to prevent any degradation of the beeswax's natural properties. A candy or deep-fry thermometer can be invaluable here to monitor the temperature closely, ensuring it stays around 145°F (63°C), which is sufficient to keep the beeswax melted without reaching a boiling point.

During this reheating phase, any remaining water content in the beeswax will start to evaporate, a process that can be aided by slightly increasing the temperature but never exceeding 160°F (71°C) to avoid burning. As the water evaporates, impurities that were not caught during the initial straining will begin to surface. Skimming these off with a stainless steel spoon or ladle requires patience and a steady hand to avoid removing too much of the beeswax itself. This step may need to be repeated several times, depending on the amount of debris present.

Once the reheating and skimming process is deemed sufficient, the next crucial step is to facilitate the separation of beeswax from any remaining water. This can be achieved by allowing the melted beeswax to cool slowly. As it cools, any water will separate and settle at the bottom of the container, while the purified beeswax solidifies on top. This separation is critical for ensuring the finished beeswax is not only clean but also dry, as any residual moisture can lead to mold growth or spoilage when the beeswax is stored.

To expedite cooling without causing the beeswax to harden too quickly, which could trap impurities or water, placing the container in a larger basin filled with cold water can be effective. Stirring the beeswax gently as it cools can also help in achieving a uniform texture and temperature throughout, minimizing the risk of crystallization or uneven hardening. Once the beeswax has solidified to a point where it is firm yet slightly pliable, it can be carefully removed from the container. If done correctly, the beeswax will form a solid block that can be lifted out, leaving any residual water behind.

The final step in the purification process involves melting the beeswax one last time to pour into prepared molds. This not only allows for the beeswax to be shaped into convenient sizes and forms for storage and use but also provides an opportunity for any final impurities to be removed. Molds should be chosen based on the intended use of the beeswax; silicone baking molds are popular for their ease of use and versatility, but any mold that can withstand the temperature of melted beeswax and release the solidified wax easily will suffice. Prior to pouring, the molds can be lightly greased with a neutral oil to ensure the beeswax releases smoothly once solidified.

Pouring the melted beeswax into molds should be done slowly and carefully to avoid introducing air bubbles or splashing, which could create imperfections in the final product. Once filled, the molds are set aside in a safe, undisturbed area to cool and harden completely. This may take several hours depending on the size of the molds and the ambient temperature. After the beeswax has fully solidified, it can be gently removed from the molds. At this stage, the beeswax is ready for use or can be wrapped in parchment paper and stored in a cool, dry place until needed.

By adhering to these detailed steps for collecting, purifying, and preparing beeswax, beekeepers and enthusiasts ensure that the final product is of the highest quality, free from impurities and moisture, and ready for a wide range of applications, from candle making to cosmetic formulations. This meticulous approach underscores the sustainable and resourceful practices integral to beekeeping, allowing for the full appreciation and utilization of beeswax as a valuable byproduct of the hive.

Crafting with Beeswax

Crafting with beeswax extends beyond the basic steps of collecting and purifying this natural resource; it involves a creative process that transforms beeswax into functional and decorative items. When delving into the crafting of candles, the selection of wicks becomes paramount. For beeswax candles, cotton wicks are recommended due to their natural composition, which ensures a cleaner burn. The thickness of the wick should be chosen based on the diameter of the candle: a general rule is to use a thicker wick for larger candles to ensure an even burn. Pre-waxing the wick, by dipping it in melted beeswax and then straightening it, can make the wick more rigid and easier to position in the center of the candle mold.

In the realm of cosmetics, when creating lip balms or lotions, the ratio of beeswax to oils (carrier and essential) plays a crucial role in determining the consistency of the final product. A starting point is a 1:4 ratio of beeswax to carrier oils for lip balms, which can be adjusted based on personal preference for a softer or firmer balm. Essential oils, added for their therapeutic properties and fragrance, should be used sparingly; typically, a few drops per ounce of base mixture are sufficient. It's essential to ensure that the essential oils are added to the mixture after it has been removed from the heat source to preserve their therapeutic qualities.

For household items, such as furniture polish, combining beeswax with mineral oil or olive oil can create a natural and effective polish. A 1:3 ratio of beeswax to oil, gently heated until fully blended, can be applied with a soft cloth to wood surfaces. This not only nourishes the wood but also provides a protective layer against moisture. The polish should be allowed to penetrate the wood for a few minutes before buffing it to a shine with a clean, dry cloth.

The versatility of beeswax is also showcased in the creation of natural lubricants for drawers and windows. A small block of beeswax, rubbed directly onto the tracks or surfaces, can reduce friction and facilitate smoother movement. This method, while simple, highlights the multifaceted uses of beeswax in everyday household maintenance.

When engaging in beeswax crafts, safety precautions should be observed, particularly when melting beeswax. Utilizing a double boiler setup can prevent direct heat and reduce the risk of igniting the beeswax. Additionally, working in a well-ventilated area is advised to avoid inhalation of fumes. The use of heat-resistant gloves can provide protection from hot containers and beeswax, ensuring a safer crafting experience.

The crafting journey with beeswax is one of exploration and innovation, where the natural qualities of beeswax are harnessed to create items that are both beautiful and beneficial. Through the careful selection of materials, adherence to safety practices, and attention to detail, crafters can transform beeswax into a myriad of products that celebrate the essence of beekeeping and the natural world.

Melting and Molding Beeswax

Transitioning from the purification process, the next phase involves melting and molding beeswax into various shapes, a procedure that requires precision and attention to detail. The melting of beeswax is a delicate operation that necessitates a controlled environment to preserve its natural qualities and prevent degradation. Utilizing a double boiler system is the most effective method for melting beeswax. This system consists of a larger pot filled with water and a smaller pot or heat-resistant bowl placed inside, containing the beeswax. The water in the outer pot heats the inner pot evenly, ensuring that the beeswax melts at a consistent temperature without direct contact with the heat source. It's important to maintain the water at a gentle simmer to avoid overheating, with the optimal melting point of beeswax being around 144 to 147°F (62 to 64°C).

Once the beeswax is completely melted, the next step is to prepare it for molding. At this stage, any desired additives should be incorporated into the molten beeswax. These can include colorants for creating colored beeswax or essential oils for scented beeswax products. When adding essential oils, it's crucial to wait until the beeswax temperature lowers slightly to around 135°F (57°C) to prevent the evaporation of the oils' volatile compounds. Stirring the mixture gently ensures an even distribution of the additives throughout the beeswax.

Choosing the right molds is essential for achieving the desired shapes and sizes of the beeswax products. Silicone molds are highly recommended due to their flexibility, non-stick properties, and the ease with which beeswax items can be released once solidified. These molds come in a variety of shapes, from simple blocks and bars to intricate designs suitable for decorative items. Before pouring the beeswax, it's beneficial to lightly coat the molds with a thin layer of oil, such as vegetable oil or mineral oil, to facilitate an even easier release of the finished beeswax product.

Pouring the molten beeswax into molds must be done cautiously to avoid air bubbles and ensure a smooth surface on the finished product. Filling the molds to the desired level, it's advisable to pour slowly and steadily from one corner, allowing the beeswax to spread naturally. If air bubbles do form, lightly tapping the sides of the mold can help release them. After pouring, the beeswax should be allowed to cool and solidify undisturbed. The cooling process can take several hours, depending on the size of the molds and the thickness of the beeswax. Cooling at room temperature is preferred, as rapid cooling methods can cause the beeswax to crack or warp.

Once the beeswax has fully solidified, removing it from the molds is the final step. This should be done with care to preserve the integrity of the shapes. Inverting the silicone mold and gently pressing on the bottom of each cavity usually releases the beeswax smoothly. If resistance is encountered, placing the mold in the refrigerator for a few minutes can firm up the beeswax, making it easier to demold without distortion.

For those looking to create beeswax sheets for rolling candles or other applications, pouring the molten beeswax onto a flat, oil-coated surface or into a shallow pan with edges can produce thin sheets. Once cooled, these sheets can be cut to size and used as desired. The versatility of beeswax molding allows for a wide range of creative and practical applications, from custom candles and ornaments to lip balm tubes and cosmetic containers.

Adhering to these detailed techniques for melting and molding beeswax ensures that the inherent beauty and utility of beeswax are fully realized. Whether for personal use, gifting, or commercial endeavors, the ability to transform raw beeswax into a myriad of shapes and forms opens up endless possibilities for beekeepers and crafters alike. With patience and practice, mastering these techniques can lead to the production of high-quality, handcrafted beeswax products that embody the essence of natural beekeeping.

Selling Beeswax Products for Extra Income

Turning your beeswax into a source of income involves more than just creating appealing products; it requires strategic marketing, understanding your target market, and utilizing various sales channels to reach potential customers. The first step in selling beeswax products is to identify who your potential buyers are. This could range from individuals interested in natural and eco-friendly products, such as homemade cosmetics and candles, to local businesses looking for quality ingredients for their own products. Knowing your audience will guide your marketing efforts and product development.

Once you have identified your target market, the next step is to create a compelling brand story. This includes packaging your beeswax products attractively and sustainably, which appeals to eco-conscious consumers. Consider using recycled or biodegradable materials for packaging and highlighting the natural and sustainable aspects of your products. Your brand story should also communicate the benefits of beeswax and the unique qualities of your products, whether it's the natural fragrance of your candles or the moisturizing properties of your lip balms.

In terms of pricing, it's important to conduct market research to understand the going rate for similar beeswax products. Pricing should cover your costs and time, while also remaining competitive. Keep in mind that handmade and natural products can often command a higher price due to their perceived value and quality.

Selling your beeswax products can be done through multiple channels. Online marketplaces like Etsy or your own e-commerce website can reach a wide audience and allow you to sell directly to consumers. Social media platforms, such as Instagram and Facebook, are powerful tools for building a brand presence and engaging with potential customers. High-quality photos and regular updates about your products can attract followers and convert them into buyers.

Local farmers' markets, craft fairs, and eco-friendly stores are excellent venues for selling beeswax products. These settings allow customers to see and smell your products firsthand, which can be a significant selling point for beeswax items. Building relationships with local shop owners can lead to your products being stocked in stores, providing another revenue stream.

To diversify your beekeeping income, consider offering a range of products that cater to different needs and preferences. In addition to candles and lip balms, you could expand into making soaps, lotions, and even beeswax wraps as an alternative to plastic cling film. Offering customizations, such as personalized labels for weddings or corporate events, can also attract a niche market willing to pay a premium for bespoke items.

Collaborating with other local artisans or businesses can open up new opportunities for cross-promotion and sales. For instance, pairing up with a local potter to create unique candle holders or working with a herbalist to develop a line of scented beeswax products can enhance your product offering and appeal to a broader audience.

Finally, consider offering workshops or classes on making beeswax products. This not only provides an additional income stream but also raises awareness of your products and brand. Workshops can be a great way to engage with the community, share your knowledge and passion for beekeeping, and inspire others to appreciate the value of beeswax.

CHAPTER 35: POLLEN

Pollen, the fine powdery substance often yellow in color, is a crucial component for the survival of bees and the continuation of plant species. It serves as the primary source of protein for bees, aiding in their growth and development. The process of collecting pollen is as fascinating as it is vital, involving several steps and tools that beekeepers can use to ensure successful harvests without harming the hive or reducing its pollination efficiency.

To collect pollen, beekeepers often use a pollen trap, a device that fits at the entrance of a beehive. As bees enter the hive, they pass through the trap, which gently brushes some pollen off their legs and into a collection drawer below. It's essential to select a pollen trap that is appropriately sized for your hive to minimize stress on the bees. The trap should be installed in such a way that it allows for easy removal and cleaning, ensuring that no bees are harmed during the process.

When installing a pollen trap, timing is critical. It should be placed during periods of high pollen flow, which typically occurs in the spring and early summer. This maximizes collection while ensuring the bees have enough pollen left for their needs. Beekeepers should monitor the trap daily, removing the collected pollen and checking for any signs of blockage or distress among the hive inhabitants.

Once collected, pollen must be dried to prevent mold growth. This can be done by spreading the pollen out on a screen in a warm, dry area with good air circulation. A food dehydrator set at a low temperature (around 95°F or 35°C) can also be used to dry pollen efficiently. The drying process is crucial; it should reduce the moisture content of the pollen to about 4% to 8%, making it safe for storage.

For storing dried pollen, airtight containers are recommended to keep out moisture and pests. Glass jars with tight-fitting lids or vacuum-sealed bags work well for this purpose. Store the containers in a cool, dark place to preserve the pollen's nutritional value. If done correctly, dried pollen can be stored for several months.

Incorporating collected pollen into recipes or using it as a dietary supplement involves careful consideration of dosage and potential allergies. Start with small amounts to ensure there are no adverse reactions. Pollen can be sprinkled over food, blended into smoothies, or mixed with other ingredients to create energy bars or supplements.

For beekeepers interested in selling pollen, understanding market demand and regulatory requirements is essential. Pollen is considered a specialty health product, and its packaging should comply with local health and safety standards. Labeling should provide clear information on the origin, processing methods, and suggested uses of the pollen. Marketing can focus on the health benefits of pollen, targeting consumers interested in natural dietary supplements and superfoods.

Nutritional Benefits of Pollen

Bee pollen stands out as a powerhouse of nutrition, densely packed with vitamins, minerals, proteins, carbohydrates, and antioxidants, making it an exceptional dietary supplement. Each granule of bee pollen is a microscopic bundle of energy and health benefits, containing over 250 biologically active substances, including vitamins A, B, C, D, E, and K, minerals such as magnesium, potassium, calcium, zinc, iron, and selenium, as well as enzymes and co-enzymes necessary for digesting food. The protein content in bee pollen is more than that found in any animal source, with approximately half of its protein being in the form of free amino acids, ready for immediate use by the body.

To incorporate bee pollen into your diet, start by selecting high-quality, fresh pollen. It should be brightly colored, indicating its freshness and nutritional value. Begin with a small dose to ensure no allergic reactions occur, gradually increasing from a few granules a day to a teaspoon or more, depending on individual tolerance and nutritional needs. Bee pollen can be consumed directly or sprinkled over breakfast cereals, mixed into yogurt, or blended into smoothies. Its slightly sweet and floral taste enhances the flavor profile of many dishes, making it a versatile addition to daily meals.

For those looking to boost their intake of antioxidants, bee pollen offers flavonoids, which are known for their anti-inflammatory and immune-boosting properties. These compounds help combat oxidative stress in the body, reducing the risk of chronic diseases such as heart disease and cancer. The presence of rutin, a specific type of flavonoid in bee pollen, strengthens blood vessels and improves circulation, offering cardiovascular benefits.

Athletes and individuals with active lifestyles may find bee pollen particularly beneficial due to its high energy content and ability to enhance stamina and reduce recovery times after intense physical activity. The natural sugars present in bee pollen, fructose, and glucose, provide a quick energy boost, while its protein content supports muscle repair and growth.

For digestive health, bee pollen contains enzymes that aid in the breakdown of food and absorption of nutrients, promoting healthy gut flora and preventing digestive disorders. Its antimicrobial properties also help protect the body from bacterial infections.

To ensure the maximum nutritional benefits, store bee pollen properly in airtight containers in the refrigerator or freezer. This preserves its vitamin and enzyme content, preventing degradation. When ready to use, simply measure out the desired amount, allowing it to come to room temperature before adding it to meals or consuming it directly.

In summary, bee pollen is a nutrient-dense superfood that supports overall health and well-being. Its wide range of vitamins, minerals, and antioxidants makes it an excellent dietary supplement for enhancing energy levels, boosting the immune system, supporting cardiovascular health, and improving digestive function. By incorporating bee pollen into your daily diet, you can take advantage of its numerous health benefits, contributing to a balanced and nutritious lifestyle.

Pollen Collection Without Hive Stress

Ensuring the well-being of the hive while collecting pollen necessitates a delicate balance, as the primary objective is to harvest without compromising the bees' essential activities or causing undue stress. The approach to pollen collection should be methodical, emphasizing the health of the bee colony and the efficiency of the collection process. To achieve this, beekeepers must employ techniques and tools designed to minimize intrusion and maintain the natural behavior of bees.

The use of a pollen trap, as previously mentioned, is a common method for collecting pollen. However, the management of the trap plays a crucial role in reducing stress on the hive. It is advisable to install the pollen trap for short periods, preferably during times of peak pollen flow. This strategy prevents excessive depletion of pollen resources that bees need for nourishment. A recommended practice is to place the trap for a couple of days, then remove it for a week, allowing bees unrestricted access to bring pollen into the hive. This intermittent trapping method helps maintain the nutritional balance within the hive and supports the health and productivity of the colony.

Another critical aspect of stress-free pollen collection is the careful monitoring of the hive's health and productivity following the installation of a pollen trap. Beekeepers should observe the behavior of the bees and the overall vigor of the hive, looking for any signs of decreased activity or distress. If any negative impacts are noticed, it may be necessary to adjust the pollen collection routine, either by reducing the frequency of trapping or by seeking alternative methods that are less intrusive to the bees.

In addition to mechanical collection methods, fostering an environment that supports the natural pollen-gathering behavior of bees is essential. This can be achieved by planting a diversity of pollen-rich flowers and plants in the vicinity of the hives. Selecting plant species that bloom at different times of the year ensures a consistent pollen supply, reducing the need for artificial collection methods and supporting the nutritional needs of the hive. Creating such an environment not only benefits the bees by providing them with a varied diet but also enhances the overall ecosystem around the hive, promoting biodiversity.

Furthermore, educating oneself on the nuances of bee behavior and the ecological factors that influence pollen availability is invaluable. Understanding weather patterns, seasonal changes, and the specific needs of your bee species can guide the timing and methods of pollen collection. This knowledge allows beekeepers to harmonize their practices with the natural cycles of their environment, ensuring that pollen collection is conducted in a way that is sustainable and aligned with the well-being of the bee colony.

Adopting a holistic approach to beekeeping, where the health of the hive is the foremost consideration, ultimately leads to more successful and sustainable pollen collection practices. By carefully managing the collection process, utilizing non-intrusive methods, and fostering a supportive environment, beekeepers can harvest pollen efficiently while ensuring the vitality and productivity of their hives. This balanced approach not only benefits the immediate health of the bee colony but also contributes to the broader goal of sustainable beekeeping, preserving these essential pollinators for future generations.

Storing and Using Pollen in Recipes

Storing bee pollen correctly is crucial for preserving its nutritional value and ensuring it remains safe for consumption over time. The process begins with drying the pollen, as previously mentioned, to remove any excess moisture that could lead to mold growth. Once dried, the pollen should be transferred to airtight containers. Glass jars with tight-sealing lids are ideal because they do not impart any unwanted flavors and prevent air exchange, which could degrade the pollen. These containers should then be placed in the refrigerator or freezer. The cold environment slows down the degradation of vitamins, enzymes, and other bioactive compounds in the pollen. Refrigeration can keep pollen fresh for up to a year, while freezing can extend its shelf life to two years or more. It's important to label each container with the date of storage to keep track of its freshness.

When integrating bee pollen into recipes, it's essential to consider the pollen's flavor profile and the nutritional benefits you aim to achieve. Bee pollen has a slightly sweet and floral taste, which can complement a variety of dishes. For breakfast, sprinkle a teaspoon of bee pollen over oatmeal, yogurt, or smoothie bowls to add a nutrient boost. Bee pollen can also be blended into smoothies directly, providing a rich source of proteins, vitamins, and minerals in a convenient and digestible form. When baking, bee pollen can be incorporated into the dough of bread, muffins, or energy bars, offering an additional layer of flavor and increasing the nutritional value of these foods. It's advisable to add bee pollen to baked goods at the end of the mixing process to minimize heat exposure, which can degrade some of its delicate nutrients.

For savory dishes, bee pollen can serve as a garnish on salads, soups, and even main dishes like grilled or roasted meats and vegetables. Its subtle sweetness can balance the flavors in savory applications, adding a unique depth that enhances the overall dish. When using bee pollen in cooking, it's important to add it after the cooking process or use it at temperatures below 118°F (48°C) to preserve its nutritional integrity. High heat can destroy some of the beneficial enzymes and vitamins present in bee pollen.

Incorporating bee pollen into beverages offers another avenue for enjoying its health benefits. It can be dissolved in warm (not hot) teas or blended into cold drinks, such as juices and smoothies. For an energy boost, mix bee pollen with a little water or juice to create a paste and consume it directly before workouts or during periods of low energy.

Finally, when crafting recipes with bee pollen, start with small quantities and gradually increase to suit your taste preferences and dietary needs. Given its potent nutritional profile, a little goes a long way, and it's wise to monitor your body's response, especially if you have pollen allergies. By storing bee pollen properly and integrating it thoughtfully into your diet, you can harness the myriad health benefits it offers, from enhanced energy and immunity to improved digestion and cardiovascular health, making it a valuable addition to a holistic wellness regimen.

PART 9: COOKING WITH HONEY

CHAPTER 36: SWEET RECIPES

Honey cakes, a traditional dessert with roots in various cultures, offer a rich canvas to explore the culinary versatility of honey. When crafting a honey cake, the choice of honey is paramount; each variety imparts a unique flavor profile to the cake. For a nuanced taste, opt for clover or acacia honey for their light and floral notes. If a more robust flavor is desired, wildflower or buckwheat honey provides depth and character. The process begins with creaming together butter and sugar until the mixture is light and fluffy, ensuring that the sugar is well dissolved into the butter for a smooth texture. Gradually add eggs, one at a time, to the mixture, ensuring each is fully incorporated before adding the next, to achieve a uniform and aerated batter.

The dry ingredients, typically a combination of all-purpose flour, baking powder, and a pinch of salt, should be sifted together to prevent clumps and ensure an even distribution throughout the batter. Alternating the dry ingredients with milk, begin to gently fold them into the butter mixture. This step should be done delicately to preserve the air incorporated during the creaming process, which contributes to the cake's lightness. The star ingredient, honey, is then warmed slightly to reduce viscosity and facilitate easier blending with the other ingredients. Stir the liquid honey into the batter until just combined to avoid overmixing, which can result in a denser cake.

For a moist and flavorful honey cake, consider adding spices such as cinnamon, nutmeg, or allspice, which complement the sweetness of the honey. A small amount of orange zest can also enhance the cake's aroma and add a subtle citrus note that balances the honey's richness. Pour the batter into a prepared pan, typically lined with parchment paper and greased to prevent sticking, and bake in a preheated oven at 350°F (177°C) until a toothpick inserted into the center comes out clean. The baking time may vary depending on the oven and the size of the pan, so monitoring the cake towards the end of the suggested baking time is crucial.

Upon removal from the oven, allow the cake to cool in the pan for a few minutes before transferring it to a wire rack to cool completely. This resting period helps the cake set and makes it easier to handle. For an added touch of elegance and flavor, a glaze made from honey thinned with a bit of warm water or citrus juice can be brushed over the top of the cake. This not only adds a glossy finish but also an extra layer of honey flavor.

Honey cookies, another delightful way to showcase honey in baking, can range from soft and chewy to crisp and golden. The key to exceptional honey cookies lies in the balance of sweetness and texture. Starting with a base of creamed butter and sugar, honey is added to the mixture to provide moisture and flavor. The inclusion of an egg helps to bind the ingredients together, while a dash of vanilla extract adds depth to the cookie's aroma. The dry ingredients, a sifted blend of flour, baking soda, and a pinch of salt, are gradually mixed into the wet ingredients until a cohesive dough forms.

Chilling the dough for at least an hour before baking is a crucial step that prevents the cookies from spreading too much in the oven, resulting in a more desirable texture. Once chilled, the dough can be portioned into balls and placed on a baking sheet, leaving enough space between each to allow for expansion. Baking at 375°F (190°C) until the edges are lightly browned but the centers remain soft ensures a perfect honey cookie with a crisp exterior and a tender interior. For an added touch, a sprinkle of coarse sugar or a light drizzle of honey on top of each cookie before baking adds texture and enhances the honey flavor.

Through the careful selection of ingredients and adherence to specific baking techniques, both honey cakes and honey cookies can be transformed into exquisite treats that celebrate the natural sweetness and complexity of honey. These recipes not only provide a delightful culinary experience but also honor the labor of bees and the art of beekeeping, connecting the pleasures of the table with the health and sustainability of the environment.

Honey Cakes: History, Culture, and Baking Techniques

To ensure your honey cake retains moisture and achieves a rich flavor, the incorporation of sour cream or yogurt into the batter is recommended. These ingredients introduce acidity, which not only tenderizes the gluten in the flour but also balances the sweetness of the honey. For every cup of flour used, adding a quarter cup of sour cream or yogurt will suffice. This ratio helps to maintain the cake's structure without compromising its moistness.

Selecting the right type of flour is crucial for the texture of your honey cake. While all-purpose flour is commonly used, substituting half of it with cake flour can make the cake lighter and more tender. Cake flour has a lower protein content than all-purpose flour, which means less gluten is formed during mixing, resulting in a softer cake crumb.

When it comes to mixing the batter, the process of folding must be gentle to avoid deflating the air bubbles created during the creaming of butter and sugar. Use a spatula to cut through the batter, gently turning it over from the bottom of the bowl to the top. This technique ensures ingredients are evenly mixed without overworking the batter.

Baking temperature and time are pivotal for a perfectly moist honey cake. A lower baking temperature, around 325°F (163°C), is preferable as it allows the cake to bake evenly without drying out. The cake should be baked until it pulls away slightly from the edges of the pan and a skewer inserted into the center comes out clean. Depending on the size and depth of the pan, this could take anywhere from 30 to 50 minutes. Avoid opening the oven door frequently during baking, as this can cause temperature fluctuations that affect the cake's rise and texture.

Once baked, allowing the cake to cool in the pan on a wire rack for about 10 minutes helps in retaining its moisture. Afterward, carefully remove the cake from the pan to cool completely on the rack. Cooling it too quickly in a drafty area can cause the cake to dry out.

For an enhanced flavor, brushing the warm cake with a honey syrup—made by simmering honey with water—adds moisture and a deeper honey taste. This step should be done while the cake is still warm, allowing it to absorb the syrup more effectively.

Honey Cookies and Pies: Traditional and Modern

Transitioning from the classic honey cake to the realm of honey cookies and pies, we delve into a variety of recipes that incorporate both traditional elements and modern twists. Honey cookies, a staple in many cultures, can be adapted to suit contemporary palates with the addition of ingredients such as ginger or almond flour, offering a gluten-free option. To achieve the perfect texture, mix the honey with softened butter and sugar until creamy before adding the dry ingredients. For a modern twist, incorporate a pinch of sea salt or a dash of espresso powder to enhance the honey's natural flavors. The dough's consistency is key; it should be pliable yet firm enough to roll into balls or cut into shapes. Baking at 375°F (190°C) for 8-10 minutes yields cookies that are golden around the edges but still soft in the center.

Pies, another versatile canvas for honey, allow for endless creativity. A honey-infused filling can range from a simple custard base to a more complex blend of fruits and spices. For a traditional honey pie, combine honey with eggs, butter, and a touch of flour to thicken the mixture, then pour into a pre-baked pie crust and bake until set. Modern variations might include layering honey with slices of pear or fig, adding a sprinkle of thyme or rosemary for an aromatic touch. The crust, whether a classic pâte brisée or a crumbly graham cracker base, should complement the filling's sweetness and texture. Pre-baking the crust ensures it remains crisp beneath the moist filling.

For those seeking a contemporary take, consider deconstructing the classic honey pie into individual tartlets, using mini tart pans for an elegant presentation. This approach allows for experimenting with different crusts and fillings, such as a chocolate ganache topped with a honey-infused whipped cream. The key to a successful honey pie or tartlet lies in the balance of flavors; the honey should enhance, not overpower, the other ingredients.

Incorporating honey into cookies and pies not only adds a natural sweetness but also a depth of flavor that refined sugars cannot replicate. Through experimenting with various ingredients and techniques, bakers can create desserts that pay homage to traditional recipes while embracing the innovation of modern cuisine.

Honey Mousse and Custards

Moving on to the delicate art of creating **honey mousse** and custards, these desserts are celebrated for their smooth texture and the way they beautifully incorporate the nuanced flavors of honey. When preparing honey mousse, the first step involves selecting a high-quality, floral honey that will impart a distinct but not overpowering sweetness. A ratio of three tablespoons of honey per cup of heavy cream ensures the perfect balance of flavor.

Begin by chilling a metal mixing bowl and beaters in the freezer for at least 15 minutes. This helps the heavy cream whip more efficiently. Pour one cup of heavy cream into the chilled bowl and whip until soft peaks form. Gradually add the honey, continuing to whip until stiff peaks form, being careful not to overbeat as this can cause the mixture to become grainy. For a lighter mousse, fold in two tablespoons of Greek yogurt to the whipped cream and honey mixture, adding a slight tang that complements the sweetness.

For honey custards, the key is to cook them gently in a **water bath** to achieve a silky, smooth texture without curdling. Start by heating two cups of whole milk or a mixture of milk and cream for a richer custard, just until it begins to simmer. Whisk together four large egg yolks and a quarter cup of honey, then gradually temper the eggs by adding the hot milk in a slow, steady stream, whisking constantly. Strain this mixture into ramekins to remove any possible bits of cooked egg.

Place the ramekins in a baking dish and fill the dish with hot water halfway up the sides of the ramekins. Bake in a preheated oven at 325°F (163°C) for 25 to 30 minutes, or until the custards are just set but still slightly wobbly in the center. Cooling the custards slowly and then chilling them in the refrigerator for at least four hours results in the ideal texture.

Whether opting for the airy lightness of honey mousse or the creamy indulgence of custard, both desserts offer a delightful way to showcase the complex flavors of honey, making them a perfect finish to any meal.

Pairing Honey with Traditional Desserts

When pairing honey with traditional desserts to elevate flavors, consider the unique characteristics of different honey varieties and how they complement or enhance the dessert's existing flavors. For example, **clover honey**, known for its mild, floral sweetness, pairs wonderfully with light desserts such as angel food cake or panna cotta. Its subtle flavor doesn't overpower the dessert but rather adds a layer of complexity.

For richer, denser desserts like chocolate cake or brownies, **buckwheat honey** with its robust, molasses-like flavor, can introduce a depth that enhances the chocolate's intensity. The key is to drizzle the honey over the dessert or incorporate it into the glaze or frosting, allowing the honey's flavor to meld with the dessert without becoming the dominant note.

When working with fruit-based desserts, such as apple pie or berry tarts, **orange blossom honey** is an excellent choice. Its citrus notes brighten the fruit flavors, creating a harmonious balance between sweet and tart. Apply the honey in a light glaze over the fruit before baking to lock in moisture and add a subtle sweetness that complements the natural sugars of the fruit.

For creamy desserts like cheesecake or custards, **acacia honey** is ideal due to its light, vanilla-like sweetness. It can be mixed into the dessert batter or used as a topping to add a hint of sweetness without altering the dessert's intended flavor profile.

Incorporating honey into traditional desserts not only adds sweetness but also complexity and depth. Experiment with different honey varieties to find the perfect match for your dessert, considering both the flavor of the honey and the dessert itself. Remember, the goal is to enhance and elevate the dessert's flavors, not to overwhelm them. Use honey sparingly to start, adjusting the amount based on taste and the strength of the honey's flavor. This thoughtful approach to pairing honey with desserts will bring a new level of sophistication and delight to your culinary creations.

CHAPTER 37: SAVORY HONEY RECIPES

Crafting marinades and glazes with honey for meats and vegetables introduces a delightful balance of sweetness and complexity to savory dishes. When preparing a honey marinade, the key is to balance the honey's sweetness with acidic components like lemon juice or vinegar and savory elements such as soy sauce or garlic. For a basic honey marinade suitable for chicken or pork, combine 1/4 cup of honey with 1/4 cup of soy sauce, 2 tablespoons of olive oil, the juice of one lemon, and 2 minced garlic cloves. This combination not only tenderizes the meat but also infuses it with flavor. Marinate the meat for at least 30 minutes, or for a more profound flavor, leave it in the refrigerator overnight.

Creating salad dressings and toppings that balance honey's sweetness with acidity is another avenue to explore. A simple vinaigrette can be elevated with the addition of honey. Mix together 3 tablespoons of extra virgin olive oil, 1 tablespoon of apple cider vinegar, 1 teaspoon of honey, a pinch of salt, and freshly ground black pepper to taste. This dressing works wonderfully on a mixed green salad or drizzled over roasted vegetables, adding a hint of sweetness that complements the natural flavors of the ingredients.

Incorporating honey into main dishes like roasted or grilled proteins can transform a simple meal into something special. For a honey-glazed salmon, whisk together 2 tablespoons of honey with 1 tablespoon of Dijon mustard and a splash of soy sauce. Brush this glaze over the salmon fillets before and during baking or grilling. The honey not only adds a rich glaze to the surface but also caramelizes slightly, providing a beautiful color and a delicious contrast to the salmon's natural flavors.

Enhancing soups and purees with a touch of honey for depth and flavor can introduce an unexpected but welcome sweetness that rounds out the dish. For a carrot soup, after sautéing onions and carrots, and simmering them until tender, blend the mixture until smooth. Stir in a tablespoon of honey to the puree to enhance the carrots' natural sweetness. This subtle addition can make a significant difference in the overall flavor profile of the soup, adding a layer of complexity that elevates the dish.

When using honey in savory recipes, it's essential to consider the type of honey and its flavor profile. Lighter honeys, such as clover or acacia, tend to work well in dressings and lighter dishes, providing sweetness without overwhelming the other flavors. In contrast, darker honeys, like buckwheat, can stand up to the robust flavors of grilled meats and roasted vegetables, adding depth and richness.

Experimenting with honey in savory dishes offers endless possibilities to enhance and transform everyday meals. Whether used as a glaze, in a marinade, dressing, or as a subtle addition to soups and purees, honey brings a unique sweetness that complements a wide range of flavors. By balancing honey's sweetness with acidic and savory elements, you can create dishes that are not only delicious but also beautifully balanced.

Honey Marinades and Glazes for Meats and Vegetables

Moving beyond the foundational honey marinades and glazes, let's delve into the nuances of crafting more complex and nuanced flavors that can elevate your culinary creations. One sophisticated approach involves the integration of spices and herbs into your honey-based marinades and glazes. For instance, combining honey with smoked paprika, garlic powder, and a touch of cayenne pepper can create a marinade that imparts a sweet, smoky, and slightly spicy flavor to chicken or pork. The key here is to balance the sweetness of the honey with the intensity of the spices, ensuring that no single flavor overpowers the others. A ratio of 2 tablespoons of honey to 1 teaspoon of each spice, mixed with 1/4 cup of olive oil and 2 tablespoons of apple cider vinegar, provides a well-rounded marinade.

For vegetables, a glaze that accentuates their natural flavors can be particularly effective. A simple yet impactful glaze can be made by whisking together honey, balsamic vinegar, and a hint of Dijon mustard. This combination, using 3 tablespoons of honey, 2 tablespoons of balsamic vinegar, and 1 teaspoon of Dijon mustard, works wonders on root vegetables like carrots or parsnips, enhancing their sweetness while adding a tangy depth. Brushing this glaze on the vegetables before roasting can transform them into a caramelized, flavorful side dish that complements a wide range of main courses.

When applying honey glazes to meats or vegetables, it's crucial to consider the cooking method and timing. Glazes should generally be applied during the last few minutes of grilling or roasting to prevent the honey from burning, which can introduce a bitter flavor. For marinades, allowing the meat to soak for several hours, or even overnight, in the refrigerator maximizes flavor absorption and tenderizes the meat, thanks to the mild acidity of ingredients like vinegar or lemon juice in the marinade.

Experimentation is encouraged when it comes to crafting honey marinades and glazes. Adjusting the proportions of ingredients or adding new elements like citrus zest, fresh herbs, or exotic spices can lead to delightful discoveries. Each adjustment not only enhances the flavor profile of the dish but also contributes to a broader understanding of how honey's sweetness interacts with other ingredients, creating a harmonious balance on the palate.

Balancing Honey's Sweetness in Dressings

For a more complex and layered salad dressing that harmonizes the sweetness of honey with the zesty tang of citrus, consider crafting a honey-lime vinaigrette. Begin with the juice of two fresh limes, ensuring you extract as much juice as possible, which should yield approximately 1/4 cup of lime juice. To this, add 2 tablespoons of honey, preferably a lighter variety such as clover or orange blossom for its mild sweetness that won't overpower the citrus notes. Whisk in 1/2 cup of extra virgin olive oil to emulsify the mixture, creating a smooth and cohesive dressing. The oil's richness balances the acidity of the lime and the sweetness of the honey, resulting in a dressing that's both refreshing and satisfying. Season with a quarter teaspoon of fine sea salt and a pinch of freshly ground black pepper to enhance the flavors. This dressing is particularly effective when drizzled over a salad featuring avocado, grilled chicken, or shrimp, as the honey-lime combination complements the ingredients' natural flavors without overwhelming them.

For those seeking a dressing with a bit more complexity, incorporating herbs such as cilantro or basil can introduce a new flavor dimension. Finely chop a tablespoon of fresh cilantro or basil leaves and whisk them into the dressing. The herbs not only add a pop of color but also a fresh, aromatic quality that pairs beautifully with the honey and lime. This variant is excellent for dressing tropical fruit salads or mixed greens that feature ingredients like mango, pineapple, or even slices of orange.

When preparing these dressings, always taste and adjust the seasoning as necessary. Depending on the honey's sweetness or the lime's acidity, you may find the need to add a little more honey to soften the tang or an extra squeeze of lime juice to brighten the dressing. The key to a perfectly balanced honey-based salad dressing lies in the careful calibration of sweetness and acidity, achieved through tasting and tweaking the ingredients to suit your palate.

Honey in Main Dishes

When incorporating honey into main dishes, particularly those involving roasted or grilled proteins, the approach should be both strategic and thoughtful to enhance the dish's flavor without overpowering the natural taste of the protein. Honey, with its natural sweetness and ability to caramelize under heat, offers a unique opportunity to create a rich, flavorful crust on meats while keeping the interior moist and tender. To achieve this, one must consider the type of protein, the cooking method, and the complementary flavors that will harmonize with honey's distinct profile.

For instance, when preparing a honey-glazed roasted chicken, start by selecting a high-quality, free-range chicken to ensure the best flavor and texture. Create a glaze by combining 1/3 cup of honey with 2 tablespoons of melted butter, 1 tablespoon of apple cider vinegar, 1 minced garlic clove, and a teaspoon of fresh thyme leaves. This blend not only brings out the sweetness of the honey but also introduces acidity, fat, and aromatics to balance the overall flavor profile. Before applying the glaze, season the chicken liberally with salt and pepper, and lightly coat it with olive oil to help the skin crisp up during roasting. Brush the honey glaze onto the chicken during the last 20 minutes of cooking to avoid burning the sugars in the honey. The oven should be preheated to 375°F, and the chicken roasted until it reaches an internal temperature of 165°F, ensuring it's cooked through but remains juicy.

For grilled meats, such as a honey-marinated flank steak, the marinade plays a crucial role in tenderizing and flavoring the meat. Combine 1/4 cup of honey with 1/4 cup of soy sauce, 3 tablespoons of balsamic vinegar, 2 minced garlic cloves, and 1 tablespoon of grated ginger. This marinade offers a balance of sweet, salty, acidic, and pungent flavors that penetrate the meat, enhancing its natural taste. Marinate the steak for at least 4 hours, or overnight for deeper flavor infusion. When grilling, ensure the grill is preheated to a high temperature to sear the steak quickly, locking in juices and creating a caramelized exterior from the honey. Cook to the desired doneness, typically 5-7 minutes per side for medium-rare, depending on the steak's thickness. Rest the steak for 5 minutes before slicing against the grain to ensure tenderness.

In both examples, the use of honey not only contributes sweetness but also aids in the browning and caramelization process, providing a visually appealing and deliciously glazed exterior. The key to success lies in balancing honey with other ingredients to complement the protein's flavor without overwhelming it. Additionally, monitoring cooking times and temperatures closely ensures that the honey enhances rather than detracts from the final dish. By following these detailed steps and recommendations, cooks can effectively incorporate honey into a variety of main dishes, elevating the taste and presentation of roasted or grilled proteins.

Enhancing Soups with Honey

When enhancing soups and purees with honey for added depth and flavor, selecting the right type of honey is crucial. Opt for a mild variety like clover or acacia honey, which won't overpower the natural flavors of the ingredients. Begin by preparing your base soup or puree, ensuring it's seasoned to your preference minus the usual amount of sugar or sweetener you might add. For a standard pot of soup (about 4-6 servings), stir in one tablespoon of honey, adding it slowly and tasting as you go. This method allows the honey's natural sweetness to meld with the soup's flavors without dominating the dish.

Incorporating honey into vegetable purees, such as carrot or sweet potato, can subtly enhance the vegetables' inherent sweetness. After cooking and blending your chosen vegetables into a smooth puree, add a teaspoon of honey for every two cups of puree. Mix thoroughly and adjust according to taste, ensuring the honey complements rather than overwhelms the puree.

For soups with a spicier profile, such as a spicy pumpkin or butternut squash soup, honey can serve as a balancing agent. Its sweetness can temper the heat from spices like cayenne pepper or chili, creating a harmonious flavor profile. In these cases, a slightly stronger honey, such as buckwheat, can stand up to the bold flavors without getting lost. Add the honey incrementally, starting with a teaspoon, taste, and then adjust as necessary.

Remember, the addition of honey to soups and purees is not just about adding sweetness but also about introducing complexity and a smooth finish to the dish. Honey's nuanced flavors can round out the taste, adding an extra layer of richness that sugar or other sweeteners cannot provide. As with all cooking, the key to success lies in balancing the ingredients. Always taste your dish after incorporating honey and adjust the seasoning accordingly to achieve the perfect blend of flavors.

CHAPTER 38: HONEY BEVERAGES

Crafting honey-based cocktails requires a nuanced understanding of how the sweetness of honey can complement and enhance the flavors of various spirits and mixers. To create a balanced and appealing cocktail, it's essential to select the right type of honey and to adjust the sweetness to suit the overall flavor profile of the drink. Here, we'll explore the steps and considerations for making two classic honey-infused cocktails: the Honey Bourbon Sour and the Honey Bee Martini.

Honey Bourbon Sour: This cocktail marries the rich warmth of bourbon with the smooth sweetness of honey, balanced by the bright acidity of fresh lemon juice. Begin by preparing a honey syrup, which will integrate more seamlessly into the cocktail than raw honey. Combine equal parts honey and hot water, stirring until the honey is fully dissolved. Allow the syrup to cool before use.

For one cocktail, in a shaker, combine 2 ounces of bourbon, 3/4 ounce of the cooled honey syrup, and 3/4 ounce of fresh lemon juice. The bourbon should be of good quality, with a smooth profile that won't overpower the delicate sweetness of the honey. Add ice to the shaker, cover, and shake vigorously for about 15 seconds, until well chilled. Strain the mixture into a rocks glass filled with ice. Garnish with a lemon twist or a cherry, adding a visual and aromatic appeal to the drink. The key to a perfect Honey Bourbon Sour is the balance between the sweet honey syrup, the sour lemon, and the bourbon, adjusting the honey syrup quantity according to taste.

Honey Bee Martini: This cocktail offers a sophisticated blend of gin, honey, and lemon, creating a smooth and slightly sweet martini with a refreshing edge. Start by making a honey syrup using the method described above. In a mixing glass, combine 2 ounces of gin, 1/2 ounce of honey syrup, and 1/2 ounce of fresh lemon juice. The gin should have a balanced botanical profile that complements rather than competes with the honey's flavor. Fill the mixing glass with ice, then stir for about 30 seconds until the mixture is well chilled. Strain the cocktail into a chilled martini glass. Garnish with a twist of lemon peel, expressing the oils over the drink to add a fragrant citrus note.

When crafting these cocktails, it's important to taste and adjust the sweetness as needed. The type of honey used can significantly impact the flavor of the drink. For a lighter, floral note, opt for clover or orange blossom honey. For a deeper, richer sweetness, buckwheat or wildflower honey may be preferred. Additionally, the freshness and quality of the lemon juice are crucial, as it provides a counterbalance to the honey's sweetness. Always use freshly squeezed lemon juice rather than bottled for the best flavor.

In conclusion, honey can transform a simple cocktail into a complex and intriguing drink. By carefully selecting the type of honey and balancing its sweetness with the other ingredients, you can create delicious honey-based cocktails that highlight the versatility and appeal of honey as a cocktail ingredient. Whether you're crafting a Honey Bourbon Sour, a Honey Bee Martini, or experimenting with your own creations, the key is to balance the flavors to achieve a harmonious and enjoyable drink.

The Art of Making Mead

Mead, often referred to as "honey wine," is a fermented beverage made from honey, water, and yeast. Its creation is a fascinating process that combines the art of brewing with the natural sweetness of honey. To craft mead, one must first select the **type of honey**. The flavor profile of the mead will vary significantly based on the honey used - for example, clover honey will produce a light and delicate mead, while buckwheat honey will result in a darker, richer flavor.

The basic steps to make mead begin with **sanitization**. Ensuring all equipment is thoroughly sanitized is crucial to prevent contamination. This includes fermenters, spoons, airlocks, and any other tools that will come into contact with the mead.

The next step is to prepare the **must**, which is the mixture that will be fermented. This involves diluting honey in water to the desired concentration. A general guideline is to use about 2 to 3 pounds of honey per gallon of water for a standard mead. However, the exact ratio can be adjusted depending on whether a dry, semi-sweet, or sweet mead is desired. The must should be mixed thoroughly until the honey is completely dissolved.

After preparing the must, **yeast is added**. The choice of yeast strain can also influence the flavor and character of the mead. Wine yeasts, champagne yeasts, or even specific mead yeasts are common choices. The yeast is typically rehydrated according to the manufacturer's instructions before being pitched into the must.

Once the yeast is added, the must is transferred to a **fermentation vessel** and sealed with an airlock. The fermentation process can take anywhere from a few weeks to several months, depending on the desired outcome and fermentation conditions. It's important to keep the fermenting mead in a location with a stable temperature, ideally between 55°F and 75°F, to ensure a steady fermentation process.

Throughout fermentation, it's essential to monitor the progress by checking the **specific gravity** with a hydrometer. This measurement will indicate when fermentation has ceased and the mead is ready to be **racked**. Racking involves transferring the mead from one vessel to another to separate it from the yeast sediment that has settled at the bottom. This process may be repeated multiple times to clarify the mead.

After fermentation and racking, the mead may still contain suspended particles. To achieve a clearer beverage, **fining agents** can be used to help these particles settle. Common fining agents include bentonite clay, isinglass, and gelatin. Each agent works differently, so selecting one that matches the specific needs of the mead is important.

The final step before enjoying the mead is **bottling**. The mead is carefully transferred into bottles, which are then sealed with corks or caps. It is often beneficial to allow the bottled mead to **age**. Aging can significantly improve the flavor of the mead, with some meads benefiting from several years of aging.

Throughout the mead-making process, patience and attention to detail are key. From selecting the right honey to monitoring fermentation and aging the bottled mead, each step contributes to the creation of a unique and flavorful beverage. Whether a novice or an experienced brewer, making mead offers an opportunity to engage with an ancient tradition and enjoy the rich, complex flavors that result from fermenting honey.

Sweetening Teas with Honey

Sweetening teas and herbal infusions with honey not only enhances the flavor but also adds a nuanced sweetness that sugar cannot replicate. The process begins with selecting the right type of honey, as its flavor profile can significantly influence the taste of the tea. For a light, floral infusion, clover or acacia honey is ideal due to its mild taste. For a more robust tea, such as black or chai, a darker honey like buckwheat provides a rich, molasses-like sweetness that complements the strong flavors of the tea.

To sweeten tea with honey, start by preparing the tea or herbal infusion as you normally would. Boil water to the appropriate temperature for the type of tea you're making; for example, green tea requires cooler water (around 175°F) to prevent bitterness, while black tea can handle boiling water (212°F). Steep the tea according to its recommended time, which usually ranges from 3 to 5 minutes for most teas.

Once the tea is steeped, stir in the honey while the tea is still hot. This ensures the honey dissolves completely, distributing its flavor evenly throughout the drink. Begin with a teaspoon of honey per 8-ounce cup of tea, then adjust according to taste. It's crucial to add honey to taste because the sweetness and flavor intensity of honey can vary widely.

For those who prefer iced tea or cold herbal infusions, dissolve the honey in a small amount of hot water to create a syrup before adding it to the cold beverage. This extra step helps the honey blend smoothly into the drink without clumping.

Experimenting with different honey varieties can lead to delightful discoveries in flavor pairing. For instance, orange blossom honey can add a citrusy note to Earl Grey tea, while a lavender-infused honey might enhance the floral notes of chamomile tea. The key is to balance the sweetness and flavor of the honey with the tea's profile, creating a harmonious blend that elevates the drinking experience.

Honey-Based Cocktails

Continuing with our exploration of honey-based cocktails, let's delve into crafting a refreshing **Honey Lavender Lemonade Cocktail** and a warming **Spiced Honey Toddy**. These drinks showcase honey's versatility in both cold and hot beverages, offering unique flavors that can be enjoyed year-round.

Honey Lavender Lemonade Cocktail: This cocktail combines the floral notes of lavender with the natural sweetness of honey, creating a refreshing and aromatic drink. Start by making a lavender honey syrup. Combine 1 cup water, 1 cup honey, and 2 tablespoons dried lavender flowers in a saucepan. Bring to a simmer, stirring until the honey dissolves. Remove from heat, cover, and let steep for 30 minutes. Strain the syrup and cool it in the refrigerator.

For the cocktail, in a pitcher, mix ¾ cup of the lavender honey syrup with 1 cup fresh lemon juice and 4 cups of cold water. Adjust the sweetness or tartness according to taste. To serve, fill a glass with ice, add 1 ½ ounces of vodka, and top with the lavender lemonade. Stir well and garnish with a lemon slice and a sprig of lavender. The choice of vodka is crucial; a neutral, high-quality vodka will let the honey and lavender flavors shine.

Spiced Honey Toddy: Perfect for chilly evenings, this toddy uses honey to add sweetness and depth to the traditional warm drink. Begin by boiling 1 cup of water and adding 1 black tea bag, steeping for about 5 minutes. In a mug, combine the hot tea with 2 ounces of bourbon, 1 tablespoon of honey, and a squeeze of lemon juice. Stir until the honey is fully dissolved. Add a pinch of ground cinnamon and cloves to introduce a subtle spice that complements the honey's richness. Garnish with a lemon wedge or cinnamon stick. The bourbon selected should have a smooth, oak-aged flavor that melds well with the spiciness and the sweetness of the honey.

Both the **Honey Lavender Lemonade Cocktail** and the **Spiced Honey Toddy** highlight honey's ability to harmonize with other ingredients, creating beverages that are both comforting and refreshing. By adjusting the amount of honey syrup or honey used, you can tailor the sweetness to your preference, making these cocktails adaptable to different tastes and occasions.

Natural Energy Drinks with Honey

Creating natural energy drinks with honey as the central ingredient is a fantastic way to harness the natural benefits of this sweet substance while avoiding processed sugars. Honey is not only a healthier alternative but also provides a quick energy boost, making it perfect for these recipes.

Honey Citrus Boost: Combine **1 cup of fresh orange juice** (preferably cold-pressed for optimal nutrients) with **1 tablespoon of raw honey**, **a pinch of sea salt** (to replenish electrolytes), and **a few drops of lemon juice** for an extra zing. Shake or stir well to ensure the honey dissolves completely. This drink is ideal for morning energy, providing vitamin C for immune support and honey for a steady supply of energy.

Green Tea Honey Energizer: Brew **1 cup of green tea** and let it cool to room temperature. Green tea is known for its antioxidant properties and a gentle caffeine boost. Mix in **1 tablespoon of honey** to sweeten and add a slice of **fresh ginger** for its anti-inflammatory benefits. This drink is perfect for a mid-afternoon pick-me-up, offering hydration, antioxidants, and a natural energy boost without the jitters commonly associated with coffee.

Honey Watermelon Refresher: Blend **2 cups of cubed watermelon** with **1 tablespoon of honey, a handful of mint leaves**, and **the juice of half a lime**. Watermelon is hydrating and rich in lycopene, while mint and lime offer a refreshing taste and digestive benefits. This light and energizing drink is perfect for hot days or after a workout to replenish fluids and energy.

For each recipe, ensure that the honey is fully dissolved in the liquid to provide a smooth texture and uniform sweetness. Adjust the amount of honey based on personal preference and the natural sweetness of the other ingredients. Using raw, unprocessed honey will offer the most health benefits, including antioxidants and enzymes that are beneficial for energy and overall wellness. These natural energy drinks with honey are not only delicious but also a healthier way to stay energized throughout the day.

CHAPTER 39: PRESERVING RECIPES

Preserving the natural flavor and properties of honey in cooking requires a meticulous approach to both preparation and storage. When incorporating honey into dishes that will be preserved, it's crucial to understand how heat affects honey's natural enzymes and flavor profile. For instance, when creating honey-infused jams or jellies, opt for a low and slow cooking method. This technique ensures that the honey's flavors meld harmoniously with the fruits without losing its aromatic qualities. A digital thermometer can be invaluable here, allowing you to monitor the temperature closely, ensuring it does not exceed 104°F, the point at which honey's beneficial enzymes begin to degrade.

In terms of storage, honey's natural acidity and low moisture content make it an excellent preservative by itself. However, when mixed into recipes, the shelf life can be affected by additional ingredients. To preserve honey-based dishes, consider vacuum sealing if possible, as this method significantly extends the longevity by removing air that can lead to oxidation. For liquid concoctions like syrups or dressings, sterilized glass bottles are ideal. Ensure they are heated before filling to prevent the glass from breaking and to kill any potential bacteria. Once filled, sealing the bottles while they are still hot can create a vacuum seal as they cool, further preserving the contents.

For solid honey-infused foods, such as energy bars or granola, oven drying at a low temperature can help remove excess moisture, which is a key factor in food preservation. Spread the mixture thinly on a baking sheet lined with parchment paper and dry at the lowest oven setting, ideally below 150°F. This process not only preserves the food but also concentrates the flavors, including the honey's sweetness and aroma.

Freezing is another effective method for preserving honey's qualities in dishes. While pure honey does not freeze solid due to its low water content, honey-infused dishes can be frozen to maintain their quality over time. Use airtight containers or freezer bags to prevent freezer burn and flavor transfer. When ready to use, thawing in the refrigerator overnight is recommended to maintain texture and flavor integrity.

Remember, the key to successfully preserving recipes with honey lies in gentle handling, minimal heat exposure, and airtight storage. By following these guidelines, you can enjoy the delightful taste and health benefits of honey in your cooking for months to come.

Preserving Honey's Natural Flavor in Cooking

When incorporating honey into baked goods, understanding the balance between honey's natural sweetness and the desired outcome of your recipe is crucial. For instance, when substituting sugar with honey in recipes, it's essential to recognize that honey is sweeter than granulated sugar. A general rule is to use three-quarters of a cup of honey for every cup of sugar the recipe calls for. Additionally, because honey adds liquid to your mixture, reducing other liquid ingredients by a quarter cup for every cup of honey used can help maintain the correct consistency. It's also beneficial to add a pinch of baking soda (about one-fourth of a teaspoon) for every cup of honey to neutralize its acidity, which can affect the rise of baked goods.

When cooking with honey, low and slow is the mantra for preserving its natural flavors and health benefits. High temperatures can cause honey to caramelize rapidly, potentially leading to a bitter taste and the loss of many of its beneficial enzymes. Therefore, when adding honey to dishes that require cooking, it's advisable to do so in the last few minutes of cooking time, especially for sauces or glazes. This method ensures that the honey retains its flavor and nutritional profile. For example, when preparing a honey glaze for roasted vegetables, brush the honey onto the vegetables in the final 5-10 minutes of roasting. This approach allows the honey to gently caramelize without burning, imparting a rich, sweet flavor to the dish.

In salad dressings, honey serves as a natural emulsifier, helping to blend and thicken the dressing while adding a touch of sweetness. To fully incorporate honey into a vinaigrette, first, mix the honey with the acidic components (like vinegar or lemon juice) until completely dissolved. Then, gradually whisk in the oil to create a stable emulsion. This technique ensures that the honey is evenly distributed throughout the dressing, providing a consistent flavor and texture.

For cold dishes or uncooked recipes, such as smoothies or yogurt parfaits, using honey in its raw form is the best way to enjoy its full flavor and benefits. Stirring honey into cold or room-temperature ingredients allows it to retain its natural enzymes and antioxidants. When adding honey to thicker cold mixtures, gently warming the honey to a pourable consistency can make it easier to blend smoothly without overheating and degrading its quality. A gentle warm bath for the honey container, keeping water temperatures below 95°F, can achieve this without compromising the honey's integrity.

Finally, when marinating meats with honey, combining the honey with acidic ingredients like vinegar or citrus juice can help tenderize the meat while infusing it with flavor. For effective marination, ensure the honey is thoroughly mixed with the other marinade ingredients before adding the meat. This step guarantees that the honey's moisture-attracting properties are evenly distributed, allowing the meat to stay moist and flavorful during cooking. Marinating for several hours, or even overnight, in the refrigerator, not only maximizes flavor absorption but also minimizes the risk of the honey burning during the cooking process, as the meat will cook more evenly and retain more moisture.

By adopting these specific techniques, you can skillfully preserve the natural flavor and beneficial properties of honey in a wide range of culinary applications, from sweet to savory dishes, enhancing both the taste and nutritional value of your meals.

Storing Honey-Based Dishes

When storing sweet and savory dishes made with honey, it's essential to consider the best methods to maintain their freshness, flavor, and texture. Honey, as a natural preservative, can extend the shelf life of many dishes when stored correctly. Here are detailed methods for storing these honey-infused creations:

Refrigeration: For dishes that contain dairy, eggs, or cooked fruit and vegetables, refrigeration is necessary. Place the dish in an airtight container to prevent the absorption of odors from other foods. Glass containers with tight-fitting lids are ideal as they do not impart any unwanted flavors into the food. If the dish is a liquid or semi-liquid, such as a honey-based dressing or marinade, ensure the container is leak-proof. Solid foods, like honey cakes or glazed meats, should be wrapped tightly in plastic wrap or aluminum foil before placing them in a container to maintain moisture levels and prevent them from drying out.

Freezer: Freezing is an excellent option for long-term storage of honey-infused baked goods, such as bread, muffins, or cookies. Wrap these items individually in plastic wrap and then again with aluminum foil to protect against freezer burn and flavor transfer. Label each item with the date to keep track of how long it has been stored. To thaw, remove the item from the freezer and let it come to room temperature while still wrapped to prevent condensation from affecting its texture.

Pantry Storage: Some honey-infused items, such as granolas or energy bars, can be stored in the pantry if they are properly sealed. Use airtight containers made of glass or BPA-free plastic to keep these items fresh. Ensure that the pantry is cool, dry, and away from direct sunlight, as heat and humidity can cause the honey to ferment or the food to spoil more quickly.

Vacuum Sealing: For both sweet and savory dehydrated or dried honey-infused foods, vacuum sealing can significantly extend shelf life by removing air and preventing oxidation. This method is particularly useful for items like jerky or dried fruits. Once vacuum-sealed, these items can be stored in the pantry, refrigerator, or freezer, depending on their specific storage requirements.

Silica Gel Packets: When storing items that may be prone to moisture, such as dried honey candies or crystallized honey, including a small silica gel packet in the container can help absorb any excess moisture and keep the items dry. Ensure that these packets are food-grade and do not come into direct contact with the food.

Labeling and Rotation: Regardless of the storage method chosen, labeling each container with the contents and the date stored is crucial for effective rotation and usage. This practice helps prevent waste and ensures that items are used while they are at their peak quality.

By adhering to these storage methods, you can enjoy the delightful taste of honey-infused dishes for longer periods, ensuring that their quality remains consistent. Whether refrigerating a batch of honey lemon chicken or freezing a honey-sweetened cake, proper storage techniques are key to preserving the integrity and flavor of your culinary creations.

Personal Honey Recipe Book Ideas

Creating a personal honey recipe book serves as a heartfelt keepsake or a unique gift, encapsulating the sweetness of your culinary journey with honey. Begin by selecting a **high-quality, durable notebook** or journal that can withstand frequent use in the kitchen. Opt for acid-free pages to ensure your recipes endure over time without yellowing or degradation.

For the content, start with a **personal introduction**. Share your journey into the world of honey, including what drew you to it, your favorite experiences, and the significance of the recipes you've chosen to include. This personal touch transforms the book from a mere collection of recipes into a story, your story, about the role of honey in your life and kitchen.

Organize the recipes into sections for easy navigation. Categories might include **Breakfasts**, **Desserts**, **Savory Dishes**, and **Beverages**. Within each section, provide a **table of contents** with page numbers. Consider adding **tabs or color-coded sections** for quick reference.

Each recipe should be detailed with **clear, step-by-step instructions** and include the **quantity of honey** needed, specifying whether raw or processed honey is preferred. Highlight any **personal tips** you've discovered that enhance the dish, such as the type of honey that best complements certain ingredients or adjustments for altitude or humidity that might affect the outcome.

Incorporate **high-quality photos** of the dishes, ideally ones you've taken yourself. These visuals not only make the book more engaging but also guide readers in their expectations of the final product.

Leave space at the end of each recipe for **notes** where you or the gift recipient can jot down personal variations, outcomes, or memories associated with the dish. This interactive element encourages ongoing engagement with the book.

Finally, consider adding a section on the **health benefits of honey**, backed by research or consultations with nutrition experts. This educates readers on why honey is not just a sweetener but a beneficial addition to their diet.

Personalizing the cover with a **custom title** and perhaps a **dedication page** if the book is a gift, adds the finishing touch to this compilation, making it a treasured item for anyone passionate about cooking with honey.

CHAPTER 40: GASTRONOMIC PAIRINGS

Pairing honey with different types of cheese creates a symphony of flavors that can elevate any dining experience. The natural sweetness of honey complements the salty, tangy, or creamy textures of various cheeses, providing a balance that is both sophisticated and delightful. Here are specific pairings that showcase the versatility of honey in gastronomic combinations:

Soft Cheeses: For soft cheeses like Brie or Camembert, a drizzle of **clover honey** enhances their creamy texture without overpowering their delicate flavors. The mild, floral notes of clover honey blend seamlessly with the buttery richness of these cheeses, making for an exquisite appetizer when served on a warm baguette slice or a crisp cracker.

Blue Cheese: The bold, pungent flavors of blue cheese such as Gorgonzola or Roquefort are perfectly balanced by the robust, slightly earthy taste of **buckwheat honey**. This darker, stronger honey cuts through the intensity of the blue cheese, creating a pairing that is both bold and harmonious. Serve this combination on a pear slice or a walnut cracker to add depth and texture.

Aged Cheeses: Hard, aged cheeses like Parmigiano-Reggiano or aged Gouda pair wonderfully with **acacia honey**. The light, vanilla-scented sweetness of acacia honey complements the nutty, crystalline texture of aged cheeses, highlighting their complex flavors without competing with them. This pairing works well as a sophisticated dessert course or a refined snack.

Goat Cheese: The tangy sharpness of fresh goat cheese is beautifully offset by the floral, citrusy notes of **orange blossom honey**. This combination brings out the best in both the cheese and the honey, creating a fresh, lively flavor profile that is ideal for spring and summer dishes. Spread the goat cheese on a toasted crostini and top with a drizzle of honey for a simple yet elegant appetizer.

Ricotta: For a sweeter pairing, fresh ricotta cheese drizzled with **wildflower honey** offers a dessert-like experience that highlights the soft, mild qualities of the cheese with the complex, varied sweetness of the honey. This pairing can be enhanced with a sprinkle of cinnamon or fresh berries for added flavor and texture, serving as a delightful end to any meal.

When pairing honey with cheese, consider the intensity and flavor profile of both the honey and the cheese to ensure a balanced combination. Experimenting with different types of honey can reveal surprising and delightful pairings that can transform ordinary cheese into a gourmet experience. Additionally, incorporating elements such as fresh fruits, nuts, or artisan breads can add dimension and contrast to the pairings, making each bite a unique exploration of flavors and textures.

Incorporating honey into main dishes like roasted or grilled proteins can introduce a subtle sweetness that enhances the natural flavors of the meat. For instance, a glaze made from **manuka honey** and Dijon mustard can be brushed onto chicken or pork before roasting, creating a caramelized, flavorful crust that is both savory and slightly sweet. The antimicrobial properties of manuka honey also add a healthful aspect to the dish, making it not only delicious but beneficial.

Enhancing soups and purees with a touch of honey can add depth and complexity to these dishes. A swirl of **lavender honey** in a butternut squash soup introduces a hint of floral sweetness that complements the natural richness of the squash. This subtle addition can elevate a simple soup to a dish of refined flavors suitable for any occasion.

By understanding the unique characteristics of different types of honey and how they interact with various foods, you can create gastronomic pairings that are both innovative and deeply satisfying. Whether used in sweet or savory dishes, as a complement to cheese, or as a key ingredient in marinades and glazes, honey's versatility makes it an indispensable tool in the culinary world.

Honey and Cheese Pairings

Pairing honey with cheese is an art that combines the sweetness of honey with the rich flavors of cheese, creating a delightful taste experience. When exploring these combinations, it's essential to consider the texture, flavor, and intensity of both the honey and the cheese to achieve a harmonious balance. For instance, a strong, pungent cheese requires a honey with a bold flavor to match, while a mild cheese pairs better with a lighter, more delicate honey.

One exceptional pairing is the combination of Manchego cheese with orange blossom honey. Manchego, a firm Spanish cheese with a slightly nutty flavor, pairs beautifully with the floral notes of orange blossom honey. The honey's citrus undertones enhance the cheese's rich taste, making it an ideal choice for a sophisticated cheese board. To serve, slice Manchego thinly and drizzle it with a small amount of honey, allowing the flavors to meld together.

Another exquisite pairing is fresh chevre (goat cheese) with lavender honey. The creamy, tangy nature of chevre is perfectly complemented by the sweet, floral essence of lavender honey. This pairing is not only a feast for the palate but also for the eyes, making it a perfect addition to a brunch menu or a dessert cheese plate. Spread the chevre on a crisp cracker and top with a dollop of lavender honey for a simple yet elegant treat.

For those who enjoy a more robust flavor profile, pairing blue cheese with chestnut honey offers a rich, intense experience. The bold, earthy flavors of blue cheese, such as Stilton or Danish Blue, are balanced by the deep, slightly bitter taste of chestnut honey. This pairing is ideal for those who appreciate complex flavors and textures. Serve crumbled blue cheese alongside a small bowl of chestnut honey for dipping, allowing guests to adjust the sweetness to their taste.

When selecting honey and cheese pairings, always consider the intensity and flavor profile of each component. Aim for balance, allowing both the honey and the cheese to shine without one overpowering the other. Experiment with different combinations to discover personal favorites and introduce guests to the delightful world of honey and cheese pairings.

Pairing Honey with Fruits for Balanced Flavors

Pairing honey with fresh and dried fruits opens a realm of culinary possibilities, each combination offering a unique balance of flavors that can enhance a variety of dishes. The natural sweetness of honey complements the tartness or sweetness of fruits, creating a harmonious blend that can be used in breakfasts, desserts, salads, and even savory dishes. To achieve the best flavor pairings, it's essential to consider the type of honey and its flavor profile alongside the fruit's characteristics.

For fresh fruits, consider their juiciness and acidity. Fruits like strawberries, peaches, and melons have a high water content and a delicate sweetness that pairs well with light, floral honeys such as clover or orange blossom. The subtle sweetness of these honeys doesn't overpower the fruit's natural flavors but rather enhances them. Drizzling honey over a fresh fruit salad or blending it into a fruit puree for a dessert sauce can elevate the taste experience.

When pairing honey with citrus fruits like oranges, grapefruits, or lemons, opt for a honey with a robust flavor, such as wildflower or buckwheat honey. The strong, bold flavors of these honeys can stand up to the acidity and tanginess of the citrus, creating a balanced and invigorating taste. A honey-citrus glaze can be an excellent addition to baked goods or a refreshing dressing for a citrus-based salad.

Dried fruits offer a concentrated sweetness and chewy texture that pairs beautifully with richer, darker honeys. Figs, apricots, and raisins, with their intense sweetness and slight tartness, are complemented by the molasses-like qualities of buckwheat or chestnut honey. Incorporating these combinations into oatmeal, yogurt, or cheese platters can introduce a complex flavor profile that is both satisfying and sophisticated.

For a tropical twist, pairing honey with exotic fruits such as mango, pineapple, or papaya can create vibrant and flavorful dishes. The exotic, bold flavors of these fruits are enhanced by the sweetness of acacia or tupelo honey, offering a taste of the tropics in every bite. These pairings work well in smoothies, tropical fruit salads, or as toppings for ice cream and other desserts.

Incorporating nuts and seeds with honey and fruit pairings adds texture and depth. For example, a salad of mixed greens, sliced pears, walnuts, and a drizzle of sage honey creates a delightful interplay of flavors and textures, with the honey acting as a bridge that ties all the components together. Similarly, a dessert of grilled peaches, a sprinkle of crushed almonds, and a generous drizzle of lavender honey can offer a satisfying crunch alongside the smooth sweetness of the fruit and honey.

When experimenting with honey and fruit pairings, it's crucial to taste the honey and fruit together to ensure their flavors harmonize. The goal is to achieve a balance where neither the honey nor the fruit dominates but rather complements each other to create a more complex and enjoyable flavor profile. Additionally, consider the visual appeal of these pairings; the vibrant colors of fruits combined with the golden hues of honey can make any dish visually stunning as well as delicious.

Ultimately, the versatility of honey makes it an invaluable ingredient in the kitchen, capable of transforming simple fruit dishes into extraordinary culinary creations. By carefully selecting and pairing different types of honey with various fruits, both fresh and dried, you can explore an endless array of flavors and textures, enhancing your cooking and offering delightful taste experiences to your family and guests.

PART 10: CRAFTING WITH HIVE PRODUCTS

CHAPTER 41: BEESWAX CANDLES

Adding natural essences to your beeswax candles not only enhances their aesthetic appeal but also imbues them with therapeutic benefits, transforming your crafting from a simple hobby into an artisanal craft. Essential oils, extracted from plants, flowers, and fruits, offer a natural way to scent your candles, providing a range of aromatherapeutic properties that can relax, energize, or uplift the mood of any room. When selecting essential oils for your beeswax candles, consider the following guidelines to ensure safety, effectiveness, and the best aromatic experience.

Choose pure, high-quality essential oils for your candles. Synthetic fragrances or low-quality oils may not provide the same natural benefits and can even emit harmful chemicals when burned. Look for oils that are certified organic, 100% pure, and free from additives or diluents.

Consider the scent throw of your chosen essential oils. Some oils, like citrus (lemon, orange, grapefruit) and mint (peppermint, spearmint), have a strong scent throw, meaning their aroma spreads well and is easily noticeable. Others, such as lavender, rose, and eucalyptus, offer a medium scent throw that is perceptible and pleasant without being overwhelming. For a subtle scent throw, consider using woody (cedarwood, sandalwood) or earthy oils (vetiver, patchouli), which provide a mild fragrance that gently permeates the room.

Calculate the correct amount of essential oil to use in your beeswax candles. A general guideline is to use about 6 to 10 drops of essential oil per ounce of melted beeswax. This ratio can be adjusted based on the strength of the oil's fragrance and personal preference. However, it's important to not exceed a 10% concentration of essential oil to wax, as too much oil can affect the candle's burn quality and safety.

Incorporate the essential oils properly into the melted beeswax. Add the essential oils after you've removed the beeswax from the heat source, just before pouring it into your molds or containers. Stir the oil evenly into the wax for at least 30 seconds to ensure a uniform distribution. This method helps preserve the integrity of the essential oils' fragrance and therapeutic properties, which can be diminished by prolonged exposure to high heat.

Test your scented beeswax candles before making a large batch. Pour a small amount of your scented wax into a candle mold and let it set. Once the candle is ready, light it in a controlled environment to observe its burning quality and scent throw. This test allows you to adjust the amount of essential oil if necessary and ensures that the wick size is appropriate for the candle's diameter.

Incorporating essential oils into beeswax candles offers a natural and enjoyable way to enhance your living space. By following these guidelines, you can create beautifully scented candles that not only light up your home but also provide the therapeutic benefits of aromatherapy. Whether you're crafting for personal use or considering a small craft business, the addition of natural essences to your beeswax candles can set your products apart, offering a unique and desirable feature for potential customers.

Traditional Beeswax Candle Techniques

Once you have selected your essential oils and prepared your beeswax, the next step is to gather the necessary tools and materials for candle making. This includes a double boiler for melting the beeswax, a thermometer to monitor the wax temperature, candle molds or containers, wicks that are appropriate for the size of the candles you are making, a pouring jug, and any decorations or colorants you wish to add to your candles.

Begin by cutting the beeswax into small pieces to allow it to melt more evenly and quickly. Set up your double boiler by filling the bottom pot with water and placing the beeswax in the top pot. Heat the water to a gentle boil, allowing the steam to melt the beeswax. It's crucial to never leave melting beeswax unattended, as it is flammable at high temperatures. Use the thermometer to monitor the wax's temperature, which should be around 145-170°F (63-77°C) for optimal pouring.

While the beeswax is melting, prepare your candle molds or containers by securing the wicks in place. If using molds, you can thread the wick through the hole at the bottom of the mold, securing it with a wick holder or a piece of tape. For containers, center the wick and hold it in place with a wick bar or a clothespin resting on the container's rim.

Once the beeswax has melted completely and reached the correct temperature, remove it from the heat and let it cool slightly. This is the time to add any colorants or decorations to your wax. If you're adding essential oils for scent, wait until the wax has cooled to about 120-130°F (49-54°C) before stirring them in. This prevents the essential oils from evaporating too quickly.

Carefully pour the melted beeswax into your prepared molds or containers, ensuring not to disturb the placement of the wicks. Fill each mold or container to the desired level, leaving a small space at the top. Allow the candles to cool and solidify completely, which can take several hours or overnight, depending on the size of the candles.

After the candles have cooled, if you used molds, gently remove the candles by pulling on the wick or tapping the mold on a hard surface to release them. For container candles, trim the wick to about ¼ inch above the wax. This is an important step as it helps ensure a clean and safe burn.

Finally, before lighting your beeswax candles, let them cure for at least 24 hours. This waiting period allows the wax and any added scents to stabilize, resulting in a better scent throw and burning experience.

Creating beeswax candles is not only a rewarding craft but also a way to bring the natural beauty and warmth of beeswax into your home. With practice, you can experiment with different shapes, scents, and colors to create unique candles for personal use or gifts. Remember, the key to successful beeswax candle making is patience and attention to detail, ensuring each candle burns cleanly and safely.

Decorative Candle Ideas

To enhance the aesthetic appeal of your beeswax candles, consider incorporating a variety of decorative elements that can transform them from simple sources of light to intricate pieces of art. Here are some specific ideas and techniques for decorating your candles:

Embedding Items: You can embed small, non-flammable items into the sides of candles as they set. Consider using dried flowers, cinnamon sticks, or coffee beans for a rustic look. Place these items against the sides of your mold or container before pouring in the melted wax. Ensure they are evenly spaced and securely positioned.

Layering Colors: Create visually striking candles by layering different colored waxes. After pouring the first layer and allowing it to partially set, add another layer of beeswax dyed with natural colorants like turmeric for yellow, beetroot for pink, or spirulina for green. Allow each layer to cool completely before adding the next to achieve clean, distinct layers.

Surface Decorating: Once your candle has set, you can decorate its surface with a fine-tipped paintbrush and natural dyes. Painting bees, flowers, or abstract patterns on the surface can add a personalized touch. For best results, choose dyes that adhere well to beeswax and practice your design on paper first.

Texturing: Add texture to your candles by manipulating the wax as it cools. Techniques such as tapping the mold with a spoon, using a heat gun to create a ripple effect, or pressing objects into the surface of the candle can create unique textures. Experiment with different tools and methods to find what works best for your design vision.

Wrapping: Wrap your finished candles with natural materials like twine, burlap, or thin strips of leather. This not only adds to the aesthetic but also provides an easy way to attach labels or charms. Secure the material with a small drop of melted wax or a decorative knot.

By employing these decorating techniques, you can elevate the appearance of your beeswax candles, making them not only a source of natural light but also a reflection of personal style and creativity. Each method allows for customization, ensuring your candles are as unique as they are beautiful.

Creating Therapeutic Scented Candles

Creating therapeutic scented candles involves the careful selection and incorporation of natural essences into the beeswax. These essences, derived from plants, flowers, and herbs, are not just about adding fragrance; they are chosen for their therapeutic properties, which can range from relaxing the mind to energizing the body. To begin, it's essential to understand the properties of various essential oils and how they can influence mood and health. For instance, lavender is renowned for its calming and soothing effects, making it ideal for stress relief, while peppermint can invigorate the senses and enhance focus.

When adding these natural essences to your beeswax candles, timing and proportion are crucial. Essential oils should be added to the melted beeswax once it has cooled to approximately 120-130°F (49-54°C) to prevent the evaporation of the volatile compounds responsible for their therapeutic effects. The general guideline for the amount of essential oil to add is about 6 to 10 drops per ounce (30 ml) of melted beeswax, but this can vary depending on the oil's potency and the desired fragrance strength.

Thoroughly mix the essential oil into the beeswax to ensure even distribution. This step is vital for achieving a consistent scent throughout the candle. Once the essential oil is well incorporated, proceed to pour the wax into your prepared molds or containers, following the previously outlined steps for securing the wick and allowing the candles to cool and solidify.

Using natural essences in beeswax candles not only imbues them with delightful fragrances but also turns them into therapeutic tools that can enhance well-being. Whether placed in a living room, bathroom, or bedroom, these scented candles can create an ambiance that promotes relaxation, invigoration, or focus, depending on the chosen essences. Moreover, by selecting high-quality, pure essential oils and integrating them thoughtfully into your candle-making process, you craft not only a source of light but also a natural, health-promoting element for any space.

CHAPTER 42: NATURAL COSMETICS

Transitioning from the art of candle making to the realm of **natural cosmetics**, we delve into the multifaceted uses of beeswax, propolis, and honey in skincare. These hive products are celebrated for their moisturizing, antibacterial, and healing properties, making them ideal ingredients for homemade cosmetics.

Beeswax is a natural emulsifier and thickener, which when combined with oils and butters, creates protective, water-resistant barriers on the skin without clogging pores. For crafting a basic **beeswax lip balm**, you'll need beeswax pellets, coconut oil, shea butter, and essential oils for scent. Melt one part beeswax with two parts coconut oil and shea butter in a double boiler, stirring continuously until smooth. Remove from heat and quickly stir in a few drops of your chosen essential oil, such as peppermint or lavender, for their therapeutic benefits. Pour the mixture into small containers or lip balm tubes and let it solidify.

Propolis, known for its antimicrobial qualities, can be infused into oils to create healing salves or added to creams for acne-prone skin. To prepare a **propolis salve**, gently heat a carrier oil like olive or almond oil in a double boiler and add propolis extract. The general guideline is to use a ratio of one part propolis extract to five parts carrier oil. After the propolis has dissolved, mix in beeswax until melted and pour into tins or jars to cool. This salve can be applied to cuts, scrapes, and dry skin patches for its soothing effects.

Honey is a natural humectant, drawing moisture into the skin, and can be used in face masks, cleansers, and moisturizing lotions. A simple **honey face mask** requires just raw honey and add-ins like ground oats for exfoliation or turmeric for its anti-inflammatory properties. Apply a thin layer of raw honey, optionally mixed with your chosen ingredients, to clean, damp skin. Leave on for up to 20 minutes, then rinse with warm water for hydrated, glowing skin.

For those interested in crafting **natural moisturizing lotions**, start with a base of distilled water, aloe vera gel for its skin-soothing properties, and oils such as jojoba or sweet almond oil. Beeswax acts as the emulsifying agent, thickening the lotion and forming an emulsion when combined with the water phase. Heat the oil phase (oils and beeswax) and water phase separately in double boilers. Once both phases reach approximately 160°F (71°C), slowly pour the water phase into the oil phase, whisking vigorously. Continue to stir as the mixture cools and thickens, then add essential oils for fragrance and additional skin benefits. Pour the final product into sterilized jars.

When creating **beeswax soap**, the process involves saponification, where fats or oils mix with lye to form soap. Beeswax can be added to the oil phase to increase the hardness of the soap bars. A basic recipe might include olive oil for gentleness, coconut oil for lather, and a small percentage of beeswax. After combining the lye water and oils, including melted beeswax, blend until trace is reached, then pour into molds to set.

Each of these projects not only allows for the personalization of skincare products but also offers the satisfaction of creating beneficial items from the fruits of one's beekeeping labors. By adhering to these guidelines and experimenting with different combinations of hive products, essential oils, and natural additives, one can craft a range of cosmetics tailored to various skin types and needs.

Creams and Lotions with Beeswax

Delving deeper into the realm of natural cosmetics with beeswax as a cornerstone ingredient, let's explore the formulation of a rejuvenating beeswax face cream that caters to dry and sensitive skin types. This cream harnesses the emollient properties of beeswax, coupled with the hydrating benefits of natural oils and the soothing touch of essential oils, to create a deeply nourishing skin care product.

To commence, gather the following materials: a digital scale for precise measurements, a double boiler for gentle heating, a thermometer to monitor temperatures, a whisk for blending, and sterile containers for storing the finished product. The selection of ingredients includes 15 grams of pure beeswax, known for its ability to form a protective barrier on the skin while still allowing it to breathe; 30 milliliters of jojoba oil, chosen for its similarity to the skin's natural sebum and its ability to provide long-lasting moisture without leaving an oily residue; 60 milliliters of rosehip oil, rich in vitamins and antioxidants to aid in skin regeneration and improve elasticity; 40 milliliters of distilled water or rosewater for hydration; 10 milliliters of aloe vera gel to soothe and heal; and a choice of essential oils such as chamomile or lavender, both renowned for their calming and anti-inflammatory properties, with a recommended dosage of 5-10 drops depending on personal scent preference.

Begin the process by measuring the beeswax and jojoba oil into the upper section of the double boiler, warming them over a low heat until the beeswax has completely melted. In a separate vessel, gently heat the rosehip oil, distilled water or rosewater, and aloe vera gel until they reach the same temperature as the beeswax mixture, approximately 160°F (71°C). This temperature synchronization is crucial to prevent the emulsion from breaking.

Once both phases are adequately heated, slowly drizzle the aqueous phase (water, rosewater, and aloe vera gel) into the oil phase (beeswax and jojoba oil), continuously whisking to create a stable emulsion. As the mixture begins to cool and thicken, add the essential oils, continuing to whisk until fully incorporated. This is the moment when the cream transforms from liquid to a soft, creamy texture.

Transfer the cream into your prepared sterile containers while it is still slightly warm, as it will continue to thicken upon cooling. Allow the cream to set undisturbed, which can take several hours depending on the ambient temperature. Once cooled and solidified, the cream is ready to use.

For application, a small pea-sized amount is sufficient, gently massaged into the face and neck after cleansing. The cream's rich composition makes it an excellent night treatment, providing intense hydration and repair throughout the night.

This beeswax face cream recipe exemplifies the versatility and efficacy of beeswax in homemade cosmetics. By adjusting the types of oils and essential oils used, this base recipe can be customized to suit different skin types and concerns, making it a valuable addition to any natural skincare regimen. Through the thoughtful combination of beeswax with other natural ingredients, one can create a range of personalized, effective skincare products that nourish the skin and embrace the essence of nature's bounty.

Nourishing Lip Balms with Natural Ingredients

Choosing the right ingredients is crucial for crafting nourishing lip balms. Begin with beeswax, a natural base that not only thickens the lip balm but also provides a protective layer on the lips, shielding them from environmental factors. For a standard batch, measure out one tablespoon of beeswax pellets. This amount is ideal for creating a firm yet easily applicable balm.

Next, incorporate coconut oil, renowned for its moisturizing properties and ability to easily penetrate the skin. Use two tablespoons of coconut oil to ensure the lip balm is soft enough to spread smoothly over the lips. Coconut oil's natural sweetness also adds a pleasant taste to the lip balm.

Shea butter is another essential component, adding a luxurious creaminess and intense hydration. One tablespoon of shea butter will complement the coconut oil, enhancing the lip balm's moisturizing capabilities. Shea butter is also rich in vitamins A and E, which help in the healing and rejuvenation of the skin.

For the scent and additional therapeutic benefits, essential oils like peppermint or lavender are perfect. Peppermint oil, with its refreshing aroma, can invigorate the senses and provide a slight plumping effect to the lips. Lavender oil, on the other hand, is known for its calming properties, making it ideal for use before sleep. Add no more than five to ten drops of essential oil to the mixture to avoid irritation.

To prepare the lip balm, use a double boiler to melt the beeswax, coconut oil, and shea butter together. Stir continuously until the mixture is completely liquid and homogeneous. Remove from heat and swiftly stir in the essential oil to ensure it's well distributed throughout the mixture. Quickly pour the liquid balm into small lip balm tubes or containers. Allow them to cool and solidify at room temperature, away from direct sunlight. This process should take about an hour, after which the lip balm will be ready to use.

By following these steps and using these recommended proportions, you can create a lip balm that not only hydrates and protects but also pleases the senses with its natural fragrance.

Homemade Soap with Beeswax and Hive Products

Moving forward with the utilization of hive products in homemade cosmetics, the crafting of soap enriched with beeswax and other bee-derived ingredients stands out as a rewarding endeavor. The process not only yields a product that's gentle and nourishing for the skin but also capitalizes on the natural properties of these ingredients. Beeswax, in particular, contributes to the hardness and longevity of the soap bars, while honey acts as a natural humectant, and propolis offers its antimicrobial benefits.

For a basic beeswax soap recipe, you'll need the following materials and ingredients: a digital scale for precise measurements, a stainless steel pot for melting the oils and beeswax, a glass or high-quality plastic pitcher for mixing the lye solution, safety gear including gloves and goggles, a stick blender for emulsification, silicone soap molds for shaping your bars, and a thermometer to monitor temperatures.

Ingredients:

- 10 ounces of olive oil, for its gentle, moisturizing properties.
- 6 ounces of coconut oil, to provide cleansing action and lather.
- 4 ounces of sustainable palm oil, for hardness and stable lather.
- 1 ounce of beeswax, to increase bar hardness and add a protective layer to the skin.
- 4 ounces of water, distilled is preferred to ensure purity.

- 2 ounces of lye (sodium hydroxide), the saponifying agent.
- 1 tablespoon of raw honey, for its humectant properties.
- Optional: Essential oils of your choice for fragrance, 1 to 2 ounces depending on preference.

Procedure:

1. Prepare your work area ensuring it's clean and free from any distractions. Wear your safety gear before handling any materials, especially when working with lye.

2. Measure the olive, coconut, and palm oils using the digital scale and combine them in the stainless steel pot. Add the beeswax to the oils. Gently heat the mixture on low until the oils and beeswax are fully melted and combined, stirring occasionally. Use the thermometer to monitor the temperature, aiming for around 120°F (49°C).

3. In the glass pitcher, carefully measure and mix the water and lye, stirring until the lye is completely dissolved. The solution will heat up quickly, so handle with care. Allow it to cool to approximately 120°F (49°C) to match the oil mixture's temperature.

4. Once both the lye solution and the oil-wax mixture have reached the target temperature, slowly pour the lye solution into the oils, using the stick blender to mix until you reach a light trace. This means the mixture has begun to emulsify and slightly thicken, leaving a trace when drizzled over the surface.

5. At light trace, add the tablespoon of raw honey and any essential oils you've chosen. Continue to blend until fully incorporated.

6. Pour the mixture into your silicone molds, tapping them on the work surface to remove any air bubbles. Cover with plastic wrap and a towel to insulate.

7. Allow the soap to set in the molds for 24 to 48 hours until they're firm enough to unmold. Once unmolded, cure the soap bars on a rack in a cool, dry place for 4 to 6 weeks. This curing process ensures the saponification is complete and the soap is gentle for use.

This beeswax soap recipe highlights the simplicity of incorporating hive products into everyday skincare routines, offering a natural and effective alternative to commercial soaps. The addition of beeswax not only enhances the soap's texture and longevity but also imbues it with the protective qualities beeswax is known for. Similarly, honey and propolis enrich the soap with hydrating and antibacterial properties, making it suitable for a variety of skin types. Through this detailed approach, crafting homemade soap becomes an accessible and fulfilling project, allowing individuals to harness the benefits of bee products in a practical and enjoyable manner.

CHAPTER 43: DECORATIVE OBJECTS

Expanding the creative use of beeswax beyond cosmetics and candles, let's delve into the art of crafting decorative objects that not only beautify spaces but also serve as a testament to the versatility of hive products.

Beeswax, with its malleable yet sturdy nature, provides an excellent medium for creating intricate and lasting items ranging from sculptures to wall art and even functional decor like coasters and picture frames. The first step in this crafting journey involves selecting the right type of beeswax. For decorative objects, it's advisable to use pure, filtered beeswax which has a smooth texture and a pleasant, natural aroma. This ensures that the final product has a clean finish and can hold fine details without impurities.

Depending on the project, you may choose between beeswax in block form, which will need to be melted down, or pellets, which are easier to measure and melt. The melting of beeswax should be done using a double boiler system to maintain a controlled, even heat, preventing the wax from burning. Keep the temperature below 160°F to preserve the natural color and scent of the beeswax. For coloring beeswax, use specially formulated wax dyes or natural colorants like powdered herbs and spices. Integrate these colorants during the melting process, stirring gently to achieve an even hue throughout the wax. When creating beeswax sculptures, start by sketching your design on paper to serve as a blueprint. Once the beeswax is melted and colored as desired, pour it into silicone molds that match your design. Silicone molds offer flexibility and ease of demolding, making them ideal for detailed shapes. Allow the wax to cool and harden completely, which can take several hours depending on the size of the object. For added stability in larger pieces, consider embedding a wireframe within the mold before pouring the beeswax.

This technique provides internal support, ensuring that the sculpture maintains its form over time. Crafting beeswax wall art involves a slightly different approach. Begin with a wooden panel or canvas as the base. If using a canvas, apply a layer of beeswax to prime the surface, creating a smooth, waxen foundation for your artwork. Layer thin sheets of beeswax over the base, using a heat gun to gently melt and manipulate the wax into your desired pattern or image. Incorporate elements such as dried flowers, leaves, or even paper cutouts into the design by placing them on the wax and then sealing them with additional layers.

This method allows for the creation of textured, multidimensional art pieces that showcase the natural beauty of beeswax. For functional decor like coasters or picture frames, measure and cut the beeswax into the required shape and size. Use a mold for coasters to ensure uniformity and smooth edges. For picture frames, construct a base frame using cardboard or thin wood, then apply melted beeswax over it, layer by layer, until the desired thickness is achieved. Embellish these items with stamped patterns or embedded objects before the final layer hardens to add a personalized touch. In crafting with beeswax, always work in a well-ventilated area and wear protective gloves to handle hot wax safely. Remember, the beauty of working with beeswax lies in its natural variability and the unique character it brings to each crafted piece. Through experimentation with techniques and materials, you can explore the full potential of beeswax in decorative arts, creating items that are not only visually appealing but also imbued with the essence of nature.

Artistic Uses of Beeswax

Transitioning from functional to purely artistic, beeswax offers a unique medium for sculptural art, allowing the artist to explore both form and texture in their creations. To begin sculpting with beeswax, one first needs to source high-quality, pure beeswax, which can be found at beekeeping supplies or craft stores. The purity of the beeswax is crucial as it affects the malleability and the finish of the sculpture. For sculpting, beeswax can be softened using a controlled heat source, such as a heat gun or a water bath, allowing the artist to mold and shape the wax with precision. It is recommended to work on a silicone mat or a surface that can release the wax easily once it cools down.

Tools for beeswax sculpting can range from professional sculpting tools to everyday items such as spoons, knives, or even dental tools. The choice of tool depends on the detail required in the sculpture. For intricate designs, fine-tipped tools are preferable as they allow for detailed work without applying too much pressure, which could distort the beeswax. As beeswax is relatively soft compared to other sculpting materials, it is conducive to adding texture and fine details by pressing objects into the surface or by carving.

Coloring beeswax sculptures can be achieved by mixing pigmented beeswax, using either natural pigments or commercially available beeswax dyes. The pigments should be melted together with the beeswax to ensure a uniform color throughout the sculpture. Alternatively, artists can paint the finished sculpture using oil paints, which adhere well to the beeswax surface and offer a wide range of color options.

To preserve the integrity of the beeswax sculpture, it is essential to consider its placement, avoiding direct sunlight or heat sources, which could cause the beeswax to soften or melt. For added protection, a clear sealant compatible with beeswax can be applied, providing a barrier against dust and handling.

Incorporating mixed media elements such as metal, wood, or fabric into beeswax sculptures can add another dimension to the artwork. These materials can be embedded into the beeswax or attached using beeswax as an adhesive, creating a cohesive piece that highlights the versatility and natural beauty of beeswax.

Making Reusable Food Wraps

Making reusable food wraps with beeswax is a sustainable way to reduce plastic use in the kitchen while utilizing natural hive products. Begin by selecting 100% cotton fabric in your desired pattern, ensuring it is pre-washed to remove any residues that might interfere with the waxing process. Cut the fabric into various sizes depending on your needs, with common dimensions being 8x8 inches for small wraps and 12x12 inches for larger items.

To prepare the beeswax coating, you will need pure beeswax, either in pellet form or grated from a block, pine resin for added stickiness, and jojoba oil to enhance flexibility. A recommended ratio is 2 ounces of beeswax, 1.5 tablespoons of pine resin, and 1.5 tablespoons of jojoba oil for a batch that can cover approximately four medium-sized fabric pieces.

Use a double boiler to melt the beeswax, pine resin, and jojoba oil together, stirring continuously until the mixture is smooth and fully integrated. The melting process should be done slowly over low heat to prevent the wax from burning, maintaining a temperature just enough to keep the mixture liquid.

While the wax mixture is melting, preheat your oven to 200°F. Line a baking sheet with parchment paper and place one piece of cut fabric on it. Once the beeswax mixture is ready, use a dedicated paintbrush to spread the mixture evenly over the fabric. Ensure the entire surface is covered, including the edges.

Place the baking sheet in the oven for a few minutes, just until the beeswax has fully melted into the fabric, making it look saturated and slightly darker in color. Remove the baking sheet from the oven and lift the fabric using tongs. Hold it in the air for a few seconds to allow the excess wax to drip off and the fabric to cool and solidify.

Once cooled, your beeswax wrap is ready for use. It can be molded around containers or food items with the warmth of your hands. After use, wash the wrap in cool water with a mild dish soap, air dry, and store it away from heat sources. With proper care, beeswax wraps can last up to a year, after which they can be composted or used as natural fire starters, closing the loop on a sustainable kitchen practice.

Care and Storage of Decorative Items

To ensure the longevity of beeswax decorative items, it's crucial to understand and implement proper care and storage techniques. Beeswax, while durable, is susceptible to damage from high temperatures, direct sunlight, and improper handling. By following these guidelines, you can preserve the beauty and integrity of your beeswax creations for years to come.

Temperature and Sunlight: Store beeswax items in a cool, dry place away from direct sunlight. Exposure to heat can cause beeswax to soften and lose its shape, while sunlight can fade the colors and lead to a brittle texture. Ideal storage temperatures for beeswax objects are between 50°F and 70°F. If you're storing larger sculptures or functional items like coasters and picture frames, consider placing them in a cabinet or a storage box that shields them from temperature fluctuations and UV exposure.

Handling: When handling beeswax decorative pieces, ensure your hands are clean and dry to avoid transferring oils and dirt onto the wax surface. For items with intricate details or delicate parts, use soft gloves to minimize the risk of leaving fingerprints or causing damage.

Cleaning: To clean beeswax items, gently dust them with a soft, lint-free cloth. Avoid using water or cleaning agents, as moisture can penetrate the wax, leading to potential mold growth or deterioration. For deeper cleaning, a soft brush can be used to remove dust from crevices without scratching the wax.

Polishing: Over time, beeswax may develop a dull surface or bloom, a whitish film that can appear on natural beeswax. To restore the shine, lightly buff the surface with a microfiber cloth. If necessary, a small amount of mineral oil can be applied sparingly and buffed into the beeswax to enhance its natural luster. However, use oil very sparingly, as over-application can attract dust and dirt.

Repairs: Should your beeswax item suffer a crack or break, it can often be repaired by gently warming the broken edges with a hairdryer set on low heat and carefully pressing them together. Once cooled, the seam should be nearly invisible. For more significant damages, remelting and reshaping might be required, following the original crafting techniques.

By adhering to these care and storage recommendations, your beeswax decorative items will maintain their aesthetic appeal and structural integrity, serving as cherished possessions or gifts for many years.

CHAPTER 44: STARTING A SMALL CRAFT BUSINESS

Turning your passion for beekeeping and crafting with hive products into a small craft business can be a fulfilling and profitable venture. To ensure success, it's essential to focus on key aspects such as **product development**, **branding**, **marketing**, and **sales channels**. Here, we delve into these areas with detailed guidance on how to create a thriving craft business centered around bee-derived products.

Product Development: The first step is to select which products you will offer. Given the versatility of hive products, your options include beeswax candles, honey-based cosmetics, propolis tinctures, and more. For each product, conduct thorough research to understand customer needs and preferences. This might involve sampling different formulations of a beeswax lotion or testing various flavors of honey in small focus groups. Ensure that your products not only meet high-quality standards but also comply with any relevant regulations, particularly if you're creating consumables or skincare items. For instance, if you're making lip balm with propolis, verify that your production process adheres to the FDA's Good Manufacturing Practices (GMP) for cosmetics.

Branding: Your brand is what sets you apart in the marketplace. It encompasses your business name, logo, product packaging, and overall aesthetic. When developing your brand, consider what message you want to convey. Is it the sustainability of beekeeping, the health benefits of hive products, or perhaps the artisanal quality of your goods? Your branding should reflect this message consistently across all elements. For example, if sustainability is a key brand value, use eco-friendly packaging and highlight this in your marketing materials.

Marketing: Effective marketing is crucial for attracting customers to your craft business. Start by creating a strong online presence through a website and social media platforms like Instagram, Pinterest, and Facebook. These channels are ideal for showcasing your products, sharing behind-the-scenes glimpses of your crafting process, and educating your audience about the benefits of hive products. Additionally, consider email marketing to keep your customers informed about new products, promotions, and events. Engaging content that tells the story of your beekeeping journey and how it inspires your crafts can resonate well with your audience.

Sales Channels: Deciding where and how to sell your products is key to reaching your target market. Online marketplaces like Etsy and Amazon Handmade offer platforms specifically designed for crafts and handmade goods, making them excellent choices for reaching customers interested in artisanal products. Additionally, setting up your own e-commerce website gives you greater control over the customer experience and branding. Don't overlook local opportunities such as farmers' markets, craft fairs, and consignment in boutique stores. These venues can provide valuable direct feedback from customers and increase local awareness of your brand.

Legal Considerations: Before launching your business, familiarize yourself with any legal requirements, such as obtaining a business license, registering your business name, and understanding tax obligations. If you're selling products that could be ingested or applied to the skin, it's particularly important to be aware of labeling laws and product liability insurance.

Customer Service: Providing excellent customer service can set your craft business apart. This includes clear communication, prompt responses to inquiries, and addressing any issues or concerns customers may have. Encourage feedback and reviews to build trust with potential customers and improve your offerings.

By focusing on these critical areas, you can build a successful small craft business that leverages your passion for beekeeping and hive products. Remember, the key to success lies in offering high-quality products, establishing a strong brand, and effectively engaging with your customers.

Market Demand for Beeswax and Propolis Crafts

Assessing the market demand for beeswax and propolis-based crafts requires a strategic approach, focusing on consumer trends, niche markets, and the unique selling proposition of your products. Begin by conducting market research to understand the current demand for natural and sustainable products. This can be achieved through online surveys, analysis of consumer behavior on platforms like Etsy and Pinterest, and monitoring discussions on social media and relevant forums. Look for trends indicating an increased interest in eco-friendly, handmade items, and specifically, products derived from beekeeping.

Next, identify your target audience. Consider demographics such as age, gender, lifestyle, and interests. For instance, younger consumers might be drawn to the sustainability aspect of beeswax wraps, while older demographics might appreciate the health benefits of propolis-based skincare products. Understanding the needs and preferences of your target market allows you to tailor your product offerings and marketing strategies effectively.

Evaluate the competition by identifying other businesses offering similar products. Analyze their product range, pricing, branding, and customer feedback. This will help you identify gaps in the market that your products can fill, and also provide insight into what consumers value in beeswax and propolis crafts.

Develop a unique selling proposition (USP) that differentiates your products from those of competitors. This could be the local sourcing of your bee products, the handcrafted nature of your items, or a specific product line that addresses a particular need or interest. Highlighting these unique aspects in your marketing efforts will attract consumers looking for products that align with their values and interests.

Selling Crafts: Online, Markets, Direct Channels

Selling crafts online, at local markets, or through direct channels requires a strategic approach to maximize your reach and sales potential. For online sales, platforms like **Etsy** and **Amazon Handmade** are popular choices due to their large customer bases and focus on handmade goods. To list your products, take high-quality photos from multiple angles to accurately represent your crafts. Write detailed product descriptions that include materials used, dimensions, and care instructions. Utilize keywords related to beekeeping and hive products to improve search visibility. Setting competitive prices involves researching similar products on the platform, considering your material and time costs, and factoring in platform fees.

For local markets, such as **farmers' markets** or **craft fairs**, research events with high foot traffic and a match for your target audience. Prepare by creating an attractive display that tells the story of your beekeeping journey and showcases the natural beauty of your products. Have informational materials available, such as brochures or business cards, that explain the benefits of hive products and your sustainable practices. Engage with customers by sharing stories about your beekeeping experience and the crafting process, which can create a personal connection and encourage sales.

Direct sales channels can include selling through your own **website** or **social media platforms**. Building a website allows you to control the branding and customer experience fully. Include an e-commerce section with secure payment options and clear shipping and return policies. Use social media to build a community around your brand by posting engaging content regularly, including behind-the-scenes looks at your beekeeping and crafting processes, customer testimonials, and special promotions.

Regardless of the sales channel, excellent customer service is crucial. Respond promptly to inquiries, ship orders on time, and handle any issues professionally and courteously. Encouraging customers to leave reviews and feedback can build trust with potential buyers and provide valuable insights for improving your products and services.

Chapter 45: Sustainability in Crafting

Ensuring ethical and sustainable use of hive resources is paramount in crafting with beeswax and propolis. When harvesting these materials from your hives, prioritize methods that minimize waste and stress to the bee colony. For beeswax, consider using a solar wax melter, which utilizes the sun's energy to melt wax caps collected during honey extraction, avoiding the need for electrically powered melters. This method not only conserves energy but also preserves the quality of beeswax by preventing overheating, which can degrade its aromatic and binding properties.

When extracting propolis, use a propolis trap, a flexible sheet with perforations placed above the frames in the hive. Bees fill the perforations with propolis, and once the sheet is removed and chilled, the propolis can be easily extracted by flexing the sheet. This method is less invasive and allows for the collection of propolis without significant disturbance to the hive.

For crafting, select materials that complement the natural qualities of beeswax and propolis. For instance, when making candles, opt for 100% cotton wicks, which burn cleanly and are biodegradable, unlike metal-cored wicks. If coloring candles, choose natural dyes derived from plants or minerals, avoiding synthetic chemicals that release harmful substances when burned.

In creating beeswax wraps, pair the beeswax with organic cotton fabric, ensuring the fabric is GOTS (Global Organic Textile Standard) certified. This certification guarantees the cotton was grown without the use of pesticides, herbicides, or synthetic fertilizers, aligning with the sustainability ethos of beekeeping. The addition of pine resin and jojoba oil, both of which should be sourced from sustainable and organic suppliers, increases the wrap's durability and antimicrobial properties.

When packaging your crafts for sale, use materials that are either recyclable or compostable. Kraft paper, for instance, is an excellent choice for wrapping products, as it is both strong and fully biodegradable. For labeling, consider using soy-based inks on recycled paper, which offer an environmentally friendly alternative to petroleum-based inks on virgin paper.

Promoting your crafts should also reflect a commitment to sustainability. Digital marketing, through social media and email newsletters, reduces the need for physical promotional materials. When physical materials are necessary, such as business cards or brochures, opt for recycled paper and minimal designs that require less ink.

By adopting these practices, you not only ensure the sustainability of your crafting endeavors but also contribute to the broader movement towards environmental stewardship. Engaging with your customers about these practices can also inspire them to adopt more sustainable habits, creating a ripple effect that extends far beyond beekeeping and crafting.

Ethical and Sustainable Hive Resource Use

When considering the **ethical and sustainable use of hive resources**, it's crucial to adopt practices that prioritize the health and well-being of the bee colony. This means harvesting hive products in a manner that supports the colony's needs first, ensuring that the bees have sufficient stores of honey, pollen, and propolis for their own survival, especially during the winter months or periods of scarce forage.

For **honey extraction**, always leave enough honey in the hive to sustain the colony. A good rule of thumb is to leave at least 60 to 80 pounds of honey in a standard Langstroth hive going into winter, which may vary based on your local climate and the length of the winter season. Use a gentle method for extracting honey, such as a centrifugal extractor, which allows you to harvest honey without destroying the comb, so bees can reuse it, saving them significant energy and resources.

When collecting **beeswax**, opt for a solar wax melter as previously mentioned, or consider a low-temperature, water bath method that gently melts the wax while preserving its quality. This method involves placing wax cappings or old combs in a mesh bag, submerging it in warm water, and allowing the wax to melt and float to the surface, where it can be collected without the use of high heat that might degrade its properties.

For **propolis extraction**, besides using a propolis trap, ensure that you only harvest propolis during times of the year when bees are actively collecting it. This is typically during warmer months when trees and plants are secreting resins. Avoid overharvesting, which can stress the bees, as propolis plays a critical role in sealing cracks and sanitizing the hive.

In crafting with these hive products, always measure and use them judiciously to minimize waste. For example, when making beeswax candles or wraps, calculate the exact amount of beeswax needed for your project to avoid unnecessary melting of excess wax. If you do have leftovers, consider how they can be repurposed or recycled in other products, such as using small beeswax remnants to make wax melts or combining them with other ingredients to create salves or balms.

Sourcing additional materials for your crafts should also reflect a commitment to sustainability. For instance, when choosing fabrics for beeswax wraps, beyond selecting organic cotton, consider the source of the fabric. Look for suppliers who use ethical labor practices and sustainable farming techniques. Similarly, when purchasing pine resin or jojoba oil, seek out vendors who harvest these materials responsibly and sustainably.

Packaging and shipping your crafted items offer another opportunity to practice sustainability. Use recyclable or compostable packaging materials and minimize the use of plastics. For shipping, choose carriers that offer carbon-neutral shipping options or invest in carbon offsets to mitigate the environmental impact of transporting your goods.

By implementing these practices, you not only ensure the ethical and sustainable use of hive resources but also contribute to a larger ecosystem of sustainability that benefits both the environment and the communities involved in the production and consumption of these valuable bee products.

Eco-Friendly Crafting Materials

When selecting **eco-friendly and biodegradable materials** for crafting with hive products, it's essential to consider the environmental impact of each component used. For instance, when creating **beeswax wraps**, the choice of fabric is as crucial as the beeswax itself. Opt for **organic cotton muslin** or **hemp fabric**, which are not only sustainable but also ensure the fabric is breathable and has a texture conducive to the beeswax adhering properly. These fabrics should be pre-washed without the use of synthetic chemicals to maintain the organic integrity of the wrap.

For the beeswax mixture, blending with **organic jojoba oil** and **tree resin** enhances the wrap's flexibility and stickiness. The jojoba oil, being structurally similar to beeswax, ensures the mixture remains pliable without cracking, while the tree resin, preferably **pine resin**, adds tackiness, allowing the wrap to cling effectively. When sourcing pine resin, look for suppliers who harvest it sustainably, ensuring no harm to the trees or surrounding ecosystem.

In the realm of **candle making**, the type of wick used significantly affects the sustainability of the candle. **Pre-tabbed organic cotton wicks** are preferable, ensuring they are lead and zinc-free, which not only benefits the environment but also ensures a cleaner burn, reducing indoor air pollution. When considering dyes for coloring candles, natural dyes can be derived from plants, such as **beetroot for red** or **turmeric for yellow**, offering a palette of colors without resorting to synthetic alternatives.

For **packaging and presentation**, materials like **biodegradable cellophane** or **recycled paper** boxes can be used. Cellophane, made from cellulose derived from plants, provides an excellent alternative to petroleum-based plastics, offering a similar level of clarity and protection for your products while being fully compostable. When printing labels or business cards, opt for **recycled paper** and **soy or vegetable-based inks**, which offer lower toxicity and are more environmentally friendly than their petroleum-based counterparts.

In crafting **decorative objects** or **furniture finishes** with beeswax, combining it with natural oils such as **linseed or tung oil** can create durable and eco-friendly wood finishes. These oils penetrate deep into the wood, providing protection and enhancing the natural beauty of the wood grain, while the beeswax seals the surface, offering a smooth, water-resistant finish. Both linseed and tung oil are derived from natural, renewable resources, aligning with sustainable crafting practices.

For adhesives or sealants in crafting, consider making a **natural glue** from beeswax, pine resin, and a small amount of oil, which can be used in various applications from woodworking to paper crafts. This adhesive not only reduces reliance on synthetic glues but also utilizes the inherent properties of beeswax and resin for a strong, durable bond.

In summary, every choice made in the selection of materials for crafting with hive products can contribute to a more sustainable practice. By prioritizing eco-friendly and biodegradable options, crafters can minimize their environmental impact while showcasing the natural beauty and versatility of beeswax, propolis, and other hive-derived materials.

Part 11: Health and Wellness with Bee Products

CHAPTER 46: MEDICINAL PROPERTIES OF HONEY

Honey's role in **wound healing** is not only traditional but also backed by scientific research, showcasing its effectiveness as a natural antiseptic. The high sugar content of honey creates an osmotic effect, which helps to draw moisture out from the wound, inhibiting the growth of bacteria and fungi. When applying honey to a wound, it's crucial to use raw, unpasteurized honey to ensure that the beneficial enzymes and properties are intact. A thin layer of honey can be directly applied to the wound or on a dressing that is then placed on the wound, changing it every 24 to 48 hours. For deeper or more serious wounds, it's advisable to consult a healthcare professional before applying honey as a treatment.

In the realm of **allergy relief**, honey acts as a natural remedy, particularly for seasonal allergies. The theory suggests that consuming local honey may help the body acclimate to the pollen present in the environment, potentially reducing allergic reactions. For this purpose, consuming a tablespoon of local, raw honey daily, several months before the allergy season starts, can help the body build a tolerance. It's important to source honey that is local to your area to ensure that it contains the same pollen types that may trigger your allergies.

Immunity boosting properties of honey are attributed to its antioxidant and antibacterial components. Incorporating honey into your daily diet can enhance your body's natural defenses. A simple way to do this is by adding honey to warm water or tea each morning, which can help soothe the throat, boost the immune system, and start the day with a healthy dose of antioxidants. Combining honey with other immune-boosting ingredients like lemon, ginger, or turmeric can further enhance its benefits.

For **digestive health**, honey can act as a mild laxative and has anti-inflammatory properties that may soothe the digestive tract. Consuming honey with ginger or lemon in warm water can aid in digestion and alleviate symptoms of indigestion. It's also beneficial for individuals suffering from ulcers, as honey can help to reduce inflammation and irritation in the stomach lining.

When using honey for **health and wellness**, it's essential to choose high-quality, raw honey, as processing and pasteurization can diminish its medicinal properties. Always start with small quantities to ensure that you do not have an allergic reaction, especially if you are new to consuming honey. For those with diabetes or blood sugar concerns, consult with a healthcare provider before incorporating honey into your diet, as it can affect blood sugar levels.

Honey's versatility extends beyond its culinary uses, serving as a potent ingredient in natural remedies for a variety of ailments. Its multifaceted medicinal properties, from antimicrobial to anti-inflammatory effects, underscore honey's valuable role in both traditional and modern medicine. By integrating honey into your health regimen, you can harness its natural benefits to enhance wellness and support healing processes.

Honey for Wounds and Burns

Honey's antibacterial properties are primarily due to the presence of hydrogen peroxide, naturally produced by the enzyme glucose oxidase, which bees add to nectar. This compound is effective against a wide range of pathogens, including those resistant to traditional antibiotics. When selecting honey for medicinal use, opt for varieties that have been tested for their antibacterial strength, often labeled as "medical-grade" honey. Manuka honey, for instance, is renowned for its potent antibacterial properties, attributed to its high methylglyoxal content. However, for those unable to access Manuka, local raw honeys have also demonstrated significant antibacterial effects and can be a more accessible option.

To utilize honey for wound care, ensure the wound has been properly cleaned with sterile saline solution. Applying honey directly to the wound should be done with sterile gloves or a sterile spatula to prevent contamination. The amount of honey used depends on the size and depth of the wound; generally, a layer about 1/8 to 1/4 inch thick is sufficient. Cover the honey-applied area with a sterile dressing, which could be a simple gauze pad or a more advanced wound dressing designed to maintain a moist environment. Change this dressing daily, or more frequently if the wound is heavily exuding.

For burns, the cooling effect of honey can provide immediate relief, but its application should follow the initial step of cooling the burn under running tap water for at least 20 minutes. After ensuring the burn area is clean, apply a thin layer of honey before covering with a sterile dressing. The dressing should be changed at least once a day, monitoring the burn for signs of infection or delayed healing, which necessitates medical attention.

Incorporating honey into skincare for burns and wounds, beyond its immediate application, can also include using honey-infused products designed for sensitive skin. These can help maintain the skin's moisture balance and support the healing process without the stickiness of direct honey application. However, the primary treatment should still be the direct application of honey to the affected area.

The efficacy of honey in wound and burn care is not only a testament to its antimicrobial properties but also to its anti-inflammatory and antioxidant capabilities. These properties help reduce swelling, pain, and scarring, promoting a conducive environment for tissue regeneration and healing. Moreover, honey's high viscosity forms a protective barrier, preventing wound contamination and further infection.

It's crucial to monitor the healing process closely. Any signs of infection, such as increased redness, warmth, pain, or pus, should prompt immediate medical consultation. While honey is a powerful natural remedy, it is not a substitute for professional medical treatment in the case of severe or infected wounds and burns.

The use of honey in medicinal contexts underscores the importance of preserving bee populations and their natural habitats. Sustainable beekeeping practices not only ensure the availability of this valuable resource but also contribute to the health of our ecosystems. As we benefit from the healing properties of honey, it's a poignant reminder of the interconnectedness of human health and environmental stewardship.

Boosting Immunity with Hive Products

Honey's role in boosting immunity is multifaceted, leveraging its antioxidant, antibacterial, and anti-inflammatory properties to fortify the body's natural defenses. The antioxidants in honey, including flavonoids and phenolic compounds, play a critical role in protecting the body against free radicals, molecules that can cause cell damage and contribute to the development of chronic diseases. By neutralizing these harmful free radicals, honey helps maintain a healthy immune system capable of warding off infections.

To effectively incorporate honey into an immunity-boosting regimen, consider starting with a daily intake of one to two tablespoons of raw, unprocessed honey. This can be consumed directly or dissolved in warm water, herbal tea, or even spread on whole-grain toast. It's important to ensure that the honey is raw and unprocessed, as heat and filtration can reduce its beneficial properties. For an enhanced effect, combine honey with other immune-supporting ingredients such as lemon juice, which provides additional vitamin C, or ginger, known for its anti-inflammatory effects.

The antibacterial properties of honey are attributed to its hydrogen peroxide content, produced by the enzyme glucose oxidase. This enzyme, introduced by bees, acts as a natural preservative and inhibits the growth of harmful bacteria. Regular consumption of honey can support the body's ability to fight off bacterial infections, reducing the burden on the immune system.

In the context of combating seasonal allergies, honey acts as a natural antihistamine. The theory behind honey's effectiveness for allergy relief is related to its pollen content. By consuming honey produced locally, individuals are exposed to the same types of pollen that often trigger allergic reactions. Over time, this can help desensitize the body to these allergens, potentially reducing the severity of allergic responses. To maximize benefits, start consuming local honey several months before allergy season begins, gradually increasing intake to build up tolerance.

For individuals looking to harness honey's allergy-fighting potential, sourcing honey from local beekeepers is key. This ensures exposure to the same pollen types found in the surrounding environment. Begin with a small daily dose, such as a teaspoon, gradually increasing to a tablespoon or more, to allow the body to adjust and build up a tolerance to local pollen.

Moreover, incorporating honey into the diet can also support overall digestive health, an important aspect of a strong immune system. Honey's prebiotic properties can nourish beneficial gut bacteria, which play a crucial role in immune function. A healthy gut microbiome is associated with improved digestion, reduced inflammation, and a more responsive immune system.

When selecting honey for its health benefits, prioritize varieties that are not only raw and local but also produced sustainably. Sustainable beekeeping practices ensure the preservation of bee populations and their environments, which is essential for maintaining the availability of high-quality honey. Look for honey that is transparently sourced, with clear labeling about its origin and processing methods, to ensure you're getting a product that is not only beneficial for your health but also supportive of environmental conservation efforts.

Honey's Health Benefits: Scientific Research

The scientific community has dedicated considerable effort to unraveling the health benefits of honey, yielding results that support its use in various medical and dietary applications. One of the most compelling areas of research focuses on honey's antibacterial properties, particularly its effectiveness against drug-resistant bacteria strains. Studies have identified that the antimicrobial activity of honey is not solely due to its hydrogen peroxide content. Instead, factors such as its low pH, high sugar concentration, and the presence of methylglyoxal, especially in Manuka honey, play a significant role. These elements work synergistically to inhibit bacterial growth, making honey a potent agent against infections.

Research into honey's antioxidant capacity reveals that it contains a wide array of phytochemicals, including flavonoids and phenolic acids. These compounds are known for their ability to scavenge free radicals, thereby reducing oxidative stress in the body. Oxidative stress is implicated in the pathogenesis of numerous chronic conditions, such as cardiovascular diseases, diabetes, and cancer. Regular consumption of honey, particularly dark varieties which are richer in antioxidants, may contribute to the prevention of these diseases by enhancing the body's antioxidant defense system.

Honey's role in wound healing is another area that has garnered attention from the scientific community. Clinical trials have demonstrated that honey, when applied topically, can accelerate the healing process of chronic wounds and burns. This is attributed to its osmotic effect, which draws fluid away from the wound, its anti-inflammatory properties that reduce swelling and pain, and its ability to provide a protective barrier against external contaminants. Furthermore, honey stimulates the regeneration of tissue, promoting faster healing with minimal scarring.

In the realm of digestive health, research has explored honey's effect on various gastrointestinal disorders, including gastroesophageal reflux disease (GERD), gastritis, and peptic ulcers. Honey's anti-inflammatory properties can soothe the esophagus and stomach lining, reducing irritation and supporting healing. Additionally, its prebiotic components feed beneficial gut bacteria, which is crucial for maintaining a healthy microbiome and overall digestive wellness.

Honey's potential to improve cardiovascular health has also been investigated. Studies suggest that honey may have a positive effect on lipid profiles, including reducing total cholesterol, low-density lipoprotein (LDL) cholesterol, and triglycerides, while increasing high-density lipoprotein (HDL) cholesterol. This lipid-modulating effect, combined with its antioxidant properties, may help reduce the risk of heart disease.

For those incorporating honey into their health regimen, it's essential to consider the source and processing of the honey. Raw, unfiltered honey preserves the maximum level of beneficial enzymes, nutrients, and compounds. In contrast, processed honey may lose some of its therapeutic properties due to heat treatment and filtration. Therefore, selecting high-quality, raw honey from reputable sources is crucial to maximizing its health benefits.

The breadth of scientific research supporting the medicinal properties of honey underscores its value beyond just a sweetener. From its antimicrobial and anti-inflammatory effects to its antioxidant and tissue-regenerative capabilities, honey stands out as a multifunctional natural product with a wide range of applications in health and medicine. As research continues to evolve, it is likely that even more benefits of honey will be discovered, further solidifying its role in both traditional and modern therapeutic practices.

CHAPTER 47: PROPOLIS FOR HEALTH

Propolis, a resinous mixture produced by honeybees from substances collected from tree buds, sap flows, or other botanical sources, has been recognized for its medicinal properties for centuries. Bees use it to seal unwanted open spaces in the hive, but humans have found its value in health and wellness, particularly for its antibacterial, antiviral, anti-inflammatory, and antifungal properties. When considering propolis for health, it's essential to understand its composition, which varies depending on the geography and the specific trees and flowers the bees have accessed. Generally, propolis is rich in flavonoids, aromatic acids, esters, and phenolic compounds, which contribute to its potent therapeutic effects.

For those interested in incorporating propolis into their health regimen, it's crucial to start with sourcing. High-quality propolis can be obtained from reputable beekeepers or health food stores, ensuring it comes from a sustainable and chemical-free beekeeping environment. Propolis is available in various forms, including tinctures, capsules, powders, and raw chunks. Each form has its specific use and method of application or ingestion.

Tinctures, for example, are a popular way to consume propolis. They are made by dissolving propolis in alcohol or a water-alcohol mixture, which extracts its active components. A propolis tincture can be applied directly to the skin to heal wounds, burns, and acne due to its antibacterial and healing properties. For internal health benefits, such as boosting the immune system or treating sore throats, a few drops of propolis tincture can be added to water or tea. The recommended dosage varies, but starting with a low dose and gradually increasing it based on tolerance and effects is advisable.

Capsules and **powders** offer a convenient way to ingest propolis, especially for those who prefer not to taste it. These forms are beneficial for systemic effects, such as enhancing immune function or combating internal infections. The dosage should follow the manufacturer's recommendations, as concentrations can vary.

For those using **raw propolis**, it can be chewed directly or used to make a homemade tincture or extract. Chewing raw propolis allows for the slow release of its compounds but be prepared for its strong taste and sticky texture. To make a homemade tincture, finely chop the raw propolis and follow a similar process to commercial tinctures, adjusting the strength based on personal preference.

When applying propolis topically, it's important to test a small skin area first to ensure there's no allergic reaction. Despite its natural origin, some individuals may be sensitive or allergic to propolis or its components.

Incorporating propolis into daily health routines can offer numerous benefits, from improved oral health—thanks to its ability to fight bacteria and heal mouth ulcers—to enhanced respiratory health by soothing sore throats and combating infections. Its anti-inflammatory properties make it a natural remedy for various skin conditions, while its antiviral effects help in the prevention and treatment of colds, flu, and other viral infections.

For those exploring the health benefits of bee products, propolis stands out as a versatile and potent option. Its wide range of applications and benefits makes it a valuable addition to natural health practices. However, as with any supplement, it's essential to consult with a healthcare provider before adding propolis to your regimen, especially for individuals with allergies to bee products or those on medication, to avoid any potential interactions.

Propolis: Boosting Immunity and Wellness

Propolis's ability to strengthen the immune system and promote wellness lies in its complex chemical composition, which includes over 300 compounds such as flavonoids, phenolic acids, and esters. These compounds are known for their powerful **antioxidant** properties, which play a crucial role in protecting the body's cells from oxidative stress and free radical damage. Oxidative stress is a condition that can lead to chronic inflammation, a root cause of many diseases, including heart disease, arthritis, and certain cancers. By neutralizing free radicals, propolis helps reduce inflammation and supports the body's natural defense systems.

The **flavonoids** in propolis are particularly effective in modulating the immune system. They can enhance the body's response to pathogens by stimulating the production of immune cells and regulating their activity. This modulation helps the body efficiently respond to and eliminate harmful bacteria, viruses, and other foreign invaders. To leverage the immune-modulating benefits of propolis, consider incorporating propolis supplements into your daily regimen. Start with a propolis tincture by adding a few drops to water or tea. Begin with a low dose to assess tolerance and gradually increase according to the product's guidelines or a healthcare provider's recommendations.

Propolis also exhibits **antibacterial, antiviral, and antifungal properties**, making it a broad-spectrum antimicrobial agent. This is particularly beneficial for oral health, as propolis can be applied to the gums or teeth to combat bacterial infections and prevent cavities. For oral health applications, a propolis mouthwash can be made by diluting propolis tincture in water. Use this rinse once or twice daily, swishing it around the mouth for one to two minutes before spitting it out.

For respiratory health, propolis acts as a natural anti-inflammatory agent, soothing irritated throat tissues and promoting the healing of ulcers and small wounds within the mouth and throat. This makes it an effective remedy for sore throats, laryngitis, and other respiratory conditions. To soothe a sore throat, mix a propolis tincture with warm water and gargle twice daily. This can help reduce inflammation and fight infection-causing bacteria.

The **antifungal properties** of propolis are beneficial for skin health, offering a natural treatment option for conditions like athlete's foot, eczema, and psoriasis. Propolis ointment or cream can be applied directly to affected areas to reduce inflammation and eliminate fungal infections. For skin application, ensure the product is designed for topical use and follow the instructions for application frequency and amount.

In terms of **wellness**, regular consumption of propolis can contribute to overall health by supporting the immune system, reducing inflammation, and protecting against oxidative stress. This, in turn, can lead to increased energy levels, improved mood, and a reduced risk of chronic diseases. For general wellness, incorporating propolis into your daily health routine through tinctures, capsules, or powders can provide a simple yet effective way to harness its benefits. Always start with the lowest recommended dose and consult with a healthcare professional before beginning any new supplement, especially if you have existing health conditions or are taking other medications.

Lastly, the adaptogenic properties of propolis help the body manage stress more effectively. Stress can suppress the immune system and make the body more susceptible to illness. By modulating the body's stress response, propolis can help maintain immune function and protect against the negative health impacts of chronic stress. To utilize propolis for stress management, consider taking a daily supplement in capsule or tincture form, following the dosing recommendations provided on the product or by a healthcare provider.

Using Propolis in Daily Health Routines

Incorporating propolis into daily health routines can be a transformative practice, offering a multitude of benefits derived from its natural, potent properties. To effectively leverage these benefits, understanding the specific applications and methods for using propolis is crucial. For skin care, propolis can be a game-changer. Its antibacterial and healing properties make it an excellent choice for acne treatment. A propolis serum, applied directly to clean skin, can help reduce inflammation and prevent future breakouts. For making a homemade propolis serum, mix a propolis extract with a carrier oil like jojoba or sweet almond oil at a ratio of 1:4. This dilution ensures the serum is potent yet gentle enough for daily use. Apply 2-3 drops of this mixture to the face after cleansing, focusing on areas prone to acne.

For oral health, propolis has shown remarkable efficacy in reducing bacterial growth, which can lead to cavities and gum disease. A daily rinse with a propolis mouthwash can fortify oral hygiene practices. To create a homemade propolis mouthwash, dissolve 10-15 drops of propolis tincture in a cup of warm water. Use this solution to rinse the mouth thoroughly for 30 seconds, twice a day, after brushing. This not only helps in reducing the bacterial load but also in healing minor mouth ulcers and sores.

Propolis's antiviral properties make it a valuable ally during cold and flu season. To harness these benefits, a daily intake of propolis can be considered. Adding 5-10 drops of propolis tincture to a morning tea or smoothie can help boost the immune system, potentially warding off viruses. It's important to start with a lower dose to gauge the body's response, gradually increasing it to the desired amount, without exceeding the recommended dosage provided on the product label or by a healthcare professional.

For those dealing with minor cuts, scrapes, or burns, propolis can accelerate the healing process. A propolis salve or balm, applied to the affected area, can act as a natural antiseptic, promoting healing while reducing the risk of infection. To prepare a simple propolis salve at home, melt 2 tablespoons of beeswax in a double boiler, then mix in 1 tablespoon of propolis extract and 4 tablespoons of coconut oil until well combined. Pour the mixture into a small container and let it solidify. Apply a small amount to the wound as needed, covering it with a bandage if necessary.

For respiratory support, especially for those prone to sore throats or laryngitis, propolis can provide soothing relief. A propolis spray, targeted for throat use, can be created by diluting propolis tincture with water in a spray bottle, using a ratio of 1:2. Spraying this solution directly to the throat area 2-3 times a day can help alleviate discomfort and inflammation.

Integrating propolis into daily health routines offers a holistic approach to wellness, leveraging the natural efficacy of this bee product. Whether used topically, ingested, or as part of oral care, propolis provides a versatile solution to various health concerns. However, it's essential to source high-quality, pure propolis from reputable suppliers to ensure the best results. Additionally, individuals with allergies to bee products should proceed with caution and consult a healthcare provider before incorporating propolis into their regimen. With these practices, propolis can become a valuable component of daily health care, contributing to improved well-being and natural resilience against common health issues.

Propolis in Respiratory Health and Infection Control

Propolis's effectiveness in respiratory health and combating infections is rooted in its unique composition, which includes bioflavonoids, aromatic acids, and esters, each contributing to its antimicrobial and anti-inflammatory properties. These components work synergistically to combat respiratory infections by inhibiting the growth of bacteria, viruses, and fungi that can cause illnesses ranging from the common cold to more severe bronchial infections.

For individuals seeking to utilize propolis for respiratory health, the application method can significantly influence its effectiveness. Inhalation of propolis vapor, for instance, delivers the antimicrobial agents directly to the respiratory tract, where they exert their therapeutic effects most efficiently. This can be achieved by adding propolis extract to hot water and inhaling the steam. Care should be taken to maintain a safe distance to avoid steam burns, and inhalation should last for about 5 to 10 minutes, covering the head and bowl with a towel to contain the vapor.

Another effective method involves the use of propolis tinctures in a saline nasal spray. Mixing a propolis tincture with saline solution at a ratio of 1:4 creates a potent nasal spray that can help clear nasal passages and combat pathogens. This mixture can be used two to three times daily, especially during the onset of symptoms or for maintenance during peak cold and flu seasons. The saline helps to moisturize nasal passages while propolis works to reduce inflammation and fight infection.

For those suffering from sore throats or laryngitis, gargling with a propolis solution can provide immediate relief. Dissolve propolis extract in warm water (approximately 10 drops of propolis to half a cup of water) and use this solution to gargle twice a day. This direct application allows propolis to come in contact with the throat's mucous membranes, offering anti-inflammatory and antimicrobial benefits, reducing pain, and speeding up recovery.

It's important to note that while propolis is a powerful natural remedy, its efficacy can be influenced by the quality of the propolis used. High-quality, ethically sourced propolis should be sought, ensuring it is free from contaminants and has been harvested sustainably. When purchasing propolis products, look for reputable suppliers who provide detailed information about the source and composition of their propolis, as well as instructions for safe and effective use.

Moreover, individuals considering propolis for respiratory health should be aware of potential allergies. As a bee product, propolis can cause reactions in individuals allergic to bee stings or other bee products. A patch test or starting with small doses can help mitigate the risk of allergic reactions. Consulting with a healthcare professional before beginning any new treatment, especially for individuals with pre-existing health conditions or those taking other medications, is always advisable to ensure safety and compatibility.

In the context of respiratory health, the integration of propolis into daily health routines offers a multifaceted approach to prevention and treatment. Its natural antiviral, antibacterial, and antifungal properties make it a valuable ally against a wide range of respiratory ailments, providing a complementary or alternative option for those seeking holistic remedies. Through careful application and consideration of individual health needs and potential allergies, propolis can be a safe and effective component of respiratory health management and overall wellness strategies.

CHAPTER 48: ROYAL JELLY AND VITALITY

Royal jelly, a milky secretion produced by worker honeybees, is renowned for its potential to promote vitality and enhance overall health. This substance serves as the exclusive food for the queen bee throughout her life, contributing to her extended lifespan and fertility compared to other bees. The unique composition of royal jelly includes water, proteins, sugars, fats, vitamins, minerals, and trace elements, making it a dense source of nutrients. Among its components, 10-Hydroxy-2-decenoic acid (10-HDA) is a fatty acid unique to royal jelly, believed to be a primary source of its health benefits.

For individuals looking to incorporate royal jelly into their diet for its health-promoting properties, it's available in various forms including fresh royal jelly, capsules, and powders. Fresh royal jelly is the most potent form but requires refrigeration and has a limited shelf life. Capsules and powders offer more convenience and a longer shelf life, making them suitable for daily supplementation. When starting with royal jelly supplementation, it's advisable to begin with a small dose to assess tolerance, as some individuals may experience allergic reactions. The recommended starting dose can vary, but a general guideline is to consume approximately 300 to 500 mg of royal jelly daily, gradually increasing as tolerated.

Incorporating royal jelly into one's diet can be done in several ways. For those using fresh royal jelly, it can be taken directly or mixed into a smoothie or yogurt to mask its strong taste. Capsules and powders can be taken according to the dosage instructions on the packaging, usually with water or another liquid. Consistency is key when supplementing with royal jelly, as its benefits are best observed with regular, long-term use.

The nutritional profile of royal jelly includes B-complex vitamins such as pantothenic acid (vitamin B5) and pyridoxine (vitamin B6), which are essential for energy metabolism and overall vitality. It also contains trace minerals like zinc and selenium, which support immune function and antioxidant defenses. The amino acids present in royal jelly, including all eight essential amino acids, contribute to its protein content, supporting tissue repair and growth.

Research into the health benefits of royal jelly has highlighted its potential in several areas, including immune system support, anti-inflammatory effects, and improved cognitive function. Its antimicrobial properties also make it beneficial for skin health when applied topically as part of a skincare routine. For topical use, royal jelly can be found in various cosmetic products or mixed with a carrier oil for a homemade facial serum.

Athletes and individuals with active lifestyles may find royal jelly particularly beneficial for enhancing physical performance and recovery. The natural sugars and proteins in royal jelly provide a source of energy and nutrients essential for muscle repair. Additionally, the anti-inflammatory properties of royal jelly can help reduce muscle soreness and accelerate recovery after intense physical activity.

For those interested in the anti-aging properties of royal jelly, its antioxidants are crucial in combating oxidative stress, one of the primary causes of aging at the cellular level. Regular supplementation with royal jelly can contribute to healthier skin, reduced signs of aging, and improved overall vitality.

When purchasing royal jelly, it's important to choose products from reputable suppliers that provide detailed information about the source and processing methods. High-quality royal jelly products should be free from additives and contaminants, ensuring the best possible health benefits. As with any dietary supplement, consulting with a healthcare professional before starting royal jelly is recommended, especially for individuals with existing health conditions or allergies to bee products.

Royal Jelly: Nutritional Energy Booster

Royal jelly's reputation as an **energy booster** is not just folklore; it's grounded in its unique nutritional profile. This creamy substance is rich in **vitamins**, **minerals**, and **antioxidants**, which collectively contribute to its energizing properties. Understanding how to harness these benefits requires a closer look at its components and how they interact with the body's systems.

One of the key elements in royal jelly that supports energy levels is the spectrum of **B vitamins** it contains. These vitamins are critical for converting food into energy, a process essential for maintaining vitality. For instance, **vitamin B5 (pantothenic acid)** is known for its role in synthesizing coenzyme A, which is involved in the metabolic pathways that release energy from fats, carbohydrates, and proteins. Incorporating royal jelly into your diet can thus help ensure that your body has the necessary components for efficient energy production.

Moreover, royal jelly contains a unique fatty acid known as **10-Hydroxy-2-decenoic acid (10-HDA)**, exclusive to this substance and believed to play a role in enhancing energy and physical performance. While the exact mechanisms of how 10-HDA influences energy levels are still being studied, its presence in royal jelly is thought to contribute to improved stamina and fatigue resistance.

To effectively incorporate royal jelly as an energy booster, consider starting with a daily intake of **300 to 500 mg** in capsule or powder form. This dosage can provide a concentrated source of its energizing nutrients without the need for large quantities of fresh royal jelly, which has a more potent flavor and shorter shelf life. For those who prefer the fresh variety, mixing a quarter teaspoon of royal jelly with a bit of honey or blending it into a morning smoothie can mask its strong taste while delivering its benefits.

Athletes and individuals with demanding lifestyles might find royal jelly particularly beneficial for sustaining energy levels throughout the day. Its combination of **natural sugars**, **proteins**, and **bioactive compounds** not only supports immediate energy needs but also contributes to endurance and recovery. The anti-inflammatory properties of royal jelly can further aid in reducing post-exercise muscle soreness, allowing for more consistent training and activity levels.

For optimal absorption and efficacy, royal jelly should be consumed on an empty stomach, preferably in the morning. This timing ensures that the nutrients are efficiently absorbed and utilized by the body, providing a natural energy boost to start the day. Additionally, maintaining a regular supplementation schedule can enhance the cumulative benefits of royal jelly, including its supportive effects on energy metabolism and physical performance.

When selecting royal jelly products, it's crucial to opt for those from **reputable sources** that offer pure, high-quality formulations. Look for products with clear labeling on dosage and ingredients to ensure you are receiving an effective and safe supplement. As with any dietary supplement, individuals with allergies to bee products or those with pre-existing health conditions should consult a healthcare professional before adding royal jelly to their regimen.

By understanding the nutritional benefits of royal jelly and incorporating it into daily health routines, individuals can leverage this natural bee product as an effective energy booster. Its rich composition not only supports immediate energy needs but also contributes to long-term vitality and wellness, making it a valuable addition to a balanced lifestyle.

Incorporating Royal Jelly into Diets and Skincare

Incorporating royal jelly into daily diets and skincare routines offers a multifaceted approach to harnessing its health and wellness benefits. For dietary incorporation, one effective method is to blend royal jelly into morning beverages. A practical recipe involves adding a quarter teaspoon of royal jelly to a smoothie composed of half a cup of blueberries, one banana, a tablespoon of ground flaxseed, and a cup of almond milk. This combination not only masks the distinctive taste of royal jelly but also ensures a nutrient-rich start to the day, leveraging the antioxidant properties of blueberries and the omega-3 fatty acids from flaxseed.

Another dietary application is integrating royal jelly into oatmeal or cereal. Warm a bowl of oatmeal and stir in up to half a teaspoon of royal jelly until it dissolves completely. Enhance the flavor and nutritional value by topping with sliced almonds and a drizzle of honey. This method provides a hearty, energizing breakfast option, with the almonds adding a crunch and a source of healthy fats.

For individuals looking to incorporate royal jelly into their skincare routine, creating a homemade facial mask offers a direct application method. Combine one teaspoon of royal jelly with one tablespoon of organic yogurt and a teaspoon of raw honey in a clean bowl. Apply this mixture to the face and neck, leaving it on for 15-20 minutes before rinsing with warm water. This mask harnesses the antimicrobial and skin-soothing properties of honey, the probiotics from yogurt, and the nourishing effects of royal jelly, making it an excellent treatment for promoting skin hydration and elasticity.

Royal jelly can also be used as a component in a natural lip balm formulation. Melt two tablespoons of beeswax over a double boiler, then mix in two tablespoons of coconut oil and a teaspoon of royal jelly until the mixture is well combined. Pour into small containers and allow to cool completely. This lip balm utilizes the moisturizing benefits of coconut oil, the protective barrier of beeswax, and the healing properties of royal jelly to help soothe and repair chapped lips.

For those interested in incorporating royal jelly into their daily health routines, it's crucial to source this product from reputable suppliers known for their sustainable and ethical practices. This ensures the royal jelly is of high quality, free from contaminants, and produced in a manner that supports bee health and biodiversity. Always store royal jelly according to the manufacturer's instructions, typically in a cool, dark place or refrigerated to preserve its potency.

When beginning to use royal jelly, especially for dietary purposes, start with small amounts to monitor for any adverse reactions, as some individuals may have sensitivities to bee products. Gradually increasing the intake or usage allows the body to adapt and minimizes the risk of allergic reactions. Regular, consistent use is key to observing the potential health and wellness benefits of royal jelly, from improved energy and vitality to enhanced skin health.

Myths and Realities of Royal Jelly Properties

Royal jelly has been surrounded by a halo of mystery and attributed almost magical properties in various cultures, leading to a mix of myths and scientific realities that warrant clarification. For instance, royal jelly is often touted as a **panacea** for a wide range of health issues, from reversing aging to boosting fertility. While royal jelly does contain a unique blend of beneficial nutrients, it's crucial to dissect these claims with a scientific lens to understand what it can genuinely offer.

Myth: Royal Jelly Can Cure Any Disease

One common misconception is that royal jelly can cure a multitude of diseases, including chronic conditions like cancer. While royal jelly possesses **antioxidant** and **anti-inflammatory** properties, there is no scientific evidence to support the claim that it can cure diseases. Its beneficial components can support overall health and may complement conventional treatments, but it should not be seen as a standalone cure.

Reality: Supports Immune System and Reduces Inflammation

Research has shown that royal jelly can enhance the immune system's function and has anti-inflammatory effects. These properties stem from its bioactive compounds, such as 10-HDA and flavonoids, which can help the body fight against certain pathogens and reduce inflammation, potentially alleviating symptoms of conditions like arthritis.

Myth: Enhances Fertility to Miraculous Levels

Another widespread belief is that royal jelly significantly enhances fertility in both men and women. While some studies suggest that royal jelly can improve sperm quality and ovarian function due to its rich nutritional content, it's not a guaranteed fertility booster for everyone. Fertility can be influenced by numerous factors, and relying solely on royal jelly for improvement is not advisable.

Reality: Nutritional Support for Reproductive Health

The reality is that royal jelly provides a spectrum of vitamins, minerals, and fatty acids, such as vitamin B5 and 10-HDA, which play roles in supporting general reproductive health. It may contribute to better reproductive outcomes when used as part of a broader approach to health and wellness.

Myth: Reverses the Aging Process

Many claims suggest that royal jelly can reverse aging, citing its antioxidant properties. While antioxidants are known to combat oxidative stress — a contributor to aging — there is no evidence that royal jelly can reverse the aging process. It may help improve skin health and vitality, giving the appearance of a more youthful complexion, but it does not alter the biological aging process.

Reality: Supports Skin Health and Vitality

Royal jelly's genuine benefit to skin health lies in its amino acids, vitamins, and minerals, which can support skin renewal and repair. Its antimicrobial and anti-inflammatory properties may also benefit skin conditions, making it a valuable addition to skincare routines for enhancing skin hydration and elasticity.

Myth: A Miracle Weight Loss Solution

Some sources claim that royal jelly can lead to significant weight loss. However, there's no scientific backing for royal jelly as a miracle weight loss supplement. Its potential metabolic benefits are more likely to support overall health rather than directly causing weight loss.

Reality: May Aid in Metabolism Regulation

The reality is that the nutritional components of royal jelly, such as B vitamins, can assist in metabolism regulation, potentially aiding in weight management when combined with a healthy diet and regular exercise. It's the holistic lifestyle changes that primarily drive weight loss, with royal jelly possibly playing a supportive role.

In summary, while royal jelly is a nutrient-rich substance with several health benefits, it's essential to approach claims about its properties with a critical eye. Understanding the difference between myth and reality can help individuals make informed decisions about incorporating royal jelly into their health and wellness routines. Always consider consulting healthcare professionals when adding supplements to your diet, especially if you have existing health conditions or concerns.

CHAPTER 49: BEE POLLEN AND NUTRITION

Bee pollen is often hailed as a **superfood** due to its dense nutritional profile, packed with proteins, vitamins, minerals, and antioxidants. It's collected by bees from flowering plants and bound with nectar and bee saliva. Understanding how to integrate bee pollen into a daily diet can offer numerous health benefits, including enhanced energy, improved immunity, and support for digestive health. To maximize these benefits, it's crucial to know how to select, store, and consume bee pollen properly.

When selecting bee pollen, look for fresh granules that are brightly colored. This indicates that the pollen is not only fresh but also of high quality. The color of bee pollen can vary, reflecting the diverse sources of flowers from which it is harvested. Store bee pollen in airtight containers in the refrigerator to preserve its nutritional value and prevent spoilage. Freezing bee pollen is also an option for long-term storage without losing its potency.

Incorporating bee pollen into the diet can be simple and versatile. Start with a small amount, such as a quarter teaspoon per day, and gradually increase to up to one or two tablespoons, depending on individual tolerance and nutritional needs. This gradual increase helps prevent potential allergic reactions, especially for those who are new to consuming bee products.

One of the simplest ways to consume bee pollen is by sprinkling it over breakfast foods like oatmeal, yogurt, or smoothies. The granules add a slight crunch and a mild sweetness, enhancing the flavor profile of these dishes. For a nutrient-packed smoothie, blend a tablespoon of bee pollen with a cup of almond milk, a handful of spinach, half a banana, and a tablespoon of flaxseed. This combination provides a balanced mix of proteins, healthy fats, and carbohydrates, along with the added benefits of bee pollen.

Bee pollen can also be used in homemade granola bars or energy balls. Mix bee pollen granules with oats, nuts, seeds, and honey, then form into bars or balls. These make for a convenient, nutritious snack that's perfect for on-the-go energy.

For those interested in exploring the culinary uses of bee pollen further, it can be incorporated into salad dressings or used as a topping for desserts, adding a unique flavor and nutritional boost. When using bee pollen in recipes, it's best to add it after cooking to preserve its delicate vitamins and enzymes that can be destroyed by heat.

Beyond its dietary uses, bee pollen has been studied for its potential health benefits. It's rich in bioflavonoids, which are known for their antioxidant properties, helping to protect the body against free radicals and support cardiovascular health. The amino acids in bee pollen are essential for muscle repair and growth, making it a valuable supplement for athletes or those with active lifestyles.

Despite its benefits, it's important to note that bee pollen may cause allergic reactions in some individuals, especially those who are allergic to bees or other bee products. It's advisable to start with a very small amount to test for any adverse reactions. Pregnant women and individuals on medication should consult a healthcare professional before adding bee pollen to their diet.

In summary, bee pollen is a versatile and nutrient-rich supplement that can enhance a balanced diet. With careful selection and proper storage, it can be easily incorporated into daily meals, offering a natural boost to health and wellness.

Bee Pollen: A Nutrient-Dense Superfood

Bee pollen's status as a superfood is well-earned, given its comprehensive nutritional profile that includes a rich array of proteins, vitamins, minerals, and antioxidants. This dense nutritional makeup makes bee pollen an exceptional supplement for enhancing daily dietary intake. To leverage the full spectrum of benefits offered by bee pollen, it's essential to understand the specifics of its nutritional components and how they contribute to health and wellness.

Proteins in bee pollen are highly bioavailable, meaning they can be easily utilized by the body. This makes bee pollen an excellent protein source for vegetarians or individuals looking to boost their protein intake without relying heavily on meat products. The amino acids present in bee pollen, including all eight essential amino acids that the body cannot synthesize on its own, are critical for muscle repair, growth, and general bodily functions. Including bee pollen in post-workout smoothies or snacks can aid in muscle recovery and strength building.

The vitamin content in bee pollen is equally impressive, featuring a wide range of B vitamins, including B1 (thiamine), B2 (riboflavin), B3 (niacin), B5 (pantothenic acid), B6 (pyridoxine), and B12 (cobalamin), which play vital roles in energy production and the functioning of the nervous system. Additionally, bee pollen contains vitamins C, D, E, and K, each contributing to various aspects of health, from immune function and skin health to blood clotting and antioxidant protection.

Minerals found in bee pollen include calcium, magnesium, iron, zinc, selenium, and more, which are essential for bone health, oxygen transport, immune function, and wound healing, among other physiological processes. The trace minerals in bee pollen, although required in smaller amounts, are crucial for maintaining balance and supporting biochemical reactions within the body.

Antioxidants in bee pollen, such as flavonoids and carotenoids, offer protection against free radicals, molecules that can cause cellular damage and contribute to chronic diseases. By neutralizing free radicals, the antioxidants in bee pollen support cardiovascular health, reduce inflammation, and may lower the risk of certain diseases.

To incorporate bee pollen into the diet effectively, it can be sprinkled over salads, blended into smoothies, mixed into yogurt, or even added to homemade bread and other baked goods. For those looking to enhance their diet with bee pollen, starting with a small daily dose and gradually increasing it allows the body to adjust and reduces the risk of allergic reactions. It's also advisable to source bee pollen from reputable suppliers to ensure purity and avoid contaminants.

Given the diverse applications and health benefits of bee pollen, it stands out as a multifaceted supplement that can enrich a balanced diet. Whether used for its protein content, vitamins, minerals, antioxidants, or all of the above, bee pollen offers a natural, nutrient-rich option for individuals seeking to optimize their health and wellness through diet.

Bee Pollen Recipes and Meal Ideas

Bee pollen, with its rich nutritional profile, can be creatively incorporated into a variety of recipes and meal ideas to enhance both flavor and health benefits. Here are specific ways to include bee pollen in your daily meals:

Morning Bee Pollen Smoothie: Start your day with a smoothie by blending 1 cup of fresh spinach, 1 banana, ½ cup of blueberries, 1 tablespoon of bee pollen, 1 cup of almond milk, and a handful of ice. This smoothie packs a punch of antioxidants, vitamins, and minerals, providing an energizing start to your day. The bee pollen not only adds a slight sweetness but also boosts the nutritional value with its protein content.

Bee Pollen Yogurt Parfait: Layer Greek yogurt with mixed berries, a drizzle of raw honey, and a sprinkle of bee pollen for a nutritious breakfast or snack. The bee pollen adds texture and a boost of vitamins and minerals to the parfait. Use about 1 teaspoon of bee pollen per serving for a balanced flavor and nutrient profile.

Salad Topping: Enhance the nutritional value of your salads by sprinkling bee pollen over the top before serving. Combine mixed greens, sliced avocado, cherry tomatoes, and cucumber in a bowl. Dress the salad with olive oil, lemon juice, salt, and pepper, then add a tablespoon of bee pollen for an extra layer of flavor and a burst of energy. The pollen's slight sweetness pairs well with the fresh, crisp ingredients of the salad.

Bee Pollen Energy Balls: Mix 1 cup of oats, ½ cup of peanut butter, ⅓ cup of honey, 1 tablespoon of bee pollen, and ¼ cup of mini chocolate chips in a bowl. Roll the mixture into balls and refrigerate until firm. These energy balls make a convenient and healthy snack, with bee pollen providing a subtle crunch and a nutritional upgrade.

Homemade Granola with Bee Pollen: Combine 2 cups of rolled oats, ½ cup of chopped nuts, ¼ cup of seeds (such as pumpkin or sunflower), ¼ cup of melted coconut oil, ¼ cup of honey, and 1 teaspoon of vanilla extract in a mixing bowl. Spread the mixture on a baking sheet and bake at 300°F for 20-25 minutes, stirring halfway through. Once cooled, stir in 2 tablespoons of bee pollen. This granola can be enjoyed with milk, yogurt, or as a crunchy snack.

Bee Pollen Dressing: Create a unique salad dressing by whisking together 3 tablespoons of olive oil, 1 tablespoon of apple cider vinegar, 1 teaspoon of Dijon mustard, 1 teaspoon of honey, and 1 teaspoon of bee pollen. This dressing combines the health benefits of bee pollen with the tangy flavors of mustard and vinegar, making it a perfect complement to any salad.

Dessert Garnish: Finish off a dessert with a sprinkle of bee pollen for a nutritious touch. Whether it's over a slice of lemon cake, a bowl of fresh fruit, or a scoop of vanilla ice cream, a light dusting of bee pollen adds a beautiful color, a hint of sweetness, and a boost of vitamins and minerals.

When incorporating bee pollen into recipes, it's important to add it after cooking or baking to preserve its delicate nutrients. Bee pollen's versatility in both sweet and savory dishes makes it an excellent supplement to enhance the nutritional value of meals while adding unique flavors and textures. Always start with a small amount to ensure tolerance and gradually increase to the desired serving size.

Allergy Precautions for Pollen in Diets

When introducing bee pollen into your diet, especially for the first time, it's paramount to proceed with caution due to potential allergic reactions. Bee pollen is a potent allergen for some individuals, and reactions can range from mild to severe. Therefore, it's advisable to follow a step-by-step approach to safely incorporate bee pollen into your nutritional regimen.

Initial Test: Begin with a very small quantity of bee pollen, such as a few granules, to test for any adverse reactions. Place these granules on your tongue and wait for at least 24 hours. If you experience no adverse reactions, such as itching, swelling, or difficulty breathing, you can gradually increase the amount over several days.

Gradual Increment: After passing the initial test without any allergic reactions, slowly increase your intake. Add an additional few granules each day, carefully monitoring your body's response. This gradual increase not only helps in identifying potential allergies but also allows your digestive system to acclimate to the new supplement.

Consult Healthcare Professionals: If you have a history of allergies, particularly to pollen or bee products, consulting with an allergist or healthcare professional before introducing bee pollen is crucial. They may recommend specific tests or precautions to ensure your safety.

Quality Matters: Ensure the bee pollen you consume is sourced from reputable suppliers. High-quality bee pollen is less likely to contain contaminants that could trigger allergies or adverse reactions. Organic and locally sourced bee pollen may also reduce the risk of exposure to pesticides and other harmful substances.

Storage and Handling: Proper storage of bee pollen is essential to preserve its nutritional value and minimize the risk of spoilage, which could potentially increase its allergenic properties. Store bee pollen in airtight containers in the refrigerator or freezer to maintain its freshness and potency.

Incorporation into Diet: Once you have safely determined that you can consume bee pollen without allergic reactions, incorporate it into your diet in small, consistent amounts. Bee pollen can be sprinkled over breakfast cereals, blended into smoothies, or mixed into yogurt. Its addition should complement a balanced diet, enhancing the intake of vitamins, minerals, and antioxidants.

Monitoring: Continuously monitor your body's response to bee pollen, especially during the initial phases of incorporation. If you notice any signs of an allergic reaction, such as hives, respiratory distress, or gastrointestinal discomfort, discontinue use immediately and consult a healthcare professional.

Special Populations: Pregnant women, nursing mothers, and individuals on medication should exercise additional caution when adding bee pollen to their diets. In such cases, it's advisable to seek guidance from a healthcare provider to avoid potential risks to health or interactions with medications.

CHAPTER 50: HOLISTIC BEE PRODUCT APPROACHES

Incorporating bee products into holistic health practices extends beyond dietary additions, touching on various aspects of wellness including skin care, wound healing, and immune system support. Each product, from honey to propolis, offers unique benefits that can be harnessed in different ways to promote health and well-being.

Honey, renowned for its wound-healing properties, can be applied topically to minor cuts and burns to expedite healing. Its natural antibacterial and anti-inflammatory properties help to prevent infection while soothing the affected area. For a homemade wound salve, mix equal parts of honey and coconut oil, apply a thin layer over the wound, and cover with a clean bandage. This mixture combines the antimicrobial properties of honey with the moisturizing benefits of coconut oil, creating an effective barrier against bacteria and promoting the healing process.

Propolis, another powerful bee product, has been shown to boost the immune system and fight against various infections. To harness its benefits, propolis can be consumed in tincture form. Add 10-20 drops of propolis tincture to a glass of water or tea and drink once daily. This method helps to introduce propolis into the body in a controlled manner, allowing for the gradual accumulation of its immune-boosting properties. Ensure the tincture is sourced from a reputable supplier to guarantee purity and potency.

Royal Jelly is often touted for its skin health and anti-aging properties. For a simple yet effective royal jelly face mask, mix one teaspoon of royal jelly with one tablespoon of aloe vera gel. Apply the mixture to a clean face, leave it on for 20 minutes, then rinse with warm water. This mask leverages the anti-inflammatory and moisturizing properties of both ingredients, resulting in hydrated, smoother skin.

Bee Pollen is a versatile supplement that can be easily integrated into daily routines to enhance nutritional intake. For individuals looking to improve their digestive health, adding a tablespoon of bee pollen to morning smoothies or oatmeal can provide a significant fiber boost, along with a variety of vitamins and minerals. The enzymes present in bee pollen also assist in the breakdown and absorption of nutrients, making it an excellent supplement for overall digestive wellness.

Beeswax is commonly used in natural skincare products due to its ability to protect and repair the skin. To create a moisturizing beeswax lip balm, melt two tablespoons of beeswax pellets with two tablespoons of coconut oil and one tablespoon of shea butter in a double boiler. Once melted, remove from heat and add five drops of lavender essential oil for its soothing properties. Pour the mixture into small containers and allow to cool completely. This beeswax lip balm provides a natural barrier against harsh environmental elements while nourishing the lips.

When integrating bee products into holistic health practices, it's crucial to source these products from reputable, sustainable beekeepers to ensure they are free from contaminants and have been harvested in a manner that supports bee health. Additionally, individuals should be mindful of potential allergies and start with small doses to gauge tolerance. With thoughtful incorporation, bee products can significantly contribute to a holistic health regimen, offering natural alternatives for skin care, nutritional supplements, and immune support.

PART 12: SUSTAINABILITY AND BEES

CHAPTER 51: CHALLENGES BEES FACE TODAY

The challenges bees face today are multifaceted and complex, requiring a nuanced understanding to appreciate the full scope of threats to their survival. One of the most significant challenges is the widespread use of **pesticides**. These chemicals, designed to protect crops from pests, can have lethal effects on bees, either killing them directly or impairing their ability to navigate and forage for food. Neonicotinoids, a class of neuro-active insecticides chemically related to nicotine, have been particularly implicated in bee declines. They are systemic chemicals, meaning they are taken up by the plant and expressed in all its tissues, including nectar and pollen. This exposure can disorient bees and make it difficult for them to find their way back to the hive, a condition often referred to as Colony Collapse Disorder (CCD).

Another critical issue is **habitat loss**. As urbanization expands and agricultural practices intensify, the natural habitats that bees rely on for foraging are increasingly being destroyed. This loss of biodiversity not only reduces the availability of the diverse floral resources bees need for a balanced diet but also affects their ability to find suitable nesting sites. Moreover, monoculture practices in agriculture, where large tracts of land are devoted to a single crop, can provide abundant resources for a limited period but leave bees without food sources for the rest of the season.

Climate change also poses a significant threat to bees, affecting the timing of flowering in plants and thus the availability of food for bees. The mismatch between the blooming of flowers and the activity patterns of bees can lead to reduced food intake, weakening bee populations. Additionally, extreme weather events, such as droughts and floods, can destroy habitats, further exacerbating the challenges bees face in finding food and shelter.

The spread of **diseases and parasites** is another area of concern. The Varroa destructor mite, for example, attaches to bees and feeds on their bodily fluids, weakening them and making them more susceptible to viruses and other pathogens. The mite has been a significant factor in the decline of honeybee populations worldwide. Efforts to control these parasites with chemical treatments have had mixed results, and there is ongoing research into more sustainable and less harmful methods of control.

In response to these challenges, there is a growing movement towards **sustainable beekeeping practices**. These practices aim to minimize the use of chemicals in managing hives, promote the planting of diverse floral resources to support bee nutrition, and encourage the restoration of natural habitats to provide safe nesting sites. Additionally, there is an emphasis on monitoring and managing bee health to prevent the spread of diseases and parasites without relying on harmful pesticides.

The plight of bees today is a reflection of broader environmental issues that require concerted efforts to address. By understanding the specific challenges bees face, beekeepers and conservationists can work together to implement strategies that support bee health and ensure their vital role in our ecosystems is preserved.

Efforts to mitigate these challenges include advocating for policies that restrict the use of harmful pesticides and promote organic farming practices. By reducing reliance on chemicals that are toxic to bees, we can create a safer environment for them to thrive. Education plays a crucial role in this endeavor, as increasing public awareness about the importance of bees to our food supply and ecosystem can drive demand for bee-friendly products and practices.

Urban and suburban areas are also becoming increasingly important in the conservation of bees. Initiatives like community gardens, green roofs, and pollinator-friendly landscaping provide essential resources for bees in areas where natural habitats are scarce. These efforts not only support bee populations but also enhance the biodiversity and resilience of urban ecosystems.

Furthermore, the development of technology and scientific research offers new solutions to the problems bees face. Advances in tracking and monitoring technology enable beekeepers and researchers to better understand bee behavior, health, and environmental impacts. This data can inform more effective conservation strategies and help identify areas where intervention is needed most.

Collaboration between farmers, beekeepers, scientists, and policymakers is essential to address the multifaceted challenges bees encounter. Integrated pest management strategies that focus on biological rather than chemical controls can reduce the impact on non-target species like bees. Similarly, landscape-level planning that incorporates habitat corridors and diverse plantings can support bees' nutritional needs throughout the year.

Breeding programs aimed at enhancing bees' resilience to diseases, parasites, and environmental stressors are another promising avenue. By selecting for traits that confer resistance to specific threats, it may be possible to develop bee populations that are better equipped to survive in changing conditions.

Finally, supporting local and small-scale beekeepers can contribute to the sustainability of beekeeping practices. These beekeepers are often more invested in the health and well-being of their bees, and they play a crucial role in educating the public about the importance of bees. By purchasing local honey and other bee products, consumers can support the economy and encourage sustainable practices.

As we move forward, it is clear that the survival of bees requires a holistic approach that addresses the complex interplay of factors threatening their existence. By combining scientific innovation with traditional knowledge, fostering collaboration across sectors, and making conscious choices in our daily lives, we can contribute to a future where bees continue to play their critical role in our world.

Pesticides: Long-term Consequences for Bees

Pesticides, substances meant to prevent, destroy, or control pests, play a significant role in modern agriculture and gardening. However, their widespread use has raised concerns about the long-term health and survival of bee populations. Bees, both wild and managed, are exposed to a variety of pesticides when foraging on treated plants, through drift from nearby applications, or when contaminants are brought back to the hive. The impact of these chemicals on bees ranges from immediate lethality to more subtle, chronic effects that can weaken colonies over time.

One of the primary ways pesticides affect bees is through **acute toxicity**. Certain pesticides, particularly those in the class of neonicotinoids, have been shown to be highly toxic to bees on direct contact. Exposure to even small amounts of these substances can result in immediate death or significant impairment of the bees' navigation, foraging ability, and learning behavior. The challenge with acute toxicity is that it can lead to significant losses in bee populations in a short period, especially if the exposure coincides with a period of heavy foraging.

Another critical aspect is the **sub-lethal effects** of pesticides, which may not kill bees outright but can have profound impacts on their health and behavior. These effects include impaired learning and memory, reduced foraging efficiency, decreased fertility in queens, and weakened immune responses. Over time, these sub-lethal effects can contribute to the decline of bee colonies by reducing their ability to reproduce, gather food, and resist diseases and parasites. The complexity of these interactions makes it difficult to pinpoint the exact cause of decline in any given colony, but the cumulative evidence points to a significant role for pesticides.

Chronic exposure to pesticides is another concern for bee health. Bees can be exposed to low levels of pesticides over long periods, leading to gradual accumulation in the hive. This chronic exposure can affect not only the current generation of bees but also future generations, as contaminated pollen and nectar are stored and fed to larvae. The long-term consequences of chronic exposure include reduced colony growth, impaired development of young bees, and increased susceptibility to diseases and environmental stressors.

The **synergistic effects** of multiple pesticides present a complex challenge. Bees are often exposed to a cocktail of chemicals, including fungicides, herbicides, and insecticides, which can interact in unpredictable ways. Some combinations of pesticides can be more toxic together than any single pesticide alone, a phenomenon known as synergistic toxicity. This complexity makes it difficult to assess the full impact of pesticide use on bee health and underscores the need for a more holistic approach to pesticide regulation and use.

Understanding the specific pathways through which pesticides affect bees is crucial for developing strategies to mitigate these impacts. Research has focused on identifying the molecular and physiological mechanisms underlying pesticide toxicity in bees. This includes studying how pesticides interfere with the nervous system, hormone regulation, and gene expression. By unraveling these mechanisms, scientists aim to identify biomarkers of exposure and effect, which can be used to monitor bee health and guide interventions.

In light of these challenges, there is a growing emphasis on **integrated pest management (IPM)** strategies that reduce reliance on chemical pesticides. IPM approaches prioritize the use of biological controls, cultural practices, and mechanical methods to manage pests, turning to chemical options only as a last resort and selecting products that are less harmful to bees. Additionally, the development of new pesticides with reduced bee toxicity is an area of active research and development.

To further protect bee populations, the implementation of **buffer zones** around agricultural fields is a practical measure. These zones are planted with bee-friendly vegetation that is not treated with pesticides, providing a safe foraging area for bees. This practice not only helps to reduce direct exposure to harmful chemicals but also supports the nutritional diversity and health of bee colonies. By integrating these buffer zones, farmers and gardeners can create a more balanced ecosystem that supports both crop production and pollinator health.

The **labeling and timing of pesticide applications** also play a crucial role in minimizing the impact on bees. Applying pesticides during times when bees are less active, such as early morning or late evening, or choosing formulations that are less volatile can significantly reduce the risk of exposure. Furthermore, clear labeling that indicates the level of toxicity to bees helps beekeepers and farmers make informed decisions about pesticide use and application timing.

Education and collaboration among farmers, beekeepers, and the general public are essential for fostering a greater understanding of the importance of bees and the threats they face from pesticides. Workshops, extension services, and community programs can provide valuable information on alternative pest control methods and the benefits of adopting bee-friendly practices.

By working together, these stakeholders can develop and implement strategies that support both agricultural productivity and the health of bee populations.

Policy and regulation also have a critical role in safeguarding bee health. Governments and regulatory bodies can enact and enforce regulations that limit the use of the most harmful pesticides, promote research into safer alternatives, and support the adoption of integrated pest management practices. Policies that encourage or incentivize the creation of pollinator habitats and the use of non-chemical pest control options can further contribute to the conservation of bees.

The development of **technology and innovation** offers new tools for monitoring and protecting bee health. Advances in sensor technology, for example, allow for real-time monitoring of bee exposure to pesticides, enabling more precise management of pesticide applications. Genetic research may also lead to the development of bee strains that are more resistant to pesticides, although such approaches must be pursued with caution to avoid unintended ecological consequences.

In conclusion, addressing the impact of pesticides on bees requires a multifaceted approach that combines scientific research, sustainable agricultural practices, policy interventions, and public education. By understanding the mechanisms through which pesticides affect bees and implementing strategies to mitigate these effects, it is possible to protect these vital pollinators. As stewards of the environment, it is our collective responsibility to ensure that bees, along with other pollinators, continue to thrive for the benefit of future generations.

Climate Change Impact on Bees

Climate change has emerged as a formidable challenge for bees, altering their natural behaviors, habitats, and ultimately, their survival. The intricate balance of ecosystems that bees have thrived in for millennia is being disrupted at an unprecedented rate, affecting their ability to forage, reproduce, and maintain healthy colonies. The effects of climate change on bees are multifaceted, encompassing shifts in temperature, precipitation patterns, and the phenology of plant life, which in turn impacts the availability of nectar and pollen sources.

Rising temperatures are a significant concern, as they can lead to mismatches between the biological cycles of bees and the flowering times of plants. Bees have evolved to emerge and forage at specific times when their preferred flowers are in bloom. However, as temperatures increase, many plants are beginning to flower earlier in the year, before bees are active and ready to forage. This temporal mismatch can result in reduced food availability for bees, weakening colonies and lowering their chances of survival. To address this issue, beekeepers and conservationists can plant a diverse array of flora that blooms at different times, ensuring a steady food supply for bees throughout the changing seasons.

Furthermore, extreme weather events, such as droughts and heavy rains, are becoming more frequent and severe due to climate change. Drought conditions can desiccate flowers, diminishing nectar production and leading to food shortages for bees. On the other hand, excessive rainfall can wash away pollen and nectar, destroy bee habitats, and hinder bees' ability to forage. Beekeepers can mitigate these risks by providing supplemental feeding during times of scarcity and employing water management practices to prevent hive flooding, such as elevating hives or installing drainage systems around apiaries.

The alteration of habitats is another critical concern. As temperatures rise, the geographical ranges of many plants and animals are shifting poleward or to higher elevations, seeking cooler conditions. Consequently, bees may lose their traditional foraging grounds and struggle to adapt to new environments. This shift can also lead to the fragmentation of habitats, isolating bee populations and reducing genetic diversity, which is crucial for resilience against diseases and environmental stressors. Promoting habitat connectivity through the establishment of wildlife corridors and supporting large, contiguous areas of natural habitat can help bees navigate these changes, allowing them to follow shifting floral resources.

To combat the effects of climate change on bees, it is essential to adopt and promote sustainable beekeeping and land-use practices that enhance the resilience of bee populations. This includes reducing greenhouse gas emissions at the source by advocating for renewable energy and sustainable agriculture, which not only benefits the climate but also reduces the exposure of bees to harmful pesticides. Encouraging local and organic farming practices can support healthy bee populations by providing them with a pesticide-free environment and diverse floral resources.

Moreover, engaging in citizen science projects that track the health and behavior of bee populations can provide valuable data to researchers studying the impacts of climate change on pollinators. This data can inform conservation strategies and help identify areas where intervention is needed most urgently.

Supporting research into bee genetics and breeding programs that focus on traits such as temperature tolerance, disease resistance, and foraging efficiency can also play a role in preparing bee populations for the challenges posed by a changing climate. By selecting for these traits, it may be possible to enhance the adaptability of bees to the rapidly evolving conditions of their habitats.

In addition, public education and awareness campaigns are vital for fostering a broader understanding of the importance of bees to our ecosystems and food systems, as well as the threats they face from climate change. By raising awareness, individuals and communities can be motivated to take action in support of bees, whether through planting pollinator-friendly gardens, supporting local beekeepers, or advocating for policies that protect pollinator habitats and address climate change.

As the climate continues to change, the survival of bees hinges on our collective efforts to mitigate these impacts through thoughtful, coordinated action. By understanding the specific challenges posed by climate change and implementing targeted strategies to address them, we can help ensure that bees continue to play their critical role in pollinating the crops and wild plants that sustain life on Earth.

Diseases and Pests Threatening Hives

The Varroa mite, scientifically known as Varroa destructor, poses one of the most significant threats to honeybee colonies worldwide. These external parasitic mites attack and feed on honeybees, including the larvae and pupae, leading to weakened individuals and, ultimately, the collapse of entire colonies if not managed effectively. The Varroa mite's lifecycle and reproduction are intricately linked to that of the honeybee, making its control a complex challenge for beekeepers.

Upon infestation, Varroa mites attach themselves to the body of the bee, sucking their hemolymph (a fluid equivalent to blood in insects), which results in the transmission of various viral pathogens. Among these, the deformed wing virus is particularly devastating, often leading to bees with malformed wings who are unable to fly and, therefore, incapable of contributing to their colony's activities. The mites preferentially target drone (male) larvae for reproduction, which exacerbates the spread and impact of infestations within colonies.

Effective management of Varroa mites requires an integrated approach, combining mechanical, chemical, and biological strategies to reduce mite populations without causing undue harm to the bee colony or the environment. Mechanical methods include the use of screened bottom boards in hives, which allow mites to fall through and prevent them from reattaching to bees. This method can be enhanced by placing a sticky sheet or oil tray beneath the screen to trap the falling mites, ensuring they do not re-enter the hive.

Chemical controls involve the application of miticides, substances designed to kill mites, with careful consideration to minimize harm to bees and avoid contaminating hive products such as honey. Organic acids, like formic and oxalic acid, have been shown to be effective against Varroa mites when used appropriately. Formic acid, for example, can penetrate the capped cells where mites reproduce, offering a way to target the mite's lifecycle directly. However, the timing, dosage, and method of application are critical to ensure the safety of the bee colony. Oxalic acid, applied as a dribble or vapor, is another option, particularly useful during times when brood (bee larvae) is minimal, as it primarily targets phoretic mites (mites attached to adult bees).

Biological control methods are also being explored, including breeding for Varroa-resistant bee strains. Some bees exhibit behaviors such as hygienic behavior, where they detect and remove infested larvae from the hive, or grooming behavior, where bees remove mites from themselves and each other. Selective breeding programs aim to enhance these traits within bee populations, offering a long-term, sustainable solution to Varroa mite infestations.

Monitoring is a crucial component of Varroa mite management, allowing beekeepers to assess mite levels within their colonies and determine the need for intervention. Techniques such as the powdered sugar roll or alcohol wash can provide an estimate of mite infestation levels. These methods involve shaking a sample of bees in a container with powdered sugar or alcohol, respectively, to dislodge mites, which are then counted to estimate the infestation level within the colony.

In addition to these strategies, beekeepers can adopt practices that indirectly reduce the impact of Varroa mites on their colonies. Maintaining strong, healthy colonies through proper nutrition and management practices can help bees better withstand the effects of mite infestations. Providing a diverse array of forage sources can improve bee health and vitality, making them more resilient to pests and diseases.

Collaboration and information sharing among beekeepers, researchers, and extension services are vital to staying abreast of the latest recommendations and advances in Varroa mite management. As our understanding of Varroa mites and their interaction with honeybee colonies evolves, so too will the strategies for their control, ensuring the health and sustainability of beekeeping operations and the vital pollination services they provide.

CHAPTER 52: SUSTAINABLE BEEKEEPING PRACTICES

Adopting sustainable beekeeping practices involves a comprehensive approach that encompasses the management of hives, the surrounding environment, and the broader ecosystem. One pivotal aspect is the **selection of hive materials**. Opt for untreated, sustainably sourced wood for constructing hives to minimize the exposure of bees to harmful chemicals often found in treated woods. Cedar and pine are excellent choices due to their durability and natural resistance to rot. When painting hives, use non-toxic, water-based paints to provide a safe and healthy environment for the bees while ensuring the longevity of the wood.

The **location of beehives** plays a crucial role in sustainable beekeeping. Place hives in areas that receive morning sunlight and are protected from strong winds. This orientation helps to keep the hive warm, encouraging early foraging. Ensure there is adequate water supply nearby but not too close to prevent drowning risks for the bees. Implementing a shallow water source with landing spots, such as stones or floating wood pieces, can provide safe hydration for the bees.

Diversity in forage is another cornerstone of sustainable beekeeping. Cultivate a variety of flowering plants that bloom at different times throughout the year to offer a continuous food supply for the bees. Focus on native and heirloom plant species that are well-adapted to the local climate and require minimal water and maintenance. This not only supports the health of the bees but also promotes local biodiversity.

Natural pest management strategies are essential to reduce the reliance on chemical treatments. Introduce beneficial insects, such as ladybugs and lacewings, to control pest populations naturally. For Varroa mite management, consider mechanical methods like drone comb trapping and powdered sugar dusting, which disrupt the mite lifecycle without chemicals. Regularly monitor pest and disease levels to address issues promptly and minimize impact.

Bee breeding practices should focus on selecting traits that enhance resilience and adaptability to local conditions. Encourage genetic diversity within the apiary by sourcing queens from reputable breeders who prioritize health and sustainability. Experiment with splitting hives to naturally raise new queens, which can increase the vigor and disease resistance of the colonies.

Community engagement is vital for expanding the impact of sustainable beekeeping. Participate in local beekeeping clubs and online forums to share experiences, knowledge, and resources. Collaborate with local farmers and gardeners to create pesticide-free zones and enhance forage diversity. Educate the public about the importance of bees to our food system and environment through workshops, school programs, and open apiary days.

Record-keeping is an often-overlooked aspect of sustainable beekeeping. Maintain detailed records of hive inspections, treatments, harvests, and any other interventions. This data is invaluable for tracking the health and productivity of your colonies, understanding the impacts of different practices, and making informed decisions for future management.

Water conservation techniques should be integrated into the beekeeping operation. Collect rainwater to provide for the bees and to irrigate nearby forage plants. Use drip irrigation systems for plantings around the apiary to minimize water usage and reduce the risk of flooding the hives during heavy rains.

Finally, **sustainable honey harvesting** practices ensure that bees have enough stores for the winter. Harvest honey judiciously, leaving ample reserves for the bees, and consider timing the harvest after the main nectar flows to reduce stress on the colonies. Utilize methods like the crush and strain or use of a honey extractor to efficiently harvest honey while preserving the integrity of the comb, which can be returned to the hive or melted down for beeswax products.

By implementing these sustainable beekeeping practices, beekeepers can contribute to the health of their bees, the environment, and the broader community. These methods foster resilience in bee populations, enhance local ecosystems, and ensure the sustainability of beekeeping for future generations.

Natural Pest Control in Beekeeping

Reducing chemical treatments in beekeeping and adopting natural pest control methods is a crucial step toward sustainable beekeeping practices. Chemical treatments, while effective in the short term, can have detrimental effects on the health of bee colonies, the quality of hive products, and the environment. Natural pest control methods, on the other hand, offer a way to manage pests and diseases within the hive without the negative side effects associated with chemical use.

One effective natural pest control method is the use of essential oils, such as thyme oil or lemongrass oil. These oils can be applied to the hive in various ways, including misting over the bees or adding to their sugar water feed. Thyme oil, for example, has been shown to be effective against the Varroa mite when used in a vaporized form. It's important to note that the concentration of essential oils should be carefully controlled to avoid harming the bees. A recommended concentration is to dilute a few drops of the oil in water or sugar syrup before application.

Another natural method is the use of physical barriers or mechanical controls to manage pests. For instance, drone comb trapping exploits the Varroa mites' preference for drone brood. By inserting a frame designed to encourage the queen to lay drone eggs, beekeepers can then remove the frame and freeze it, killing the mites that have infested the drone larvae without the use of chemicals. This method requires precise timing to remove the drone comb before the drones hatch and release the mites back into the hive.

Biological control is another avenue, utilizing the natural enemies of pests to keep their populations in check. For example, introducing predatory insects that feed on Varroa mites or small hive beetle larvae can help manage these pest populations without resorting to chemical treatments. Care must be taken to ensure that the introduced predators do not become pests themselves or disrupt the ecosystem within the hive.

Integrated Pest Management (IPM) is a comprehensive approach that combines multiple strategies to manage pests in a way that minimizes risks to bees, humans, and the environment. IPM emphasizes regular monitoring of pest populations, using physical and biological controls wherever possible, and resorting to chemical treatments only as a last resort and in a targeted manner. For example, if a Varroa mite infestation exceeds a certain threshold and cannot be controlled with natural methods alone, a carefully selected and timed application of organic acids, such as formic or oxalic acid, may be used. These acids can be effective against mites while having a lower impact on bees and hive products than synthetic chemicals.

It's crucial for beekeepers to maintain accurate records of pest and disease levels, as well as the methods and treatments applied. This data helps in making informed decisions and adjusting strategies as needed. Beekeepers should also stay informed about new research and developments in natural pest control methods and be willing to adapt their practices accordingly.

By adopting natural pest control methods, beekeepers can contribute to the health and sustainability of their bee colonies and the environment. These methods, while sometimes requiring more effort and diligence than chemical treatments, offer a way to manage pests and diseases without compromising the integrity of hive products or the well-being of the bees.

Sustainable Harvesting for Colony Health

Promoting sustainable harvesting to preserve colony health involves a meticulous approach that ensures the well-being of the bee population while allowing beekeepers to benefit from the fruits of their labor. Sustainable harvesting is not just about when and how much honey or other hive products are taken; it's about understanding the intricate balance between taking from the hive and leaving enough resources for the bees themselves. This balance is critical for the colony's survival through the winter months or during periods of scarce forage.

To begin, timing the harvest is crucial. Harvesting should occur after the main nectar flows, when the bees have had ample opportunity to collect and store enough food for themselves. This period varies by region and can be determined by observing local floral sources and consulting with experienced beekeepers in the area. The goal is to ensure that the removal of honey does not deplete the hive's essential reserves.

The method of extraction plays a significant role in sustainable harvesting. Using a honey extractor that gently spins the frames allows for the honey to be removed while keeping the wax comb intact. This method is preferable because it preserves the structure of the comb, which bees can then quickly refill, reducing the energy and resources they need to rebuild. For small-scale or hobbyist beekeepers, the crush and strain method can also be utilized, though it requires the bees to expend more energy reconstructing comb. If using this method, it's vital to leave ample honey in the hive to compensate for the additional work imposed on the colony.

Leaving enough honey for the bees is another cornerstone of sustainable harvesting. As a general guideline, a strong colony needs approximately 60 to 90 pounds (27 to 40 kilograms) of honey to survive the winter, although this can vary based on local climate conditions. Beekeepers should assess the weight of their hives in late summer or early fall to ensure that this threshold is met before considering any further harvesting.

In addition to honey, beekeepers may also harvest beeswax, propolis, and pollen. However, each of these comes with its own considerations for sustainable practice. For example, when harvesting propolis, only remove excess from the hive frames and boxes, ensuring not to compromise the hive's structural integrity or the bees' ability to regulate temperature and defend against pests. Similarly, when collecting pollen, use pollen traps sparingly and for short periods to avoid depriving the bees of a crucial protein source.

Beekeepers can further support colony health during the harvesting process by practicing gentle handling of bees and frames, minimizing open-hive time to reduce stress and exposure to predators or environmental extremes. Using smoke judiciously can help calm the bees without causing undue stress or harm.

Monitoring colony health continuously is essential, especially after harvesting, to ensure that the bees are recovering well and have sufficient resources. If resources are found to be lacking, supplemental feeding with sugar syrup or pollen patties may be necessary until the bees can gather more natural forage.

Sustainable harvesting extends beyond the immediate concerns of quantity and timing; it encompasses a holistic approach to beekeeping that prioritizes the health of the bee colony. By adopting these practices, beekeepers not only ensure the longevity and productivity of their hives but also contribute to the broader ecological balance, supporting the vital role bees play in our ecosystems.

CHAPTER 53: SUPPORTING LOCAL POLLINATORS

Creating pollinator-friendly spaces involves a series of deliberate actions and choices that collectively contribute to the health and prosperity of local pollinator populations, including bees. One effective strategy is the planting of native flowers and shrubs, which serve as vital sources of nectar and pollen for bees and other pollinators. When selecting plants, prioritize species that are indigenous to your region as they are adapted to the local climate and soil conditions, requiring less water and maintenance than non-native plants. Additionally, these plants are more likely to co-evolve with the local pollinator species, providing the nutrients they need to thrive.

To maximize the benefits for pollinators, aim for a diversity of plant species that bloom at different times throughout the year, creating a continuous food supply. For example, include early bloomers like crocus and snowdrop, summer bloomers such as lavender and echinacea, and late bloomers like goldenrod and aster. This not only ensures a steady food source for bees from early spring to late fall but also adds color and vibrancy to the garden throughout the growing season.

Another critical aspect of supporting local pollinators is the provision of water. Bees, like all living creatures, need water to survive. However, their small size and flying capabilities necessitate specific considerations for water sources. Create shallow water stations by filling a shallow dish or birdbath with clean water and lining it with stones, pebbles, or floating pieces of wood to provide landing spots. This setup prevents bees from drowning while allowing them to hydrate safely. Ensure to change the water regularly to keep it clean and prevent the breeding of mosquitoes.

Shelter is another essential element for pollinators. While honeybees live in hives, many native bees and other pollinators require different types of habitats for nesting and protection from predators. Leaving areas of your garden untouched with piles of leaves, dead wood, and bare soil can provide natural nesting sites. Alternatively, installing bee hotels or nesting boxes can offer a safe haven for solitary bees and other beneficial insects. When constructing or purchasing a bee hotel, ensure it is made from natural, untreated materials and placed in a protected area away from direct rain and strong winds.

Avoiding the use of pesticides is paramount in creating a pollinator-friendly environment. Chemicals used to kill pests can also harm bees and other beneficial insects, either killing them directly or impairing their ability to forage and navigate. If pest control is necessary, opt for natural and organic methods that pose minimal risk to pollinators. For instance, encourage the presence of natural predators like birds and ladybugs, or use barriers and traps to protect plants without resorting to harmful chemicals.

Engaging with the community can amplify your efforts to support local pollinators. Share your knowledge and experiences with neighbors, friends, and local gardening clubs to encourage others to adopt pollinator-friendly practices. Participating in or organizing community projects, such as creating pollinator gardens in public spaces or schools, can have a significant positive impact on local pollinator populations. Additionally, advocating for local policies that protect pollinator habitats and restrict pesticide use can help create safer environments for bees and other pollinators on a larger scale.

By implementing these strategies, individuals can play a crucial role in supporting local pollinators. Through thoughtful plant selection, providing water and shelter, avoiding pesticides, and engaging with the community, it is possible to create environments where pollinators can flourish. These efforts not only benefit the pollinators themselves but also enhance the health of local ecosystems, agricultural productivity, and the beauty of our natural surroundings.

Creating Pollinator-Friendly Spaces

When selecting native flowers and shrubs for your garden to support pollinators, it's essential to consider the variety of plant species that bloom at different times of the year to provide a consistent source of nectar and pollen. Begin by researching plants that are native to your specific region. Your local extension office or native plant society can be an invaluable resource, offering lists of native plants suited to your climate and soil type.

For early spring bloomers, consider adding **redbud trees (Cercis canadensis)** and **wild columbine (Aquilegia canadensis)**. These early sources of nectar and pollen can be critical for bees emerging from hibernation. In the summer, plants like **purple coneflower (Echinacea purpurea)** and **black-eyed Susan (Rudbeckia hirta)** thrive in full sun and attract a wide range of pollinators. For late-season blooms, **New England aster (Symphyotrichum novae-angliae)** and **goldenrod (Solidago spp.)** provide vital resources as bees prepare for winter.

Incorporate a variety of shrubs as well, such as **buttonbush (Cephalanthus occidentalis)** and **blueberry bushes (Vaccinium spp.)**, which offer both forage and habitat for bees. When planting, group the same species in clusters to create a "target" for pollinators, making it easier for them to find and access the plants.

Consider the light requirements and water needs of each plant. Most native flowering plants and shrubs prefer full sun, defined as at least six hours of direct sunlight daily, and well-drained soil. Amend your soil with compost to improve its structure and fertility if necessary. When planting, give each plant enough space to reach its full size at maturity, which allows for adequate air circulation and reduces competition for resources.

Watering new plantings is crucial until they become established, which typically takes one to two growing seasons. A drip irrigation system or soaker hose can provide deep, infrequent watering that encourages strong root growth, as opposed to shallow, frequent watering, which can promote weak root systems susceptible to drought stress.

Mulching around your plants with organic material, like shredded bark or leaf mulch, will help retain soil moisture, suppress weeds, and add nutrients to the soil as it decomposes. Apply a 2 to 3-inch layer of mulch, being careful to keep it away from the plant stems to prevent rot.

For year-round interest and support for pollinators, incorporate evergreen shrubs and trees into your landscape. These can offer shelter during extreme weather and serve as landmarks for pollinators navigating their environment.

Finally, avoid the use of pesticides and herbicides in your garden. If pests become a problem, identify them and use targeted, pollinator-friendly control methods. Introducing beneficial insects, like ladybugs to control aphids, can be an effective strategy. Regularly inspect your plants for signs of disease and address any issues promptly to prevent spread.

Partnering with Farmers for Pollinator Networks

Partnering with farmers and gardeners to expand pollinator networks involves a strategic approach that emphasizes collaboration, education, and the implementation of bee-friendly practices. This partnership is a critical step towards enhancing the resilience and health of bee populations and, by extension, the broader ecosystem. Here's how to effectively foster these partnerships:

Identify Potential Partners: Start by reaching out to local agricultural cooperatives, gardening clubs, and community gardens. These groups often include members who are already interested in or practicing sustainable agriculture and gardening techniques. Use local social media groups, community bulletin boards, and agricultural extension offices to make initial contacts.

Educate on Best Practices: Once potential partners are identified, organize workshops or informational sessions that focus on the importance of pollinators and how agricultural practices can be modified to support them. Topics should include the creation of pollinator habitats, the reduction of pesticide use, and the planting of crop varieties and flowering plants that provide nectar and pollen throughout the growing season.

Collaborate on Plant Selection: Work together to select and plant native flowering plants that are beneficial to both bees and agricultural goals. For crops, recommend varieties that are known to be good for pollinators or that bloom at times when food might be scarce for bees. Encourage the planting of cover crops like clover or alfalfa, which are excellent sources of nectar when not in bloom.

Implement Integrated Pest Management (IPM): Educate and assist farmers and gardeners in implementing IPM strategies that minimize the impact on pollinators. This includes using biological pest control agents, mechanical pest control methods, and chemical controls as a last resort with careful consideration of timing and application methods to avoid harming bees.

Create Shared Pollinator Habitats: Encourage the development of shared spaces such as community gardens or buffer zones around agricultural fields that are dedicated to pollinator-friendly plants. These habitats can serve as crucial links between fragmented landscapes, offering bees and other pollinators safe passage and abundant resources.

Promote Water Conservation and Quality: Discuss the importance of clean, accessible water sources for pollinators. Implementing rainwater harvesting systems and creating shallow water features with landing spots can provide bees with the hydration they need without the risk of drowning.

Advocate for Policy Changes: Work together to advocate for local and regional policies that support pollinator health. This could include restrictions on certain pesticides, incentives for organic farming practices, or funding for pollinator habitat restoration projects.

Monitor and Share Successes: Establish a system for monitoring the health and success of pollinator populations in partnership areas. Use social media, newsletters, and community meetings to share successes and challenges, fostering a sense of community achievement and encouraging wider adoption of pollinator-friendly practices.

By following these detailed steps, the partnership between beekeepers, farmers, and gardeners can become a powerful force for positive change, contributing significantly to the sustainability and productivity of local ecosystems and agriculture.

CHAPTER 54: EDUCATION AND ADVOCACY

Engaging the younger generation in bee conservation is crucial for the future of beekeeping and the sustainability of our ecosystems. Schools are an ideal platform for instilling an appreciation for bees and their role in our world. **Educational programs** tailored to various age groups can make learning about bees an interactive and impactful experience. For elementary students, activities could include **bee-themed art projects** that encourage creativity while teaching about bee anatomy and the importance of pollination. Middle school students might benefit from **science experiments** that explore the behavior of bees, such as how they communicate through the waggle dance or their role in the growth of plants. High school students could engage in more complex projects, such as **building bee hotels** or **designing pollinator gardens**, which provide hands-on experience with conservation efforts.

Teacher workshops are an effective way to equip educators with the knowledge and resources to teach about bees. These workshops can cover a range of topics, from the basics of bee biology to the complexities of ecosystem interdependencies. Providing teachers with **curriculum materials**, such as lesson plans and multimedia resources, can help integrate bee education into various subjects, including science, geography, and even art.

Community outreach programs can extend education beyond the classroom, involving families and community members in bee conservation efforts. Organizing **bee awareness days**, with activities like honey tasting, demonstrations of beekeeping equipment, and talks by local beekeepers, can raise public awareness about the importance of bees. **Citizen science projects**, such as tracking local bee populations or planting native flowers, can engage people of all ages in hands-on conservation work.

To support these educational efforts, partnerships with local beekeeping associations, environmental organizations, and universities can provide valuable expertise and resources. These partnerships might facilitate **guest lectures**, **field trips** to apiaries or research labs, and **access to educational materials**. Additionally, grants and funding from governmental or environmental bodies can help schools and communities initiate and sustain bee education programs.

Incorporating **technology** into bee education can also enhance learning experiences. **Virtual reality experiences** that simulate the inside of a beehive or **apps** that teach about bee conservation can make learning interactive and accessible. **Online platforms** can serve as repositories for educational materials, enabling teachers and students to share projects and learn from each other's experiences.

By fostering an environment of learning and curiosity about bees from a young age, we can cultivate a generation that values and actively participates in the conservation of bees. Through education and advocacy, every individual can contribute to the sustainability of bee populations, ensuring a healthy planet for future generations.

The Role of Bees in Ecosystems

Creating and maintaining bee-friendly environments within urban settings presents a unique set of challenges and opportunities. Urban beekeeping has gained momentum as a means to support bee populations, enhance urban green spaces, and educate the public on the importance of pollinators. The integration of hives into city landscapes requires careful planning to ensure the health of the bees and the safety of the community. Selecting locations for urban hives demands consideration of several factors, including exposure to sunlight, noise levels, and proximity to pesticide-free green spaces. Rooftops, community gardens, and parks often serve as ideal sites, offering sufficient sunlight and a buffer from ground-level pollutants. These areas should be assessed for their ability to provide a continuous supply of flowering plants throughout the growing season, ensuring a steady source of nectar and pollen for the bees.

When establishing hives in urban areas, it's crucial to use barriers or elevated platforms to keep hives out of direct contact with pedestrians and pets. Strategic placement of hives, with the entrance facing away from foot traffic and towards green spaces, can direct bee flight patterns to minimize interactions with the public. Additionally, incorporating water sources near hives, such as shallow birdbaths with landing stones, can prevent bees from venturing into neighboring pools or drinking fountains in search of water.

Educational outreach plays a vital role in urban beekeeping, transforming public spaces into living classrooms. Informational signage near hives can inform passersby about the role of bees in urban ecosystems, the basics of beekeeping, and how they can contribute to pollinator conservation. Hosting regular workshops and open hive days can demystify beekeeping for the public, fostering a community of informed advocates for bees and biodiversity.

Urban beekeepers must navigate local ordinances and community concerns with transparency and engagement. Before installing hives, it's advisable to communicate with neighbors and local authorities, addressing any questions or concerns and outlining the benefits of urban beekeeping. This dialogue can pave the way for supportive policies and community initiatives aimed at increasing urban green spaces and reducing pesticide use.

Collaboration with local businesses and schools can amplify the impact of urban beekeeping efforts. Businesses can sponsor hives or adopt bee-friendly landscaping practices, while schools can integrate beekeeping into their science curriculum, offering students hands-on learning experiences that highlight the importance of pollinators in food production and ecosystem health.

In summary, urban beekeeping requires a multifaceted approach that combines hive management with public education and community engagement. By fostering environments where bees can thrive, urban beekeepers contribute to the resilience of local ecosystems and the global effort to protect pollinator populations. Through strategic placement of hives, community education, and collaboration with local stakeholders, urban areas can become sanctuaries for bees, offering hope for their future and the health of our planet.

Teaching Kids About Pollinators

Developing school programs focused on teaching children about pollinators involves a multi-faceted approach that integrates curriculum development, hands-on activities, and community involvement. The goal is to provide students with a comprehensive understanding of pollinators' roles in ecosystems, the challenges they face, and how individuals can contribute to their conservation.

Curriculum Development: Begin by designing age-appropriate lesson plans that align with existing science and environmental education standards. For younger children, lessons can focus on the basic biology of bees and other pollinators, their role in pollination, and simple ways to support them. As students progress in age, curriculum can delve into more complex topics such as the impact of pesticides, habitat loss, and climate change on pollinator populations. Incorporate multimedia resources, such as videos and interactive websites, to cater to different learning styles and keep students engaged.

Hands-on Activities: Practical, hands-on experiences are crucial for fostering a deeper connection between students and pollinators. Activities can include planting pollinator-friendly gardens on school grounds, which serve as outdoor classrooms for observing pollinators in action and learning about plant-pollinator relationships. Crafting bee hotels from natural or recycled materials teaches students about habitat needs. Setting up a school beehive, with professional guidance, offers an immersive experience into the world of beekeeping and honey production.

Community Involvement: Extend learning beyond the classroom by involving the local community. Invite local beekeepers, entomologists, or environmental scientists to speak with students about their work and the importance of pollinators. Organize field trips to botanical gardens, nature reserves, or working farms where students can see pollination and conservation efforts firsthand. Encourage students to participate in citizen science projects, such as tracking local pollinator species or contributing to national databases, to provide them with a sense of contribution to larger conservation efforts.

Educational Materials: Equip teachers with a toolkit of resources to effectively teach about pollinators. This can include lesson guides, reference materials, and access to online platforms for further research. Provide materials for students to take home, such as informational brochures or seeds for pollinator-friendly plants, to encourage family discussions and activities related to pollinator conservation.

Evaluation and Feedback: Implement mechanisms for evaluating the effectiveness of the program and gathering feedback from students, teachers, and community participants. Use this feedback to refine and expand the program, ensuring it remains relevant, engaging, and informative. Surveys, quizzes, and group discussions can help assess students' knowledge gains and attitudes towards pollinators and conservation.

PART 13: ADVANCED BEEKEEPING TECHNIQUES

Chapter 55: Queen Rearing

Evaluating the quality of a queen bee is paramount for maintaining a healthy and productive hive. A queen's egg-laying patterns and the behavior of her colony offer critical insights into her vitality and the overall hive health. When assessing a queen, beekeepers should observe the brood pattern on the comb. A strong queen will lay eggs in a compact, solid pattern, with few missed cells. This indicates that she is fertile and laying consistently. Eggs should be centered at the bottom of each cell, which suggests that the queen is healthy and the hive is well-managed.

The behavior of the colony can also indicate the queen's quality. A content and productive colony, with workers busily foraging, tending to the brood, and performing hive maintenance, often reflects the presence of a robust queen. Conversely, signs of aggression, frequent swarming tendencies, or a lack of cohesion within the hive may point to issues with the queen's pheromones and her ability to unify the colony.

Introducing a new queen to an existing hive requires careful planning and execution to ensure colony stability. Before introduction, the new queen should be kept in a queen cage with a candy plug, allowing the worker bees to gradually become accustomed to her scent. This cage is then placed between the frames in the brood chamber, with the candy plug facing upwards to prevent any attendant bees that die from blocking the exit. Over several days, the worker bees will eat through the candy to release the queen into the hive. During this period, it's crucial to monitor the hive's behavior for signs of acceptance or rejection, such as the workers clustering around the cage in a calm manner versus aggressively attempting to sting through the cage.

Successful queen introduction is often followed by a noticeable increase in hive activity and an improvement in brood patterns. However, beekeepers should continue to monitor the hive, especially in the first few weeks after introduction, to ensure the new queen is accepted and begins laying eggs effectively. If the queen is rejected or fails to start laying, further intervention may be necessary, such as re-queening or examining the hive for underlying health issues.

In summary, queen rearing and introduction are critical components of advanced beekeeping techniques. By carefully selecting, evaluating, and introducing queens, beekeepers can maintain productive and healthy hives. This process requires a deep understanding of bee behavior, meticulous observation, and timely intervention to address any issues that arise during the queen's introduction. Through these practices, beekeepers can ensure the long-term success and sustainability of their beekeeping endeavors.

Queen's Role in Hive Health and Productivity

The queen bee's role within the hive extends far beyond her reproductive duties; she is the nucleus around which the entire colony operates. Her health, vitality, and genetic makeup are paramount to the hive's productivity and survival. The queen's pheromones, for instance, play a crucial role in regulating the social structure of the hive. These chemical signals help to suppress the development of ovaries in worker bees, ensuring that the queen remains the sole egg-layer within the colony. Additionally, her pheromones promote cohesion and harmony among the colony's members, guiding behaviors such as foraging, nursing, and defending the hive.

The process of rearing a new queen, whether for replacing an aging queen or for establishing new colonies, requires precision and attention to detail. Selecting a larva from a strong, healthy colony is the first step. The larva should be no older than three days to ensure its developmental potential as a queen. Beekeepers then transfer the selected larva into a queen cell cup, which is placed in a specially prepared queen rearing colony. This colony must be made queenless a day before introducing the queen cell cups to stimulate the worker bees' instincts to rear a new queen.

Feeding the developing queen larvae is a critical aspect of queen rearing. A diet of royal jelly is essential for the larvae to develop into queens. Worker bees in the queen rearing colony are responsible for this, but the beekeeper can enhance the process by ensuring the colony is well-fed with sugar syrup and pollen to stimulate royal jelly production. The environment within the rearing colony must be closely monitored; temperature and humidity levels are vital for the larvae's development. A stable temperature around 95°F (35°C) and high humidity promote optimal growth conditions.

Once the new queens emerge, they must undertake mating flights to fertilize their eggs. This critical phase requires careful planning from the beekeeper to ensure the safety and success of the mating flights. Mating yards, areas where drones are intentionally concentrated, can improve the genetic diversity and quality of the queens' offspring. After successful mating, the young queens return to their hives, where they begin their egg-laying duties. Beekeepers must monitor these new queens closely, observing their laying patterns and the health of the brood to assess the queens' quality and the potential need for further intervention.

Throughout her life, which can span several years, the queen's productivity will naturally decline. Beekeepers must vigilantly monitor her performance and the overall health of the colony, ready to rear or introduce a new queen when necessary. This cyclical process of queen rearing and introduction ensures the hive's longevity, productivity, and genetic strength.

The critical role of the queen in hive health and productivity underscores the importance of meticulous queen rearing practices. By understanding the queen's influence on the colony and mastering the techniques of queen selection, rearing, and introduction, beekeepers can enhance their hives' resilience, productivity, and contribution to the ecosystem. Through diligent practice and observation, beekeepers play a pivotal role in the perpetuation and health of bee populations, embodying the essence of advanced beekeeping techniques.

Raising Queen Bees: Grafting to Natural Selection

Grafting is a precise technique in queen rearing that involves transferring bee larvae by hand from their original worker cells into artificial queen cups. This method allows beekeepers to select larvae from colonies with desirable traits, such as gentleness, productivity, or disease resistance. The process begins with the preparation of a starter colony, which is essentially a strong, queenless hive that will readily accept and nurture the grafted larvae due to their instinct to raise a new queen in the absence of one.

To perform grafting, a beekeeper needs a grafting tool, which is a small, delicate instrument designed to scoop up the tiny larvae without harm. The ideal larva for grafting is no older than three days, as its developmental stage is perfectly suited for queen rearing. The beekeeper carefully transfers each selected larva into a queen cell cup, which is then placed in a cell bar holder. These holders are inserted into the starter colony. The worker bees in the starter colony, recognizing the need for a new queen, will begin to feed the grafted larvae with copious amounts of royal jelly, initiating their development into queen bees.

Temperature and humidity control within the starter colony is critical during this period. The beekeeper must ensure that the environment remains stable, with a temperature around 95°F

(35°C) and high humidity, to support the larvae's growth. After approximately 10 days, the queen cells are ready to be transferred to mating nucs or queenless hives where the emerged queens can mate and begin their egg-laying duties.

Natural selection, on the other hand, allows the bees to choose their own queen without human intervention. This method involves creating conditions that prompt the colony to raise its own queen, typically by making a hive temporarily queenless or by simulating swarming conditions. In this scenario, the bees select several young larvae and begin to feed them royal jelly, creating what are known as emergency queen cells. The first queen to emerge will usually destroy the other queen cells, establishing herself as the new queen of the colony.

Natural selection is less labor-intensive than grafting and can produce queens that are well-adapted to their local environment, as the bees choose the larvae based on their own criteria. However, this method offers less control over the genetic traits of the new queens, which may be a drawback for beekeepers focused on specific breeding goals.

Both grafting and natural selection have their places in advanced beekeeping practices. Grafting allows for more precise control over the genetics and traits of the future queens, making it an excellent choice for beekeepers looking to improve or maintain specific qualities within their colonies. Natural selection, while more unpredictable, can yield strong, locally adapted queens that contribute to the genetic diversity and resilience of bee populations. Depending on the beekeeper's objectives, experience level, and the specific needs of their apiary, either method—or a combination of both—can be effectively employed to rear new queens, ensuring the health, productivity, and sustainability of the hive.

Evaluating Queen Quality

Evaluating the quality of a queen bee is a nuanced process that hinges on observing her egg-laying patterns and the resultant behavior of the colony. A high-quality queen is characterized by her ability to lay eggs in a consistent and orderly fashion. Ideally, she fills the cells of the brood chamber in a compact pattern, starting from the center and working her way outwards. This pattern is crucial as it ensures that the brood is kept warm and fed efficiently by the worker bees. The presence of a solid brood pattern with minimal empty cells indicates that the queen is fertile and healthy, laying eggs at an optimal rate.

The eggs themselves offer clues to the queen's quality. Each egg should be centered at the bottom of its cell, a sign of a queen's precision and care in laying. Multiple eggs in a single cell can be a sign of a failing queen or, in some cases, laying workers, which can occur if the queen's pheromone production is insufficient to suppress the ovaries of the worker bees.

Beyond the brood pattern, the behavior of the colony provides insights into the queen's effectiveness. A strong, cohesive colony with high levels of activity, especially in foraging and brood care, often reflects the presence of a robust queen. Worker bees in such a colony are typically more productive and exhibit less aggressive behavior, attributed to the queen's pheromones promoting social harmony within the hive.

Conversely, signs of a poor-quality queen include spotty brood patterns with numerous empty cells, indicating irregular egg-laying. Such patterns can lead to a weakened colony structure, as the worker bees must spread their resources thin to care for the brood. Additionally, a decline in the queen's pheromone production can lead to increased aggression or unrest within the colony, as the social order begins to break down.

Monitoring the queen's performance over time is essential. A decrease in brood production or a noticeable change in colony behavior can signal that the queen's vitality is waning. Beekeepers must then decide whether to introduce a new queen to the colony. This decision is critical, as the timing and method of introduction can significantly impact the colony's acceptance of a new queen and, consequently, its future productivity and health.

In evaluating a queen's quality, beekeepers must also consider the broader environmental and genetic factors at play. The queen's genetics play a role in her longevity, disease resistance, and productivity, influencing the overall traits of the colony. Environmental factors, such as availability of forage and exposure to pesticides, can also affect a queen's performance and the health of her colony.

In summary, the evaluation of a queen bee's quality is a comprehensive process that requires careful observation of her egg-laying patterns and the behavior of her colony. By understanding these indicators, beekeepers can make informed decisions about queen rearing and management, ensuring the health and productivity of their hives.

Introducing a New Queen to a Hive

When introducing a new queen to an existing hive, the beekeeper's approach must be methodical and sensitive to the colony's dynamics to ensure a smooth transition and acceptance. The initial step involves isolating the new queen in a queen cage, a small, mesh enclosure that allows the hive's bees to acclimate to her scent without direct contact, minimizing the risk of immediate rejection. The cage typically contains a candy plug, a sugar-based substance that gradually dissolves over several days. This design serves a dual purpose: it provides nourishment for the queen during the introduction process and allows worker bees to slowly chew through the candy to release her, ensuring a gradual and less confrontational introduction.

Before placing the queen cage into the brood chamber, it's crucial to remove the existing queen to prevent any immediate conflict between the two queens. This action requires careful inspection of the hive frames to locate and safely remove the old queen. Once this is accomplished, the beekeeper positions the queen cage between two central brood frames, a location that ensures the new queen is immediately surrounded by the brood, where her pheromones can have the most significant impact on the colony. The positioning also encourages the worker bees to attend to and eventually accept the new queen as they care for the brood.

During the next few days, while the worker bees work to release the new queen from her cage, the beekeeper must monitor the hive for signs of unrest or rejection, such as aggressive clustering around the queen cage or attempts to block access to it. Such behaviors may indicate that the colony is not ready to accept the new queen, necessitating additional measures or a reassessment of the introduction process.

Once the queen is released, observation becomes even more critical. The beekeeper should look for signs of normalcy resuming within the hive, such as the new queen being freely allowed to move throughout the hive and begin laying eggs. Successful egg laying is one of the most definitive signs of acceptance, as it demonstrates that the worker bees have accepted her pheromones and authority as the colony's new matriarch. Regular inspections should follow in the subsequent weeks to ensure the queen's continued acceptance and to monitor the brood pattern for signs of healthy laying activity.

In cases where the new queen is rejected, evidenced by her disappearance or failure to start laying, the beekeeper must be prepared to intervene quickly. This might involve introducing another queen, potentially using a different introduction method or reassessing the colony's health and readiness for a new queen. Factors such as timing within the beekeeping season, the colony's health, and the presence of stressors such as disease or insufficient resources can all impact the success of queen introduction.

Throughout this process, maintaining a calm and minimally invasive approach is paramount. Excessive disturbance can stress the colony, undermining the introduction process. Beekeepers should wear appropriate protective gear and use smoke judiciously to calm the bees during inspections, ensuring that their actions promote a peaceful environment conducive to the new queen's acceptance.

Ultimately, the success of introducing a new queen to an existing hive hinges on a deep understanding of bee behavior, meticulous observation, and the ability to adapt strategies based on the colony's response. This nuanced approach underscores the art and science of beekeeping, where each action is taken with the health and productivity of the colony in mind.

CHAPTER 56: HIVE SPLITTING

Hive splitting is a crucial technique for beekeepers aiming to expand their apiaries, manage swarming, or rejuvenate aging colonies. It involves dividing an existing colony into two or more separate units, each with the potential to grow into a full-sized, independent hive. This process not only increases the number of hives but also reduces the risk of swarming, where a single colony splits naturally, and half the population leaves with the old queen to find a new home.

The first step in hive splitting is to assess the parent colony's strength. A strong colony, ideally with multiple frames of brood, ample worker bees, and sufficient food stores, is necessary for a successful split. The presence of healthy, capped brood and eggs indicates that the queen is laying well and the colony is robust enough to be divided.

Once a suitable colony is identified, the beekeeper must decide on the type of split: a simple split or a walk-away split. In a simple split, the beekeeper manually divides the brood and resources between the new and old hive, ensuring both have a mix of eggs, larvae, capped brood, honey, and pollen. The original queen stays with one part, while a new queen must be introduced to the other. This method requires the beekeeper to have a new queen ready, either purchased or reared in advance.

In a walk-away split, the beekeeper divides the colony and leaves the queenless part to raise its own queen from existing young larvae or eggs. This method relies on the bees' natural instinct to ensure their survival by raising a new queen. However, it takes longer for the new queen to start laying, and there's a risk the bees may not successfully raise a queen, leading to the need for intervention by introducing a new queen if the first attempt fails.

For both types of splits, it's essential to ensure that each new colony has enough resources to sustain itself. This includes frames of honey and pollen for immediate food sources and enough bees to maintain brood temperature and forage for new resources. Placing a feeder with sugar syrup in each new hive can help support the bees until they establish strong foraging patterns.

The location of the new hive is also crucial. If the new hive is placed too close to the original, there's a risk that foraging bees will return to their original hive, leaving the new colony with insufficient numbers to thrive. To prevent this, the new hive can be placed at a significant distance from the original, or the beekeeper can temporarily relocate the new hive to a different location for a few weeks before bringing it back to its permanent position.

After the split, monitoring is critical. The beekeeper should check for signs of queen acceptance in the colony with the introduced queen and signs of successful queen rearing in the walk-away split. This includes observing for new eggs and larvae approximately a month after the split, indicating that the new queen is laying and has been accepted by the colony.

Additionally, the beekeeper must manage the risk of pests and diseases, which can quickly overwhelm a weakened or stressed colony. Regular inspections to monitor the health of the split colonies, along with timely interventions for pest control and disease management, are vital to ensure the success of the new hives.

Hive splitting is an effective way to manage bee populations, prevent swarming, and increase the number of colonies in an apiary. With careful planning, attention to detail, and ongoing management, beekeepers can successfully split hives, contributing to the growth and sustainability of their beekeeping operations.

Step-by-step instructions for successfully splitting a hive.

Materials and tools

- Two beekeeping suits and veils for protection

- A smoker to calm the bees

- Hive tool for opening hives and separating frames

- A new hive setup (including bottom board, hive body or brood box, frames with foundation, inner cover, and lid)

- A queen for the new hive (purchased or raised)

- Bee brush to gently move the bees

- Optional: Queen excluder to keep the queen in the brood box

Step-by-step instructions

1. **Choose the right time**: Splitting a hive is best done in the spring when colonies are expanding rapidly and there's plenty of nectar and pollen available. This gives the new colony time to establish before winter.

2. **Prepare the new hive**: Place the new hive close to the original hive. This should be done a few days in advance to allow for any adjustments in placement based on weather or shading.

3. **Smoke the original hive**: Lightly smoke the entrance of the hive and under the lid to calm the bees. Wait a few minutes before opening the hive.

4. **Open the original hive and locate the queen**: Carefully inspect frames to find the queen. This step is crucial to ensure that each hive ends up with one queen – the original hive retains its queen, and a new queen will be introduced to the split hive.

5. **Select frames to move**: Choose 2-3 frames of brood in various stages of development (eggs, larvae, capped brood) and at least one frame of honey and pollen. Ensure these frames have bees on them. These will provide the new colony with a workforce and resources.

6. **Transfer frames to the new hive**: Gently brush any excess bees off the sides of the frames and place them into the center of the new hive body. Fill in the spaces on either side with frames of foundation.

7. **Introduce a new queen**: If you have a mated queen ready, you can introduce her to the new hive now. Place the queen cage between the frames in the center of the brood area. If the cage has a candy plug, the bees will eat through it to release the queen. If not, release her manually after a few days, ensuring the colony has accepted her.

8. **Close up both hives**: Replace the lids on both hives. Use the smoker to gently encourage any bees still outside to enter the hives.

9. **Monitor the new hive**: Check the new hive after a week to ensure the queen has been released and is laying eggs. Look for signs of normal activity and growth in the colony.

10. **Feed the new colony if necessary**: If nectar flow is weak, provide sugar syrup to help the new colony get established.

Troubleshooting

- If the new colony appears to be rejecting the queen, check for signs of aggression towards the queen cage. If aggression is observed, remove the queen and try introducing another queen after a few days.

- If the original hive seems weakened by the split, consider reducing its entrance size to help the bees defend against pests and adding a frame of brood from another strong colony to boost its numbers.

Care and maintenance

- Regularly inspect both the original and new hives for signs of disease or pests.
- Ensure both colonies have sufficient space to grow by adding additional boxes or frames as needed.
- Monitor food stores in both hives, especially the new colony, and feed as necessary to support development.

By following these detailed steps, you will successfully split a hive, helping to increase your beekeeping operation and prevent swarming by providing overcrowded bees with more space.

Avoiding Stress-Induced Absconding

Avoiding stress-induced absconding during the hive splitting process requires careful attention to several critical factors. Stress in bees can be triggered by various factors including abrupt changes in their environment, handling, or the sudden absence of their queen. To mitigate these risks, follow these detailed strategies:

Timing of the Split: Choose a period of strong nectar flow, typically in the spring or early summer, when bees are naturally more inclined to expand and the presence of abundant resources can reduce stress. Ensure that the weather forecast is favorable, with mild temperatures and no rain expected for several days post-split to allow the bees to acclimate without additional environmental stress.

Gentle Handling: When manipulating frames and moving bees, do so gently to minimize disturbance. Use a smoker lightly to calm the bees before and during the process, but avoid excessive smoke as it can disorient and stress the colony. Move slowly and steadily, giving bees time to adjust to your presence and actions.

Ensuring Resources: Both the parent and new hives should have ample resources. This includes frames of capped brood, stores of honey and pollen, and drawn comb for the queen to lay eggs immediately. If resources are scarce, supplement with sugar syrup and pollen patties to ensure the colony does not experience stress from lack of food, which can trigger absconding.

Queen Introduction: If introducing a new queen to one of the splits, use a proven method such as a queen cage with a candy plug to allow gradual acceptance. Placing the queen cage between brood frames helps integrate her scent with the colony's. Monitor the hive closely for several days to ensure that the queen is accepted and starts laying eggs. If using a walk-away split method, ensure there are eggs or very young larvae from which the bees can rear a new queen, and check back in about a week to confirm queen cells are being constructed.

Hive Location: After splitting, if one of the new hives remains in the original location, the other should be moved to a new location at least 2 miles away to prevent foraging bees from returning to the old site, which can deplete the population of the moved hive. If moving the hive a great distance is not possible, use barriers or reorient the hive entrance to confuse returning foragers and encourage them to enter the new hive.

Monitoring and Intervention: In the weeks following the split, monitor both hives for signs of stress, such as pacing at the hive entrance, reduced foraging activity, or aggression. Check for successful queen acceptance and egg laying. Be prepared to intervene if signs of queen rejection or failure are observed, which may include adding a new queen or combining the split back with another colony if it fails to thrive.

Minimizing External Stressors: Ensure the hives are protected from predators and have adequate water sources nearby. Avoid pesticide exposure by communicating with neighbors and local farmers about your beekeeping activities and the importance of notifying you before they spray.

By adhering to these detailed strategies, beekeepers can significantly reduce the risk of stress-induced absconding following a hive split. Each step, from timing to monitoring, plays a crucial role in ensuring the success of both the parent and new colonies, contributing to a healthy and expanding apiary.

CHAPTER 57: OVERWINTERING HIVES

As temperatures begin to drop signaling the approach of winter, the beekeeper's focus shifts towards ensuring their hives are well-prepared to withstand the cold months ahead. **Overwintering** hives is a critical aspect of beekeeping that requires detailed planning and execution to prevent colony losses during this dormant period. The key to successful overwintering lies in the combination of adequate food stores, proper hive insulation, and moisture control, all of which contribute to a healthy and robust colony ready to emerge in spring.

Food Stores: The first step in preparing a hive for winter is to assess and supplement the colony's food stores. A typical hive requires approximately 60 to 90 pounds of honey to survive the winter months when foraging is not possible. To ensure your bees have enough to eat, check the weight of the hive in late fall by gently lifting it from the back. If the hive feels light, or inspection reveals insufficient honey reserves, feeding the bees with a 2:1 sugar syrup can help them build up their stores. It's crucial to complete this feeding before the temperature drops too low, as bees will not take syrup in cold weather.

Hive Insulation: Proper insulation helps maintain a stable internal temperature within the hive, allowing the bee cluster to keep warm. Wrapping the hive in a black tar paper can absorb heat from the sun during the day and retain warmth. Ensure that the hive's entrance and ventilation holes are not covered, as ventilation is critical to prevent moisture buildup inside the hive. Some beekeepers also use foam insulation boards placed on top of the hive under the outer cover for additional warmth.

Moisture Control: Excess moisture is one of the biggest threats to a hive during winter. Moisture can accumulate inside the hive from the bees' respiration, leading to mold growth and chilling the cluster. To combat this, ensure good ventilation by slightly propping up the hive's outer cover or using a moisture board or quilt box on top of the frames. These devices absorb moisture and can be removed and dried periodically throughout the winter.

Pest and Disease Management: Before winter sets in, it's also vital to address any pest or disease issues within the hive. Treatments for Varroa mites should be administered in late summer or early fall to ensure the colony goes into winter as healthy as possible. A final inspection to remove any dead bees from the bottom board and check for signs of disease can help prevent problems from worsening over the winter.

Reducing Hive Entrances: To help bees defend against rodents and reduce cold drafts, the hive entrance should be reduced or equipped with a mouse guard. This not only keeps mice and rats out but also helps maintain the warmth generated by the bee cluster.

Positioning for Winter: Finally, consider the hive's orientation and location. Hives should be positioned to receive maximum sunlight exposure during the winter months, ideally with a slight forward tilt to assist in water drainage. If possible, hives can be moved to a more sheltered location to protect them from prevailing winds, or windbreaks can be set up to reduce wind chill.

By meticulously addressing each of these areas—food stores, insulation, moisture control, pest and disease management, entrance reduction, and positioning—beekeepers can significantly increase their colonies' chances of surviving the winter. This preparation allows the bees to conserve energy and emerge strong in the spring, ready for the new foraging season. Remember, the health and strength of the colony going into winter are the best predictors of a successful overwintering period. Therefore, ongoing monitoring and maintenance of the hive's condition throughout the year are essential to ensure the bees are in the best possible shape as they head into the colder months.

Winterizing Hives: Insulation and Ventilation

The process of preparing hives for winter involves a careful balance between insulation for warmth and ventilation to prevent moisture accumulation, which can be detrimental to the health of the bee colony. To achieve this balance, specific materials and techniques are recommended to ensure the hive remains a safe and conducive environment for the bees during the cold months.

For insulation, one effective material is rigid foam insulation boards, which can be cut to the size of the hive and placed around the exterior. These boards have a high insulation value and are lightweight, making them easy to install and remove as needed. When applying insulation, it's crucial to cover the sides of the hive while leaving the bottom board and the hive entrance unobstructed to allow for proper air circulation and to enable bees to leave the hive on warmer winter days for cleansing flights.

In addition to side insulation, the top of the hive should be insulated to prevent heat from escaping. A simple method is to place a piece of rigid foam insulation under the hive's outer cover. However, it's important to ensure that this top insulation does not block the hive's upper ventilation holes, which are critical for moisture control.

Ventilation can be managed by creating an upper entrance or vent, if one does not already exist. This can be done by slightly propping up the hive's outer cover with small sticks or spacers, allowing moist air to escape. Alternatively, a notched inner cover can provide a passageway for moisture to leave the hive while keeping the bees protected from direct exposure to cold air. Moisture boards or quilt boxes filled with wood shavings or moisture-absorbing materials can also be placed above the frames to wick away moisture from the cluster below.

Another key aspect of winter hive management is ensuring the entrance is reduced and protected. A reduced entrance minimizes cold air flow into the hive and helps the bees defend against pests. An entrance reducer can be made from a piece of wood or metal with a small opening, placed at the hive entrance. Additionally, to protect against rodents, a mouse guard, which can be a piece of hardware cloth with holes large enough for bees but too small for mice, should be secured over the entrance.

The hive's physical location plays a role in its winter survival. Hives should be positioned to maximize sun exposure, ideally facing south or southeast. This positioning helps the hive warm up during the day. If strong winds are a concern, creating a windbreak using bales of straw or a temporary fence can reduce wind chill on the hive.

Lastly, regular winter inspections, though minimized to reduce heat loss, are necessary to assess the health of the colony, the adequacy of food stores, and the effectiveness of insulation and ventilation measures. These inspections should be brief and conducted during warmer, sunny days to minimize disturbance and heat loss.

By following these detailed strategies for insulating and ventilating hives, beekeepers can help ensure their colonies emerge from winter strong and ready for the spring nectar flow. Proper winter preparation not only protects the bees from the cold but also from the moisture and pests that can threaten their survival during the dormant season.

Feeding Strategies for Low Activity Periods

During periods of low activity, particularly in the winter months when bees are less active and foraging opportunities are scarce, it's essential to have a strategic approach to feeding to sustain the colony's health and vitality. **Feeding strategies** must be tailored to meet the nutritional needs of the bees while considering the limitations imposed by colder weather.

Sugar Syrup: In the lead-up to winter, transitioning from a lighter 1:1 sugar-to-water ratio syrup to a heavier 2:1 ratio can provide the bees with a denser energy source. This heavier syrup mimics the consistency of honey, making it easier for bees to store and use during the cold months. It's crucial to complete this feeding before temperatures drop below 50°F, as bees will not take syrup in colder conditions. Utilize a feeder that sits inside the hive, such as a frame feeder or top feeder, to minimize exposure to cold air when bees access the syrup.

Fondant or Candy Boards: As temperatures plummet, liquid feeding becomes impractical. Fondant, or sugar candy, provides an alternative food source that bees can consume without the need to leave the cluster. Place a candy board or fondant directly above the brood chamber where the bee cluster will reside during winter. This placement ensures that bees have immediate access to food without expending valuable energy moving throughout the hive. Fondant can be made by boiling sugar and water to a specific temperature, then cooling it to form a solid that's easy for bees to digest.

Dry Sugar Method: An emergency feeding strategy involves sprinkling granulated sugar on a paper placed directly on the top bars of the uppermost box. This method, often referred to as the "mountain camp method," provides an emergency food source if the colony exhausts its stores. The bees' movement across the sugar and their respiration moisture will cause some of the sugar to clump and become more palatable for the bees.

Protein Supplements: While carbohydrates are crucial for energy, bees also require protein, especially in early spring before natural pollen sources are available. Protein patties can be placed inside the hive near the cluster. These patties, made from soybean flour or other protein sources mixed with sugar syrup, can help maintain the colony's health and stimulate brood rearing in preparation for spring. Ensure that the patties are placed close to the cluster but not directly on top, to avoid disrupting the bees' natural thermoregulation.

Water: Even in winter, bees require access to water for digestion and to dilute stored honey for consumption. On warmer days, bees may venture out to find water, but providing a water source closer to the hive can reduce the risk of cold exposure. A small, shallow water container with floating materials like cork or small twigs can prevent drowning and should be placed near the hive entrance.

Monitoring and Adjustment: Regularly monitoring the weight and activity of the hive can provide insights into the colony's food stores and health. If a hive feels light or inspection reveals low food stores, additional feeding may be necessary. Adjustments to feeding strategies should be made based on the colony's consumption rates and the availability of natural resources.

By implementing these feeding strategies during periods of low activity, beekeepers can ensure their colonies remain strong and healthy through the winter, ready to thrive in the coming spring.

Monitoring Hive Health in Winter

Monitoring hive health during the winter months is a delicate task that requires a nuanced approach to ensure the well-being of the colony without causing undue stress or disturbance. The key lies in the subtle observation of external indicators and the judicious use of technology to assess the internal condition of the hive without opening it and exposing the bees to the cold.

One effective method for monitoring hive health is through the observation of bee activity at the entrance on warmer days. Bees will often take advantage of milder temperatures to go on cleansing flights. A lack of activity during these periods can be an early warning sign of potential issues within the hive. Additionally, the presence of dead bees outside the entrance is normal to some extent, as the colony cleans out deceased members. However, an unusually high number of dead bees could indicate problems such as disease, starvation, or exposure to harsh conditions inside the hive.

Another non-intrusive technique involves tapping gently on the side of the hive and listening to the intensity of the buzz within. A strong, vigorous buzzing indicates a healthy and active colony, while a weak or absent buzz may suggest that the colony is in distress or has dwindled in numbers. This method requires experience and a good understanding of the normal sounds of a bee colony, as variations can be subtle.

The use of technology offers beekeepers advanced methods for monitoring without opening the hive. Hive scales, for instance, can be employed to track the weight of the hive over time. A steady decrease in weight during the winter can indicate that the bees are consuming their honey stores, which is expected. However, a rapid decline might suggest that the colony is consuming its stores too quickly and could face starvation before winter's end. Conversely, a sudden increase in weight could signal robbing by other bees or pests.

Temperature and humidity sensors placed inside the hive can provide valuable data on the internal conditions. Ideal temperature ranges are crucial for the survival of the colony, as bees cluster together to maintain a warm environment for the queen and brood. Excessive humidity is a sign of poor ventilation and can lead to mold growth and disease. Monitoring these parameters can alert beekeepers to potential issues that could threaten the health of the colony, allowing for early interventions that might include adjusting hive insulation or ventilation without directly disturbing the cluster.

Visual inspections should be minimized during the winter but can be conducted during warmer days if absolutely necessary. These inspections must be quick and purposeful, aimed at verifying the presence of the queen or assessing food stores without dismantling the hive or significantly reducing its temperature. The use of a thermal imaging camera can be a valuable tool in this regard, allowing beekeepers to assess the size and location of the cluster within the hive, as well as identifying potential cold spots or areas of concern without direct contact.

In conclusion, monitoring hive health in winter revolves around a balance between attentive observation and the strategic use of technology. By understanding and employing these methods, beekeepers can ensure their colonies remain strong and healthy throughout the cold months, poised for a successful start to the new season.

CHAPTER 58: MIGRATORY BEEKEEPING

Migratory beekeeping involves the relocation of bee hives to different geographical areas throughout the year to maximize honey production and provide pollination services. This practice requires detailed planning, precise timing, and careful management to ensure the health and productivity of the bee colonies during transport and in new locations. **Selecting suitable locations** is the first critical step in migratory beekeeping. Each location must offer abundant forage opportunities corresponding to the plants in bloom at that time of year. Researching floral sources, bloom times, and the nutritional needs of bees is essential. For example, almond orchards in California require pollination services in early spring, while clover fields in the Midwest might offer forage in late spring to early summer.

Preparing bees for transport involves ensuring that hives are healthy and strong enough to withstand the stress of moving. This includes performing health checks for diseases and pests, such as Varroa mites, and treating any issues well before the planned move. Hives should be well-ventilated during transport to prevent overheating, which can be achieved by using screens in place of the hive's entrance and inner cover.

Securing hives for travel is crucial to prevent damage and ensure the safety of the bees. Hives can be fastened together using hive staples or straps, and individual components like supers and frames should be secured with screws or hive locks to prevent shifting. The use of non-slip mats under hives on the transport vehicle can also prevent movement.

Transportation logistics require careful consideration of the distance, travel time, and weather conditions. Traveling during cooler hours or at night can reduce stress on the bees by minimizing temperature fluctuations and keeping the hives cooler. It's also important to plan the route in advance to minimize travel time and ensure there are suitable locations to stop if checks on the bees' well-being are necessary.

Acclimating bees to new locations upon arrival involves gradually exposing them to their new environment. This can include initially placing them in a shaded area if arriving during the day to prevent overheating and allowing the bees to orient themselves to their new surroundings. Providing water sources near the hives immediately upon arrival is crucial, especially in warmer climates or if natural water sources are not readily available.

Monitoring and management at the new site are vital to ensure the colonies adjust well and remain healthy. This includes regular inspections to monitor for pests and diseases, ensuring that bees have access to adequate forage, and providing supplemental feeding if natural forage is insufficient. The timing of pesticide applications in the surrounding area must be communicated with landowners or farmers to protect the bees from exposure.

Legal considerations and agreements with landowners are also a key aspect of migratory beekeeping. This includes obtaining permission to place hives on private property, negotiating pollination contracts, and ensuring compliance with local and state regulations regarding beekeeping and transport.

Environmental impact and sustainability practices should be considered, aiming to minimize the ecological footprint of migratory beekeeping. This includes selecting locations that support biodiversity, avoiding overgrazing floral resources with too many hives in one area, and practicing responsible beekeeping to ensure the health and vitality of bee populations.

By adhering to these detailed strategies and considerations, migratory beekeeping can be a successful endeavor that benefits both beekeepers and the agricultural crops that rely on bees for pollination. The practice requires a commitment to the well-being of the bees, a deep understanding of their needs, and a willingness to adapt and respond to the challenges of moving and establishing hives in new environments.

Benefits of Relocating Hives for Seasonal Blooms

Relocating hives to follow seasonal blooms allows beekeepers to optimize honey production and enhance the health and productivity of their bee colonies. This practice, known as migratory beekeeping, taps into the natural cycle of flowering plants, ensuring that bees have access to a continuous supply of nectar and pollen. Here, we detail the multifaceted benefits of this approach, emphasizing the strategic considerations beekeepers must account for to ensure success.

Maximized Honey Production: By moving hives to areas where specific crops or wildflowers are in bloom, beekeepers can significantly increase the quantity and variety of honey produced. For instance, moving hives to almond orchards in California during early spring can result in almond-flavored honey, while late spring relocation to orange groves in Florida can yield orange blossom honey. This strategy requires precise timing to coincide with peak bloom periods, which can vary based on local climate conditions and plant species.

Enhanced Bee Health: The diversity of pollen and nectar sources available through migratory beekeeping can improve the nutritional health of bee colonies. Different plants provide varied nutritional profiles in their pollen and nectar, contributing to a more balanced diet for bees. This diversity can bolster the immune system of bees, making them more resilient to diseases and pests.

Pollination Services: Migratory beekeeping plays a crucial role in the pollination of crops, contributing to the agricultural economy. Beekeepers can enter into pollination contracts with farmers, providing a valuable service by enhancing crop yields through efficient pollination. This symbiotic relationship benefits both parties; crops receive necessary pollination, and bees gain access to abundant forage resources.

Risk Mitigation: By spreading hives across different locations, beekeepers can mitigate risks associated with environmental factors, diseases, and pests that may affect bees in a particular area. If a disease outbreak or adverse weather event occurs in one location, not all hives will be impacted, safeguarding the beekeeper's overall operation.

Legal and Environmental Considerations: Before relocating hives, beekeepers must navigate legal requirements, such as permits for transporting bees across state lines and agreements with landowners for hive placement. It's crucial to ensure that the chosen locations are free from pesticide exposure and other environmental hazards. Moreover, beekeepers should practice responsible hive management to prevent overgrazing of floral resources, maintaining the ecological balance and supporting the health of local bee populations.

Preparation for Relocation: Successful relocation requires thorough preparation, including ensuring that hives are healthy and strong, securing hives properly for transport to prevent damage, and choosing the right time for movement to minimize stress on the bees. Upon arrival at the new location, careful monitoring and management are essential to help the bees acclimate and thrive in their new environment.

In implementing migratory beekeeping practices, beekeepers leverage the natural foraging behavior of bees, aligning it with agricultural needs and the changing seasons. This approach not only maximizes the productivity of beekeeping operations but also supports the health of bee colonies and the broader ecosystem. Beekeepers must carefully plan and execute each step of the process, from selecting suitable locations and preparing hives for transport to managing hives at the new site and navigating legal requirements. By doing so, they can enjoy the numerous benefits of migratory beekeeping while contributing to the sustainability of agriculture and the well-being of bees.

Ensuring Hive Health During Relocation

Ensuring hive health during and after relocation is paramount to the success of migratory beekeeping. This process begins with a thorough inspection of the bee colonies prior to the move. Beekeepers must check for signs of diseases such as American or European foulbrood, and for pests like Varroa mites and small hive beetles. Treatment should be administered well in advance of the relocation to allow time for the colony to recover and to prevent the spread of these issues to new locations. For Varroa mite control, options include organic acids such as formic or oxalic acid vaporization, ensuring treatments align with the life cycle of the bees to maximize effectiveness while minimizing harm to the colony.

The physical preparation of hives for transport is critical. Hives should be closed securely with screens that allow for ventilation while preventing bees from escaping during transit. Utilize ratchet straps to tightly secure hive bodies and supers together, and to the transportation platform, reducing the risk of shifting or toppling. Employ entrance reducers to limit the opening, helping to maintain the internal environment of the hive and to prevent bees from leaving the hive during transport. It's advisable to move hives during the evening or at night when bees are less active, which helps in reducing stress and keeping the colony intact.

Upon arrival at the new location, carefully position the hives with consideration for the sun's path, ensuring they receive morning sunlight which stimulates bees' activity. Hives should be slightly tilted forward to prevent water accumulation inside the hive which can lead to mold growth and negatively impact bee health. Immediate access to water is crucial; set up water stations if natural sources are not available, adding a small amount of salt or apple cider vinegar to make it more attractive to the bees, mimicking natural nectar.

Post-relocation, monitor the bees closely for signs of stress or disorientation. This includes observing for normal foraging behavior and checking if the bees are returning to the hive with pollen, which indicates successful adaptation to their new environment. Supplemental feeding may be necessary until bees adapt to the new forage sources, using a sugar syrup mixture or pollen patties to provide essential nutrients.

Regular hive inspections in the weeks following relocation are important to assess the health of the colony, the queen's egg-laying performance, and to ensure that the hive is free from pests and diseases. Adjustments to the hive's location or orientation may be needed based on the bees' behavior and environmental factors such as wind or sun exposure. Beekeepers should also communicate with local farmers and landowners about pesticide use in the area to safeguard the bees from accidental exposure.

In areas where bees are introduced for pollination services, it's essential to ensure that the forage available is sufficient to support the introduced colonies without depleting resources for native pollinators. This involves coordination with other beekeepers and land managers to prevent overcrowding of bee populations which can lead to resource competition and stress on all pollinators in the area.

The practice of relocating hives requires a commitment to ongoing education and adaptation to new challenges. Beekeepers should stay informed about local flora and fauna, weather patterns, and potential hazards to bee health in the regions they operate. Networking with local beekeeping communities can provide valuable insights and support, facilitating a smoother transition for the bees and increasing the success rate of migratory beekeeping operations.

Part 14: The Cultural Significance of Bees

CHAPTER 59: BEES IN HISTORY AND MYTHOLOGY

Bees have been revered and mythologized throughout human history, playing significant roles in various cultures' folklore, religion, and mythology. In ancient Egypt, bees were believed to be born from the tears of the sun god Ra, symbolizing birth and resurrection, and were associated with royalty and governance. This connection between bees and divinity underscores their importance in the natural world and human society. The Egyptians mastered the art of beekeeping, with honey being a prized commodity used for sweetening foods, as an offering to the gods, and for its medicinal properties.

In Greek mythology, bees, also known as the "birds of the Muses," were considered messengers between the natural world and the divine. Honey was often referred to as the nectar of the gods, a substance of divine origin that was believed to confer immortality. The Oracle at Delphi was often called the "Delphic Bee," and priestesses of Apollo were named after bees. This highlights the bee's role in prophecy and wisdom in ancient Greek culture.

The Celts also held bees in high esteem, viewing them as symbols of wisdom, and believed that bees represented the soul and had the ability to travel between worlds. Celtic lore includes stories of bees as messengers carrying news to the spirit world. Honey was used in mead production, a drink that played a central role in Celtic feasts and rituals, further emphasizing the cultural significance of bees.

In Hinduism, the god Vishnu is sometimes depicted as a blue bee resting on a lotus flower. This imagery signifies the interconnectedness of life, with bees serving as a link between the natural and spiritual realms. The lotus, symbolizing purity and enlightenment, combined with the bee, represents the soul's journey towards spiritual knowledge.

In Norse mythology, bees were believed to have been born from the carcass of the giant Ymir, and their ability to find their way home was admired, symbolizing guidance and perseverance. Mead, made from honey, was considered the drink of the gods, imbuing poets and scholars with divine inspiration and wisdom.

The Christian tradition often uses bees and honey as metaphors for diligence, purity, and the sweetness of the divine. Monasteries in the Middle Ages kept bees for honey, which was used not only as food but also in the production of mead and beeswax candles for religious ceremonies. The bee's work ethic and communal living were seen as models for human society, emphasizing cooperation and dedication to the common good.

In modern times, the symbolism of bees has evolved, but many of their associations with wisdom, immortality, and communication between realms remain. Environmental movements have adopted the bee as a symbol of ecological interdependence and the importance of preserving natural habitats. Bees continue to inspire, reminding us of our connection to the natural world and the importance of living in harmony with it.

The cultural significance of bees is vast and varied, reflecting their impact on human societies across the globe. From ancient deities and myths to modern environmental symbols, bees have been integral to our understanding of the world around us. Their role in pollination, essential for food production and biodiversity, further underscores their importance, not just in mythology, but in the very fabric of life on Earth.

CHAPTER 60: THE HONEY TRADE THROUGH THE AGES

The honey trade, a vital component of human civilization, has evolved significantly from its humble beginnings. Initially, honey was collected from wild bee colonies, a practice that dates back thousands of years. The ancient rock paintings in Spain, which depict humans foraging for honey, stand as a testament to this early interaction. As beekeeping practices developed, so did the methods of honey extraction and preservation, leading to the establishment of a structured honey trade.

In ancient Egypt, honey was not only a sweetener but also an offering to the gods and an integral part of embalming practices. The Egyptians were among the first to domesticate bees in artificial hives, made from straw and mud, facilitating the collection and storage of honey. This innovation allowed for the surplus production of honey, which became a valuable commodity for trade both within the empire and with neighboring regions. Honey was transported along the Nile and traded for other luxury goods, demonstrating its high value in ancient economies.

The trade routes of honey extended beyond Egypt, with the Greeks and Romans also recognizing its value. Honey played a crucial role in Greek and Roman cuisine, medicine, and rituals. The Romans, in particular, were known to have established apiaries in their provinces, notably in Spain and the Middle East, to ensure a steady supply of honey for their empire. The Roman legions often carried honey with them, using it as a source of energy and for wound treatment, underscoring its importance in their society.

During the Middle Ages in Europe, the honey trade flourished under the auspices of monasteries. Monks and nuns were instrumental in advancing beekeeping techniques and honey production, with monasteries acting as centers of trade. Honey, along with beeswax, was traded for goods and services, and it played a pivotal role in the economy of the time. Beeswax was highly sought after for candle making, a necessity in religious ceremonies, while honey was a staple in the diet and medicine.

The discovery of the New World opened new avenues for the honey trade. European settlers introduced honeybees to the Americas, which led to the expansion of beekeeping and honey production in the new colonies. Honey became a significant export from the Americas back to Europe, enriching the transatlantic trade network. The introduction of sugar cane to the Americas eventually led to a decline in the use of honey as a primary sweetener, but it remained an essential product in local economies and trade.

In the modern era, the honey trade has become global, with advancements in beekeeping, processing, and transportation allowing for the mass production and distribution of honey. Countries like China, Turkey, and Argentina have emerged as leading honey producers, contributing significantly to the global honey trade. The rise of organic and artisanal honey products has also opened new markets, catering to consumers' growing demand for natural and sustainably produced goods.

Technological advancements have further transformed the honey trade, with traceability and quality control becoming paramount. The implementation of blockchain technology and sophisticated tracking systems ensures the authenticity and purity of honey, protecting consumers and producers alike. These measures address concerns over adulteration and mislabeling, ensuring that the honey trade remains transparent and sustainable.

The honey trade, with its rich history and cultural significance, continues to evolve, reflecting changes in consumer preferences, technological advancements, and global trade dynamics. From the ancient Egyptians to modern-day beekeepers, the trade of honey has remained an enduring link between cultures, economies, and the natural world, showcasing the timeless value of this golden elixir.

CHAPTER 61: BEES IN MODERN CULTURE

In the realm of modern culture, bees have transcended their biological and ecological roles to become symbols of community, sustainability, and environmental activism. The rise of social media and digital platforms has allowed for a broader dissemination of information regarding the plight of bees and the critical challenges they face, including habitat destruction, pesticide use, and climate change. This increased awareness has sparked a global movement towards the conservation of bees and the promotion of bee-friendly practices among individuals and communities alike.

Documentaries and Films have played a pivotal role in educating the public about the importance of bees. These visual narratives combine compelling storytelling with scientific research to highlight the bees' role in pollination and the consequences of their decline on global food security. For instance, films such as "More Than Honey" and "Vanishing of the Bees" delve into the complex world of bees, showcasing the intricate relationship between humans and bees and urging a reevaluation of our agricultural and landscaping practices.

Social Media Campaigns have leveraged platforms like Instagram, Twitter, and Facebook to spread awareness and mobilize action. Hashtags such as #SaveTheBees and #BeeConservation have become rallying points for sharing information, bee-friendly gardening tips, and advocacy for policy changes. These campaigns often feature stunning macro photography of bees, personal stories from beekeepers, and updates from scientific research, creating a vibrant and engaged online community passionate about bee conservation.

Urban Beekeeping has emerged as a popular trend in cities around the world, reflecting a growing interest in sustainable living practices and local food production. Urban beekeepers manage hives on rooftops, balconies, and community gardens, contributing to local pollination and biodiversity while producing honey. This practice not only supports bee populations but also educates urban residents about the importance of bees and encourages the creation of green spaces in cities.

Educational Programs in schools and communities have integrated beekeeping and pollinator conservation into their curricula. These programs aim to instill a sense of stewardship for the environment in young minds through hands-on learning experiences. By participating in the planting of bee-friendly gardens, building bee hotels, and even managing school hives, students gain a deeper understanding of ecology and the critical role bees play in our ecosystem.

Art and Literature continue to draw inspiration from bees, with artists and writers using bees as motifs to explore themes of cooperation, resilience, and environmental harmony. From intricate bee-themed sculptures and murals to poetry and novels that metaphorically delve into the life of bees, these creative expressions reflect our enduring fascination with these vital pollinators and serve as a reminder of our interconnectedness with the natural world.

Consumer Choices and Market Trends have also shifted towards supporting bees, with an increasing demand for products that are sustainably sourced and produced. Honey, beeswax, and other bee-derived products are now often marketed with an emphasis on ethical production methods, including organic and biodynamic farming practices that are beneficial to bees. This shift not only supports the livelihood of bees but also promotes a broader movement towards sustainable agriculture and food systems.

In essence, bees have become emblematic of the broader environmental movement, symbolizing the urgent need for humans to adopt more sustainable practices and live in harmony with nature. Through a combination of education, activism, and lifestyle changes, modern culture continues to evolve in its relationship with bees, highlighting the critical role these pollinators play in sustaining life on Earth.

CHAPTER 62: INSPIRING CHANGE THROUGH BEES

Bees have become a powerful symbol in the fight against environmental degradation, serving as a catalyst for change and a reminder of the interconnectedness of our ecosystem. The decline in bee populations has highlighted the urgent need for sustainable practices and conservation efforts. By drawing on bees' role in our world, we can inspire significant environmental and societal changes.

Community Gardens and Bee Sanctuaries offer a practical approach to supporting bee populations while engaging communities in conservation efforts. Establishing these gardens involves selecting locations that can provide a variety of native flowering plants to ensure a continuous bloom throughout the growing season. These sanctuaries not only serve as vital food sources for bees but also create educational spaces for communities to learn about pollinators and the importance of biodiversity.

Sustainable Farming Practices are crucial in reducing the impact of agriculture on bee populations. Farmers and landowners can adopt crop rotation methods that improve soil health and reduce the need for chemical pesticides. Integrating cover crops, such as clover or alfalfa, attracts bees and other pollinators while enriching the soil. Reducing pesticide use by implementing integrated pest management (IPM) strategies that rely on biological and mechanical control methods can significantly decrease the harmful effects on bees.

Corporate Responsibility and Policy Advocacy play a significant role in promoting bee conservation. Businesses can adopt sustainability policies that support bee-friendly practices, such as reducing pesticide use in their supply chains or funding bee research. Advocating for policies that protect pollinators, support habitat restoration, and regulate pesticide use is essential. Engaging in public-private partnerships can amplify the impact of conservation efforts, leveraging the resources and reach of corporations to effect change.

Educational Outreach and Public Engagement are vital in raising awareness about the importance of bees. Schools, museums, and community centers can host workshops and events that teach about bee biology, the role of pollinators in our food system, and how individuals can contribute to their protection. Collaborating with local beekeeping associations to offer hands-on experiences with beekeeping can demystify bees and highlight their importance, fostering a new generation of bee advocates.

Innovative Research and Technology can offer new solutions to the challenges facing bees. Supporting research into bee health, genetics, and habitat requirements can lead to better conservation strategies. Technology, such as tracking systems for monitoring bee populations and health, can provide valuable data for researchers and conservationists. Developing new agricultural technologies that reduce the need for pesticides or create bee-friendly farming equipment can also contribute to bee conservation.

Individual Actions and Lifestyle Choices have a cumulative impact on bee populations. Planting bee-friendly gardens, reducing pesticide use, and supporting local beekeepers by purchasing their honey and other products are simple yet effective ways to help. Participating in citizen science projects that track pollinator populations can contribute valuable data to conservation efforts. Advocating for local and national policies that protect bees and their habitats ensures that individual actions are part of a larger movement towards sustainability.

By harnessing the cultural significance of bees and their role in our ecosystem, we can inspire a wide range of actions and policies that support bee health and biodiversity. From individual efforts to global initiatives, the movement to save bees is a testament to the power of collective action in addressing environmental challenges. Through education, advocacy, and sustainable practices, we can ensure that bees continue to thrive, supporting our food systems and ecosystems for generations to come.

PART 15: THE FUTURE OF BEEKEEPING

CHAPTER 63: TECHNOLOGICAL INNOVATIONS IN BEEKEEPING

Emerging technologies in beekeeping have revolutionized the way beekeepers manage hives, monitor bee health, and improve honey production. One of the most significant advancements is the development of **smart hive technology**. Smart hives are equipped with sensors that monitor temperature, humidity, weight, and even sound frequencies within the hive. These sensors transmit data in real-time to a beekeeper's smartphone or computer, allowing for immediate adjustments to maintain optimal hive conditions. For instance, a sudden drop in weight might indicate a decrease in honey stores, prompting the beekeeper to check for issues or supplement the bees' food source.

Another innovative tool is the **beekeeping drone**, used for aerial inspection of hives situated in difficult-to-reach areas. These drones are equipped with high-resolution cameras, enabling beekeepers to inspect the health of the colony without disturbing the bees. This is particularly useful for large apiaries where traditional inspection methods would be time-consuming and potentially stressful for the bees.

RFID (Radio Frequency Identification) technology represents a breakthrough in tracking bee movement and behavior. By attaching tiny RFID tags to bees, researchers and beekeepers can study their foraging patterns, identify changes in behavior that may indicate health issues, and monitor the effects of environmental factors on bee activity. This technology provides invaluable insights into the dynamics of bee populations and helps in the early detection of diseases or pests.

The use of **automated honey extraction systems** has streamlined the process of honey collection, making it more efficient and less labor-intensive. These systems minimize the disturbance to bees, reduce the risk of contamination, and ensure a consistent product quality. Automated extractors can be programmed to handle frames gently, extract honey, and then return the frames to the hive, all while keeping the integrity of the comb intact for the bees to refill.

Artificial intelligence (AI) and machine learning are being applied to analyze the vast amounts of data collected from smart hives and other monitoring tools. AI algorithms can predict potential problems within the hive, such as the likelihood of disease outbreak or the need for swarm control measures. This predictive capability allows beekeepers to take preemptive actions to protect their colonies.

In the realm of **pest and disease management**, innovative solutions like **thermal treatment devices** use controlled heat to eliminate Varroa mites without the use of chemicals. This method is effective, as the mites are more sensitive to temperature changes than bees. It offers a sustainable alternative to chemical treatments, which can harm bees and lead to resistance among pests.

The integration of **solar panels** into hive designs is a sustainable innovation that powers various hive technologies, including heating elements for cold climates and electronic monitoring systems. Solar-powered hives ensure that technology does not add to the carbon footprint of beekeeping operations and supports the overall health of the colony by maintaining a stable internal environment.

Finally, **online platforms and mobile apps** have created communities where beekeepers can share data, insights, and advice. These platforms offer access to a wealth of knowledge, from beginner tips to advanced techniques, and foster a collaborative approach to beekeeping. They also serve as a hub for citizen science projects, contributing to global efforts in bee conservation and research.

These technological innovations in beekeeping not only enhance the efficiency and effectiveness of hive management but also contribute to the sustainability of beekeeping practices. By adopting these technologies, beekeepers can ensure the health and productivity of their colonies, supporting the vital role bees play in our ecosystems and agriculture.

CHAPTER 64: GLOBAL TRENDS IN BEEKEEPING

As the world becomes increasingly interconnected, global trends in beekeeping reflect a blend of tradition and innovation, with sustainability at the forefront. One significant trend is the **shift towards organic beekeeping**. This approach emphasizes the use of natural substances and methods for managing hives and combating pests, avoiding synthetic chemicals that can harm bees and the environment. Organic beekeeping requires careful selection of hive locations away from industrial agriculture to minimize exposure to pesticides and GMO crops. Beekeepers adopting this method often use essential oils like thyme and lemongrass oil for pest control and feed bees with organic sugar syrup or let them forage on natural nectar sources during times of scarcity.

Another growing trend is the **integration of beekeeping with other agricultural practices**, known as agroforestry or silvopasture, depending on the specific combination. This practice involves planting crops or grazing livestock among trees, which provide a diverse array of flowering plants for bees throughout the year. This not only helps in improving the health and productivity of bees by providing them with a rich, varied diet but also enhances biodiversity and soil health. Beekeepers working within these systems may plant specific types of trees and shrubs known to be beneficial for bees, such as willow, almond, and clover species.

Urban beekeeping has also seen a surge in popularity, driven by the desire to support pollinator populations in cities and educate the public about the importance of bees. Urban beekeepers face unique challenges, such as limited space and potential conflicts with neighbors, but they also have the opportunity to tap into the rich variety of urban flora. Successful urban beekeeping requires careful hive placement to ensure bees have access to water and are protected from extreme heat, which might involve rooftop gardens or specially designed apiaries that blend into urban architecture.

The rise of **technology in beekeeping** is another critical trend. Innovations such as remote hive monitoring systems allow beekeepers to track hive health without frequent disturbances. These systems can measure temperature, humidity, hive weight, and even sound frequencies, sending alerts to the beekeeper when parameters indicate potential problems. To implement this technology, beekeepers must ensure a stable power source, which can often be achieved through solar panels, and a reliable internet connection for data transmission.

Community-supported beekeeping is gaining traction as well, where individuals can sponsor hives or bee colonies in exchange for a share of the honey produced. This model not only provides financial support to beekeepers but also builds a community of bee advocates and educates the public about the importance of bees. Beekeepers looking to adopt this model should focus on building strong relationships with their local communities through education and engagement activities, such as open days, workshops, and honey tastings.

Conservation and rehabilitation of native bee species have become a priority in many regions. Efforts include creating habitat corridors with native plants, reducing pesticide use, and rehabilitating areas affected by development or agriculture. Beekeepers can contribute by dedicating portions of their land to native habitats or participating in local and national conservation programs.

In summary, global trends in beekeeping are moving towards practices that are sustainable, technologically advanced, and community-oriented. These trends not only aim to improve the health and productivity of bee colonies but also to ensure the long-term sustainability of beekeeping as an essential part of our agricultural and ecological systems. Beekeepers adopting these trends will need to balance traditional knowledge with new technologies and methods, always with an eye towards the health of their bees and the broader environment.

Chapter 65: Educating Future Beekeepers

Educating the next generation of beekeepers involves a multifaceted approach that combines theoretical knowledge with hands-on experience. To ensure a comprehensive understanding, it's essential to start with foundational concepts of bee biology and behavior. This includes teaching about the different types of bees, their roles within the hive, and the life cycle of bees. For instance, lessons should cover the distinctions between the queen, workers, and drones, highlighting their unique contributions to the hive's functioning. Detailed diagrams and live observations can be used to illustrate the anatomy of bees, emphasizing the importance of each part in their daily activities and survival.

Introducing young beekeepers to the equipment and tools of the trade is another critical step. This should include detailed demonstrations on the use of protective gear such as veils, gloves, and suits to ensure safety while handling bees. Additionally, practical sessions on the use of smokers, hive tools, and brushes will familiarize them with the techniques for managing hives effectively. It's beneficial to include a hands-on workshop where participants can practice assembling a Langstroth hive, understanding the purpose of each component from the bottom board to the cover.

The curriculum should also encompass the environmental and ecological significance of bees, discussing pollination, the relationship between bees and flowering plants, and the impact of bees on food security. This can be complemented by field trips to local farms or gardens, where learners can observe bees in their natural habitat, fostering a deeper appreciation for their role in the ecosystem.

Hive management practices are crucial for maintaining healthy colonies, so educational programs must cover topics such as monitoring for pests and diseases, providing supplementary feeding during scarcity, and techniques for overwintering hives. Detailed case studies on challenges like Varroa mites and American foulbrood disease can offer insights into problem-solving and preventive measures. Interactive sessions that simulate hive inspections can teach students how to identify signs of a healthy hive versus indicators of stress or disease.

Sustainable beekeeping practices should form a core part of the education, emphasizing the importance of biodiversity, organic methods to control pests, and strategies to minimize the impact on native bee populations. Workshops on creating bee-friendly gardens, using native plants to support pollinators, and understanding the effects of pesticides on bee health can inspire responsible beekeeping and conservation efforts.

To prepare future beekeepers for the practical aspects of beekeeping, including honey harvesting and hive product processing, demonstrations on extracting honey, processing beeswax, and crafting products like candles or balms can be highly engaging. These activities not only teach valuable skills but also highlight the economic aspects of beekeeping, from product creation to marketing.

Incorporating modern technology into beekeeping education, such as the use of apps for monitoring hive health or platforms for community engagement, can appeal to the tech-savvy generation. Discussions on the latest innovations in beekeeping technology, like smart hives and thermal treatment for pest control, can open their minds to the possibilities of modern apiculture.

Finally, fostering a sense of community and collaboration among young beekeepers is essential. Encouraging participation in local beekeeping clubs, online forums, and citizen science projects can provide ongoing learning opportunities and support. Mentorship programs that pair experienced beekeepers with novices can offer personalized guidance, helping to bridge the gap between theoretical knowledge and practical experience.

By covering these topics in depth and providing hands-on learning experiences, educational programs can equip the next generation of beekeepers with the knowledge, skills, and passion needed to sustain and grow the beekeeping community. This holistic approach to education will ensure that future beekeepers are well-prepared to face the challenges of modern apiculture, contribute to bee conservation, and continue the vital work of supporting our ecosystems.

Printed by Libri Plureos GmbH in Hamburg, Germany